METHODS IN MOLECULAR BIOLOGY

Series Editor
John M. Walker
School of Life and Medical Sciences
University of Hertfordshire
Hatfield, Hertfordshire, UK

For further volumes:
http://www.springer.com/series/7651

For over 35 years, biological scientists have come to rely on the research protocols and methodologies in the critically acclaimed *Methods in Molecular Biology* series. The series was the first to introduce the step-by-step protocols approach that has become the standard in all biomedical protocol publishing. Each protocol is provided in readily-reproducible step-by-step fashion, opening with an introductory overview, a list of the materials and reagents needed to complete the experiment, and followed by a detailed procedure that is supported with a helpful notes section offering tips and tricks of the trade as well as troubleshooting advice. These hallmark features were introduced by series editor Dr. John Walker and constitute the key ingredient in each and every volume of the *Methods in Molecular Biology* series. Tested and trusted, comprehensive and reliable, all protocols from the series are indexed in PubMed.

Stem Cell-Based Human Embryo Models

Methods and Protocols

Edited by

Vincent Pasque

Department of Development and Regeneration, KU Leuven, Leuven, Belgium

Kursad Turksen

Ottawa, ON, Canada

Humana Press

Editors
Vincent Pasque
Department of Development
and Regeneration
KU Leuven
Leuven, Belgium

Kursad Turksen
Ottawa, ON, Canada

ISSN 1064-3745 ISSN 1940-6029 (electronic)
Methods in Molecular Biology
ISBN 978-1-0716-5355-5 ISBN 978-1-0716-5356-2 (eBook)
https://doi.org/10.1007/978-1-0716-5356-2

This Humana imprint is published by the registered company Springer Science+Business Media, LLC, part of Springer Nature.
The registered company address is: 1 New York Plaza, New York, NY 10004, U.S.A.

Preface

Over the last decade or so, our understanding of not only the pluripotency of stem cells but how to regulate and manipulate them appropriately in vitro has advanced so far as to now allow for the generation of very early embryo-like structures. These structures or "blastoids" have now in turn opened up huge, exciting, and very important opportunities for analyzing and delineating the very early steps in embryo development under controlled in vitro conditions. Being able to take full advantage of these opportunities continues to drive rapid innovations in the methods and protocols needed for studying early embryonic development.

In this volume, we have collected a series of protocols that cover diverse aspects of the topic and exemplify some of the developments and improvements in the field.

Once again, the protocols gathered here are faithful to the mission statement of the Methods in Molecular Biology series: They are well-established and described in an easy-to-follow, step-by-step fashion so as to be valuable for not only experts but also novices in the stem cell field. That goal is achieved because of the generosity of the contributors who have carefully described their protocols in this volume, and we are very grateful for their efforts.

Our thanks as well go to Dr. John Walker, the Editor-in-Chief of the Methods in Molecular Biology series, for giving us the opportunity to create this volume and for supporting us along the way.

We are also grateful to Patrick Marton, the Executive Editor of Methods in Molecular Biology and the Springer Protocols collection, for his continuous support from idea to completion of this volume.

A special thank you goes to Anna Rakovsky, Assistant Editor for Methods in Molecular Biology, for her support from the beginning to the end of this project.

We would also like to thank David C. Casey, Senior Editor of Methods in Molecular Biology, for his outstanding editorial work during the production of this volume.

Finally, we would like to thank the production crew for their work in putting together an outstanding book.

Leuven, Belgium *Vincent Pasque*
Ottawa, ON, Canada *Kursad Turksen*

Contents

Contributors

ASLI AK • *MERLN Institute for Technology-Inspired Regenerative Medicine, Maastricht University, Maastricht, The Netherlands; Department of Obstetrics & Gynecology, GROW Institute for Oncology and Reproduction, Faculty of Health, Medicine and Life Sciences, Maastricht University, Maastricht, The Netherlands*
OMAR MARTÍNEZ ALARCON • *Departamento de Fisiología y Desarrollo Celular, Instituto Nacional de Perinatología, Ciudad de México, Mexico*
CANTAS ALEV • *Institute for the Advanced Study of Human Biology (ASHBi), Kyoto University, Kyoto, Japan*
SHARON ARCURI • *Department of Veterinary Medicine, University of Sassari, Sassari, Italy; Laboratory of Biomedical Embryology and Tissue Engineering, Department of Veterinary Medicine and Animal Sciences, Centre for Stem Cell Research, Università degli Studi di Milano, Lodi, Italy*
ROCHELLE AUBRY • *Laboratory of Bioengineering and Morphogenesis, Biomechanics Section, Department of Mechanical Engineering, KU Leuven, Leuven, Belgium*
DANIELA ÁVILA-GONZÁLEZ • *Departamento de Fisiología y Desarrollo Celular, Instituto Nacional de Perinatología, Ciudad de México, Mexico*
MEIR AZAGURY • *Department of Developmental Biology and Cancer Research, The Institute for Medical Research Israel-Canada, The Hebrew University-Hadassah Medical School, Jerusalem, Israel*
FELICITAS AZPIROZ • *Research and Development, Eugin Group, Barcelona, Spain*
IVAN BEDZHOV • *Embryonic Self-Organization Research Group, Max Planck Institute for Molecular Biomedicine, Münster, Germany*
MARLOES BLOTENBURG • *Center for Integrative Genomics, University of Lausanne, Lausanne, Switzerland*
TIZIANA A. L. BREVINI • *Laboratory of Biomedical Embryology and Tissue Engineering, Department of Veterinary Medicine and Animal Sciences, Centre for Stem Cell Research, Università degli Studi di Milano, Lodi, Italy*
JOSHUA M. BRICKMAN • *Novo Nordisk Foundation Center for Stem Cell Medicine (reNEW), Department of Biomedical Sciences, University of Copenhagen, Copenhagen, Denmark*
SHRUTHI BUDDHARAJU • *Department of Molecular Biology and Genetics, Aarhus University, Aarhus, Denmark*
YOSEF BUGANIM • *Department of Developmental Biology and Cancer Research, The Institute for Medical Research Israel-Canada, The Hebrew University-Hadassah Medical School, Jerusalem, Israel*
DANIEL CARVALHO • *MERLN Institute for Technology-Inspired Regenerative Medicine, Maastricht University, Maastricht, The Netherlands*
AXEL CASTRO-ABREGO • *Departamento de Fisiología y Desarrollo Celular, Instituto Nacional de Perinatología, Ciudad de México, Mexico*
SUSANA M. CHUVA DE SOUSA LOPES • *Department of Anatomy and Embryology, Leiden University Medical Center, Leiden, The Netherlands; The Novo Nordisk Foundation Center for Stem Cell Medicine (reNEW), Leiden University Medical Center, Leiden, The Netherlands; Department for Reproductive Medicine, Ghent University Hospital, Ghent, Belgium*

SYLWIA M. CZUKIEWSKA • *Department of Anatomy and Embryology, Leiden University Medical Center, Leiden, The Netherlands; The Novo Nordisk Foundation Center for Stem Cell Medicine (reNEW), Leiden University Medical Center, Leiden, The Netherlands*
NESTOR FABIÁN DIAZ • *Departamento de Fisiología y Desarrollo Celular, Instituto Nacional de Perinatología, Ciudad de México, Mexico*
NESTOR EMMANUEL DIAZ-MARTINEZ • *Laboratorio de Reprogramación Celular y Bioingenería de Tejidos, Biotecnología Médica y Farmacéutica, Centro de Investigación y Asistencia en Tecnología y Diseño del Estado de Jalisco, Guadalajara, Mexico*
BERIL ESIN • *Department of Molecular Biology and Genetics, Aarhus University, Aarhus, Denmark*
TOSHIHIKO EZASHI • *Colorado Center for Reproductive Medicine, Lone Tree, CO, USA*
FULVIO GANDOLFI • *Department of Agricultural and Environmental Sciences – Production, Landscape, Agroenergy, Centre for Stem Cell Research, Università degli Studi di Milano, Milan, Italy*
SEHER IPEK GASSALOGLU • *Department of Genetics, Yale School of Medicine, Yale University, New Haven, CT, USA*
BORZO GHARIBI • *Quantitative Stem Cell Biology Laboratory, The Francis Crick Institute, London, UK*
STEFAN GISELBRECHT • *MERLN Institute of Technology-Inspired Regenerative Medicine, Department for Cell Biology-Inspired Tissue Engineering, Maastricht University, Maastricht, The Netherlands*
NIRAIMATHI GOVINDASAMY • *Embryonic Self-Organization Research Group, Max Planck Institute for Molecular Biomedicine, Münster, Germany*
CARLY GUILTINAN • *Department of Biomolecular Engineering, University of California, Santa Cruz, CA, USA; Institute for the Biology of Stem Cells, University of California, Santa Cruz, CA, USA; Genomics Institute, University of California, Santa Cruz, CA, USA*
GE GUO • *Living Systems Institute, University of Exeter, Exeter, UK*
SOFIANE HAMIDI • *Institute for the Advanced Study of Human Biology (ASHBi), Kyoto University, Kyoto, Japan*
IDSE HEEMSKERK • *Department of Biomedical Engineering, University of Michigan, Ann Arbor, MI, USA; Department of Computational Medicine and Bioinformatics, University of Michigan Medical School, Ann Arbor, MI, USA; Department of Cell and Developmental Biology, University of Michigan Medical School, Ann Arbor, MI, USA; Center for Cell Plasticity and Organ Design, University of Michigan Medical School, Ann Arbor, MI, USA; Department of Physics, University of Michigan, Ann Arbor, MI, USA*
ANAYANSI MOLINA HERNÁNDEZ • *Departamento de Fisiología y Desarrollo Celular, Instituto Nacional de Perinatología, Ciudad de México, Mexico*
SHU ISHIDA • *Uehiro Division for Applied Ethics, Graduate School of Humanities and Social Sciences, Hiroshima University, Higashihiroshima, Japan*
ERA JAIN • *Department of Biomedical and Chemical engineering, Syracuse University, Syracuse, NY, USA; Bioinspired Syracuse: Institute for Material and Living System, Syracuse University, Syracuse, NY, USA*
SHASHANK JAITLY • *Embryonic Self-Organization Research Group, Max Planck Institute for Molecular Biomedicine, Münster, Germany*
KEUN-TAE KIM • *Institute for the Advanced Study of Human Biology (ASHBi), Kyoto University, Kyoto, Japan*

CHIE KOBAYASHI • *Uehiro Division for Applied Ethics, Graduate School of Humanities and Social Sciences, Hiroshima University, Higashihiroshima, Japan*
SAYAKA KOZUKI • *Department of Biomolecular Engineering, University of California, Santa Cruz, CA, USA; Institute for the Biology of Stem Cells, University of California, Santa Cruz, CA, USA; Genomics Institute, University of California, Santa Cruz, CA, USA*
KUN LIU • *Colorado Center for Reproductive Medicine, Lone Tree, CO, USA*
GERRALD A. LODEWIJK • *Department of Biomolecular Engineering, University of California, Santa Cruz, CA, USA; Institute for the Biology of Stem Cells, University of California, Santa Cruz, CA, USA; Genomics Institute, University of California, Santa Cruz, CA, USA*
DORIAN G. LUIJKX • *MERLN Institute of Technology-Inspired Regenerative Medicine, Department for Cell Biology-Inspired Tissue Engineering, Maastricht University, Maastricht, The Netherlands*
SHIJIAN LYU • *Department of Molecular Biology, University of Texas Southwestern Medical Center, Dallas, TX, USA*
ZHEN MA • *Department of Biomedical and Chemical engineering, Syracuse University, Syracuse, NY, USA; Bioinspired Syracuse: Institute for Material and Living System, Syracuse University, Syracuse, NY, USA*
ALEJANDRO MARTINEZ-JUAREZ • *Departamento de Fisiología y Desarrollo Celular, Instituto Nacional de Perinatología, Ciudad de México, Mexico*
JENNIFER NICHOLS • *MRC Human Genetics Unit, Institute of Genetics and Cancer, University of Edinburgh, Western General Hospital, Edinburgh, UK*
TAKUMI OKUBO • *Kyoto University, CiRA, Kyoto, Japan*
SEIYA OURA • *Department of Molecular Biology, University of Texas Southwestern Medical Center, Dallas, TX, USA*
MONIQUE PEDROZA • *Department of Genetics, Yale School of Medicine, Yale University, New Haven, CT, USA*
DUANQING PEI • *Laboratory of Cell Fate Control, School of Life Sciences, Westlake University, Hangzhou, China*
GEORGIA PENNAROSSA • *Laboratory of Biomedical Embryology and Tissue Engineering, Department of Veterinary Medicine and Animal Sciences, Centre for Stem Cell Research, Università degli Studi di Milano, Lodi, Italy*
CARLOS A. PINZON-ARTEAGA • *Department of Molecular Biology, University of Texas Southwestern Medical Center, Dallas, TX, USA; Howard Hughes Medical Institute, Department of Cell Biology, Blavatnik Institute, Harvard Medical School, Boston, MA, USA*
MINA POPOVIC • *Research and Development, Eugin Group, Barcelona, Spain; Department for Reproductive Medicine, Ghent University Hospital, Ghent, Belgium*
WENDY PORTILLO • *Departamento de Neurobiología Conductual y Cognitiva, Instituto de Neurobiología, UNAM, Quéretaro, Mexico*
DEVILA PRIT • *Embryonic Self-Organization Research Group, Max Planck Institute for Molecular Biomedicine, Münster, Germany*
MARTIN PROKS • *Novo Nordisk Foundation Center for Stem Cell Medicine (reNEW), Department of Biomedical Sciences, University of Copenhagen, Copenhagen, Denmark*
ADRIAN RANGA • *Laboratory of Bioengineering and Morphogenesis, Biomechanics Section, Department of Mechanical Engineering, KU Leuven, Leuven, Belgium*
JONATHAN SALAZAR-ALONSO • *Departamento de Fisiología y Desarrollo Celular, Instituto Nacional de Perinatología, Ciudad de México, Mexico*

NAZMUS SALEHIN • *Novo Nordisk Foundation Center for Stem Cell Medicine (reNEW), Department of Biomedical Sciences, University of Copenhagen, Copenhagen, Denmark*
IDRIS SALMON • *Laboratory of Bioengineering and Morphogenesis, Biomechanics Section, Department of Mechanical Engineering, KU Leuven, Leuven, Belgium*
MENAKA SANGHVI • *Department of Molecular Biology, University of Texas Southwestern Medical Center, Dallas, TX, USA*
SILVIA D. M. SANTOS • *Quantitative Stem Cell Biology Laboratory, The Francis Crick Institute, London, UK*
HIROO SASAKI • *Department of Veterinary Anatomy, Graduate School of Agriculture and Life Sciences, The University of Tokyo, Tokyo, Japan*
IORI SASAKI • *Department of Veterinary Anatomy, Graduate School of Agriculture and Life Sciences, The University of Tokyo, Tokyo, Japan*
TSUTOMU SAWAI • *Uehiro Division for Applied Ethics, Graduate School of Humanities and Social Sciences, Hiroshima University, Higashihiroshima, Japan; Graduate School of Humanities and Social Sciences, Hiroshima University, Higashihiroshima, Japan; Centre for Biomedical Ethics, Yong Loo Lin School of Medicine, National University of Singapore, Singapore, Singapore*
MICHAEL PATRICK SEITZ • *Department of Biomedical and Chemical engineering, Syracuse University, Syracuse, NY, USA; Bioinspired Syracuse: Institute for Material and Living System, Syracuse University, Syracuse, NY, USA*
S. ALI SHARIATI • *Department of Biomolecular Engineering, University of California, Santa Cruz, CA, USA; Institute for the Biology of Stem Cells, University of California, Santa Cruz, CA, USA; Genomics Institute, University of California, Santa Cruz, CA, USA*
BERNA SOZEN • *Department of Genetics, Yale School of Medicine, Yale University, New Haven, CT, USA; Yale Stem Cell Center, Yale University, New Haven, CT, USA; Department of Obstetrics, Gynecology and Reproductive Sciences, Yale School of Medicine, Yale University, New Haven, CT, USA*
YASUHIRO TAKASHIMA • *Kyoto University, CiRA, Kyoto, Japan*
SETH TEAGUE • *Department of Biomedical Engineering, University of Michigan, Ann Arbor, MI, USA*
BENJAMIN R. TOPACIO • *Department of Biomolecular Engineering, University of California, Santa Cruz, CA, USA; Institute for the Biology of Stem Cells, University of California, Santa Cruz, CA, USA; Genomics Institute, University of California, Santa Cruz, CA, USA*
RON VAN GOLDE • *Department of Obstetrics & Gynecology, GROW Institute for Oncology and Reproduction, Faculty of Health, Medicine and Life Sciences, Maastricht University, Maastricht, The Netherlands; Department of Gynaecology, Women Mother Child Centre, Maastricht University Medical Centre+ (MUMC+), Maastricht, The Netherlands*
ERIK J. VRIJ • *MERLN Institute of Technology-Inspired Regenerative Medicine, Department for Cell Biology-Inspired Tissue Engineering, Maastricht University, Maastricht, The Netherlands; GROW Research Institute for Oncology and Reproduction, Department of Obstetrics & Gynaecology, Maastricht University, Maastricht, The Netherlands*
HANNING WANG • *Laboratory of Cell Fate Control, School of Life Sciences, Westlake University, Hangzhou, China*
JUN WU • *Department of Molecular Biology, University of Texas Southwestern Medical Center, Dallas, TX, USA; Hamon Center for Regenerative Science and Medicine, University of Texas Southwestern Medical Center, Dallas, TX, USA; Cecil H. and Ida*

Green Center for Reproductive Biology Sciences, Department of Obstetrics and Gynecology, Children's Research Institute, University of Texas Southwestern Medical Center, Dallas, TX, USA

JINZHU XIANG • *Laboratory of Cell Fate Control, School of Life Sciences, Westlake University, Hangzhou, China*

AYAKA YANAGID • *Department of Veterinary Anatomy, Graduate School of Agriculture and Life Sciences, The University of Tokyo, Tokyo, Japan; Stem Cell Therapy Division, Advanced Research Institute, Institute of Integrated Research, Institute of Science, Tokyo, Japan*

ZHIYUAN YU • *Department of Computational Medicine and Bioinformatics, University of Michigan Medical School, Ann Arbor, MI, USA*

YE YUAN • *Colorado Center for Reproductive Medicine, Lone Tree, CO, USA*

PETER ZELLER • *Department of Molecular Biology and Genetics, Aarhus University, Aarhus, Denmark*

MARIA CAROLINA ZIMARA • *Colorado Center for Reproductive Medicine, Lone Tree, CO, USA*

Methods in Molecular Biology (2026) 3048: 1–16
DOI 10.1007/7651_2025_668

Published online: 04 October 2025

Human Stem Cell-Based Embryo Model Research: Ethics and Regulations

Chie Kobayashi, Shu Ishida, and Tsutomu Sawai

Abstract

Given rapid advances in the life sciences, stem cell–based embryo models (SCBEMs) call for careful ethical and regulatory consideration. As with similar research related to human embryos and reproduction, SCBEMs raise important normative questions. In this chapter, we survey these questions, outline possible answers, and discuss their implications for global regulatory development. We also map major ethical issues related to SCBEM research, identify areas in need of further debate, and examine the governance frameworks required for responsible progress. We conclude with recommendations to guide SCBEM research so that it advances scientific innovation while protecting the public interest. In particular, we highlight the importance of careful planning, transparent communication, inclusive dialogue, robust governance, and ongoing empirical inquiry. Taken together, these measures can help ensure that SCBEM research proceeds responsibly, remains ethically sound, and earns public trust.

Key words Embryonic stem cells, Human reproduction, Induced pluripotent stem cells, Informed consent, Moral Status, Public trust, Responsible science, Stem cell research, Stem cell–based embryo models, Techno-moral change

1 Introduction

Since 2014, several studies have been reported on the stem cell culture methods involved in modeling early human embryonic development [1, 2]. More recently, researchers have succeeded in modeling various stages of early embryogenesis, ranging from the blastocyst stage to gastrulation, and even early organogenesis [3]. These advances have not only deepened our understanding of early human development but also hold promise for fundamental biological research and clinical applications. Although no standardized terminology has yet been adopted for the tissues created in such studies—with terms such as "embryoids," "synthetic embryos," "gastruloids," and "blastoids" in use [4]—we follows the International Society for Stem Cell Research [5] in adopting the term "stem cell–based embryo model" (SCBEM) in this chapter.

There is broad international consensus that the use of human embryos for research must be limited to either 14 days post-fertilization or the appearance of the primitive streak, whichever occurs first. Human embryo research is also subject to further restrictions depending on its purpose [6]. These restrictions, however, do not apply to SCBEMs because they are currently regarded not as embryos themselves but as *models* of parts or stages of human embryogenesis. SCBEMs are therefore expected to play key roles in both basic research on human development and applied biomedical fields [7]. Their promise lies in two key features: their morphological and functional resemblance to natural embryos, and the fact that they are not bound by the strict ethical and legal regulations governing natural embryos. Taken together, these features create opportunities to advance our scientific knowledge of early pregnancy loss, improve early pregnancy management, and aid the development of therapies for early developmental disorders [8].

Some scholars further suggest that SCBEMs can offer ethical alternatives to animal experiments and help circumvent ethical and practical concerns relating to the use of human embryos [9]. Others, by contrast, highlight that SCBEMs remain mere models, different from natural embryos [10]. Skeptics also argue that, as long as SCBEMs are created solely for research and not for reproductive purposes, they do not require particular ethical consideration at all [11]. Still others cautiously point out that, as science evolves, SCBEMs could eventually acquire levels of functionality comparable to embryos [12]. At that point, questions might arise as to whether they should be treated as equivalent to embryos.

At present, however, it remains unclear when or whether SCBEMs begin to approximate the status of natural embryos [13]. This uncertainty gives rise to a wide range of ethical, legal, and sociocultural questions, including those rooted in religious beliefs. Such questions concern not only how to evaluate ongoing developments in SCBEM research but also how these models ought to be understood and used. To ensure responsible research and innovation in SCBEM science, it is therefore essential to identify and, if possible, address normative concerns surrounding this fast-emerging domain.

This chapter surveys the ethical issues associated with SCBEM research and sets out considerations for global regulatory development. Subheading 2 classifies major ethical issues and identifies areas requiring further discussion. Subheading 3 examines governance frameworks needed for responsible development of this research field. Through this discussion, these analyses offer a structured approach to addressing normative questions ranging from realistic near-term concerns to more speculative, future-oriented challenges.

2 Ethical Issues Associated with SCBEM Research

Here, we provide a systematic overview of the ethical issues associated with SCBEM research. We begin by identifying three types of stakeholders whose interests may be affected: (a) the SCBEMs themselves, (b) closely related parties, such as donors of research materials, and (c) third parties, including society at large. This categorization intersects with another tripartite framework that classifies ethical issues in emerging science as (1) already apparent, (2) foreseeable in the near future, or (3) conceivable in the more distant future. Given the inherent uncertainties of SCBEM research, some ethical concerns are currently pressing, while others remain largely speculative. We address both in this section.

The intersection of these two dimensions produces the 3 × 3 matrix shown in Table 1, which covers nine distinct domains (a1–c3). This framework organizes normative concerns along two axes: (X) the affected parties—that is, the stakeholder on whose behalf the normative question is raised—and (Y) the urgency and immediacy of the issue. Framing the discussion in this way allows for a systematic treatment of the ethics of SCBEM research, rather than a piecemeal list of isolated concerns. The following subsections apply this framework to several of the most frequently discussed issues in the normative debate on SCBEM research.

2.1 The Moral Status of SCBEMs

One of the central ethical concerns in SCBEM research relates to the question of their moral status. According to a widely accepted view, if an entity X has moral status, then moral agents such as humans are obliged to consider X's desires, interests, and welfare *for X's own sake*, and not merely for the benefit of others [14]. Moral status is therefore directly linked to how we should treat a particular entity. If SCBEMs possess moral status, for instance, researchers have a duty to consider their welfare in the course of

Table 1
Typical ethical considerations associated with SCBEM research

	(1) Already apparent	(2) Foreseeable in the near future	(3) Conceivable in the distant future
(a) SCBEMs themselves	(a1) Moral status (e.g., potentiality and relational traits)	(a2) Moral status (e.g., more sophisticated traits)	(a3) Reproductive application (e.g., solo reproduction and multiplex parenting)
(b) Donors and related parties	(b1) Informed consent; donor rights	(b2) Benefit sharing; privacy (e.g., personal data)	(b3) Reproductive application (e.g., cloning); heritable genetic risks
(c) Society at Large	(c1) Hype; public misunderstanding	(c2) Misuse	(c3) Techno-moral shift (e.g., changes in social norms)

experimentation. This question, i.e., what we owe to SCBEMs in their own right, is a recurring concern in Domain a1 in Table 1 (the "domains" refer to those listed in Table 1 hereafter, unless otherwise specified).

To determine whether SCBEMs have moral status, we must identify the relevant criteria—i.e., the characteristics that confer such status—and assess whether SCBEMs meet them. Previous studies have highlighted criteria such as the capacity for sentience (the ability to feel pain or pleasure) and the potential to develop into human organisms [15]. Arguments based on *sentience* specifically suggest that entities capable of experiencing pain have moral status. In human embryos, however, sentience does not arise before the appearance of the primitive streak at around 14 days post-fertilization, when the nervous system begins to develop. As noted in the previous section, human embryos before this point are generally thought to lack sentience and therefore moral status.

This threshold may not apply straightforwardly to SCBEMs because they do not follow the same developmental trajectory as natural embryos. SCBEMs are not created through sperm–egg fusion and therefore lack a defined "day 0 post-fertilization," which renders the 14-day rule unworkable in this context. More significantly, researchers have also proposed that neural tissue could form in SCBEMs without passing through the primitive streak stage, suggesting that limits on SCBEM research should not be based on standard developmental milestones for human embryos [16]. Instead, they should be guided by the actual emergence of physiological features relevant to moral status, such as neural substrates linked to sentience. Advocates of this view support a more flexible, case-specific approach that accounts for developmental variation among SCBEMs.

A different line of argument appeals to *potentiality*—the idea that entities capable of developing into human beings possess moral status. In the context of the 14-day rule, Sophia McCully has argued that moral status depends on *both* biological development and whether the embryo is, or will be, situated in an environment that allows it to mature into a human organism [17]. If an embryo is destined for disposal and lacks the capacity to develop further, then it is not considered to have full moral status—i.e., moral status that warrants as strong protection as is granted to typical human beings.

If we apply this reasoning to SCBEM research, it follows that SCBEMs created solely for research and not intended to develop into human beings similarly lack full moral status. This potentiality-based view of moral status is reflected in the regulatory practices of many countries [18]. Notably, it does not rely only on intrinsic properties of SCBEMs such as sentience. The permissibility of using SCBEMs in research also depends on balancing their moral worth against competing values and interests.

Both the sentience-based and potentiality-based approaches face limitations. Sentience-based criteria might exclude entities, such as embryos or the recently deceased person's remains, that many believe deserve moral consideration despite lacking sentience. Meanwhile, the concept of "potentiality" has become increasingly unstable and ambiguous given advances in developmental biology [19], challenging its tenability as a criterion for moral status.

Alternatively, some scholars have highlighted *relational properties*. On this view, an entity may have moral status if individuals or communities regard it as valuable, or if it stands in a special relationship with them [14]. For instance, a human embryo often has a psychological tie with its parents, and human remains can retain cultural, religious, or spiritual significance for the deceased person's family and the broader community to which that person belonged.

The relational view of moral status comes in two forms: "strong" relational views, which ground moral status entirely in the pertinent relationships, and "weak" relational views, which combine relational and intrinsic characteristics. This distinction was well acknowledged by Mary Anne Warren, one of the most important philosophical figure who explored the relational account of moral status [14]. Importantly, proponents of weaker variations may admit that an entity's intrinsic traits (e.g., sentience or potentiality) can also inform moral considerations it deserves.

Applied to SCBEMs, the relational approach implies that moral status can be attributed even if an SCBEM lacks sentience or developmental potential, based on its cultural or religious significance within a specific community. Although this view has received limited attention in ethical debates about SCBEMs, it highlights that intrinsic properties alone are unlikely to settle questions of moral status. By analogy with natural human embryos, the assessment of SCBEMs' moral status should consider both intrinsic and relational features. Given their resemblance to human embryos, SCBEMs may also carry powerful symbolic meaning [20, 21].

This point is particularly important for designing governance frameworks for SCBEM that are sensitive to diverse cultural values. For instance, religious traditions often differ in their views on the moral status of human embryos [22], and it is therefore unsurprising that they may also diverge in their assessments of SCBEMs in ways that reflect their broader worldviews. At present, one might argue that debates about the moral status of SCBEMs are more grounded when they focus on their relational traits (Domain a1) rather than intrinsic properties such as sentience (Domain a2), since most concerns about SCBEMs developing sentience remain speculative and pertain to possible future scenarios.

2.2 The Interests of Donors and Closely Related Parties

In SCBEM research, donors of cells, gametes, or embryos are key stakeholders. Protecting their rights and interests is a widely acknowledged ethical priority, falling squarely within Domain b1. Unlike the moral status question, this issue concerns the potential impact of SCBEM research on donors and related parties, rather than on SCBEMs themselves.

The donor's identity and role vary depending on the source material used to generate SCBEMs. For instance, in research based on embryonic stem cells (ESCs), currently the most common approach, the donors are typically the two individuals who provide the gametes used to create the embryo. In contrast, when researchers use induced pluripotent stem cells (iPSCs), the donor is the individual who provides the somatic cells from which the iPSCs are derived.

A central ethical issue is *informed consent* [23]. Which procedures ensure that the consent is ethically valid and sufficiently informed? In SCBEM research, biological materials such as cells or embryos are often repurposed from earlier collections under broad or general consent. However, this practice raises concerns about respect for donor autonomy, especially when secondary use of the materials involves novel and ethically complex applications that donors could not have foreseen at the time of giving their "informed" consent [24]. Even if general consent has been obtained, using donors' biological materials beyond the purposes clearly mentioned at the point of gaining consent may fail to adequately respect the autonomy of those donors; for example, tissue originally donated for disease-genomics research may later be repurposed to construct SCBEMs, which the donor did not foresee and might have declined.

To address these concerns, scholars have proposed several consent models. In the adjacent field of brain organoid research—where stem cells are used to create three-dimensional neural tissue—some argue that specific, *project-based consent* is ethically preferable to general consent [25]. Another proposal is *dynamic consent*, which involves ongoing communication with donors (e.g., information-sharing and regular updates) as the research evolves [26]. Both strategies enhance respect for donor autonomy, but they might also increase logistical burdens on both researchers and donors. Researchers must recontact donors, track changes in their consent preferences over time, and update documentation and data-access controls across studies. Donors, in turn, must repeatedly review study materials and revisit their previous consent. These difficulties highlight the need for balanced, workable systems.

In addition to the consent issue, the use of SCBEMs can raise concerns about privacy and data protection, especially when sensitive personal data (e.g., genomic information) are involved. SCBEMs derived from ESCs, the relevant data pertain to the embryo's biological parents. For those derived from iPSCs, the data pertain to the somatic cell donor. Long-term storage and analysis of SCBEMs further intensify these concerns [23, 26]. Although anonymization is a common safeguard, recent and future advances in AI and omics analysis might increase the risk of re-identification, making it increasingly difficult to guarantee complete anonymity (Domain b2).

Finally, if SCBEM research leads to applications in regenerative medicine or pharmaceutical development, then the issue of *benefit sharing* becomes ethically salient—i.e., how to ensure fair and equitable distribution of profits and other positive outcomes among all stakeholders (Domain b2). Allowing a small number of research institutions or companies to capture the scientific and commercial benefits from SCBEM breakthroughs raises concerns about *distributive injustice* [26, 27]. Thus, equitable benefit-sharing mechanisms are essential to acknowledge and fairly compensate all contributors to the research, including donors. The possibility that global inequities will be reinforced is particularly troubling: donors from low- and middle-income countries or marginalized backgrounds might be excluded from benefits enjoyed by wealthy stakeholders.

2.3 Social Impact

Beyond immediate stakeholders such as donors, SCBEM research might also affect other actors, or even society at large, in various ways (Domain c1). One notable ethical challenge lies in the issue of *hype*, whereby researchers or the media exaggerate a study's scientific potential or implications (whether positive or negative). For instance, SCBEM research is expected to contribute to future therapies for early developmental disorders, but overstating its potential might lead to unrealistic expectations among patients and their families. Failure to meet such expectations can, in turn, undermine public trust in this and related research. Conversely, suggesting that SCBEM research could directly lead to human cloning (e.g., when derived from an individual's iPSCs) might provoke public alarm. This could lead to overly restrictive regulations. In both cases, hype has a significant influence on shaping public perception, policy development, and hence research funding. Thus, as in adjacent fields (e.g., brain organoid research), it is essential to avoid hype by, for instance, using accurate terminology and responsible communication practices [28].

The challenge of naming the entities generated during SCBEM research is closely related to the problem of hype. Terminology plays a crucial part in shaping public understanding and social attitudes. As of now, various terms—including "embryoids," "synthetic embryos," "gastruloids," and "blastoids"—have been proposed and used, often interchangeably [4]. However, naming is far from trivial. It carries significant ethical and social weight, especially when one considers that SCBEMs can appear novel and alien to non-experts. For instance, the term "synthetic embryo" might trigger moral unease among the public, while highly technical terms such as "gastruloid" or "blastoid" might be too obscure for non-experts. This can undermine transparency and accessibility [15].

In any event, the International Society for Stem Cell Research has recommended adopting the term "stem cell-based embryo model" (or "SCBEM") to promote international consensus. We therefore adopt that convention in this chapter. However, many researchers continue to use a variety of terms, and terminological inconsistency remains a problem [29]. This lack of standardization not only complicates public communication but also has the potential to undermine broader social acceptance of research in this field. This means that developing effective science communication strategies—including the use of clear and consistent nomenclature—is essential to ensure that the field evolves responsibly.

2.4 Reasoning and Justifications in SCBEM Research Ethics

In the preceding subsections, we discussed three classes of ethical concerns that are often raised in relation to SCBEM research, roughly corresponding to the main stakeholders affected by (un) ethical practices in that field: SCBEM themselves, closely related parties like donors, and society at large. A further concern is the forms of reasoning used to support or challenge ethical claims about SCBEM research, under conditions of uncertainty, irrespective of their substantive content. This question, in other words, concerns the general issue of how we can and should engage in principled ethical discussion.

Current international guidelines and ethical codes explicitly prohibit certain uses of SCBEMs, including (i) uses that cause or enable pain or suffering in SCBEMs, and (ii) reproductive use, particularly implantation into a uterus [5, 30]. It is, however, important to examine the precise nature of such prohibitions and their ethical justifications rather than accepting them uncritically. Are the bans categorical, rooted in some unnegotiable ethical principle? Or are they conditional prescriptions based on current scientific knowledge and prevailing social values, and thus potentially subject to future revision? Answering this two-part question is important for developing governance frameworks that evolve alongside scientific advances and shifting societal norms (see also Subheading 2.5). As we discuss below, some widely accepted ethical statements are absolute and uncompromising, while others are contingent upon the social, cultural, and scientific realities that are more or less prone to change.

Take, for instance, the prohibition against research that could cause pain to SCBEMs. The associated suffering will be ethically problematic under many prominent normative theories, whether they be consequentialist, deontological, or even virtue ethics. Typically, consequentialists will argue that such negative experiences amount to unnecessary disvalue in the world. Deontologists might state that inflicting pain on other sentient beings is a violation of ethical obligations. And virtue ethicists might consider causing pain to others to be an ethical flaw in the relevant agent's disposition. As mentioned in Subheading 2.1, if SCBEMs have the capacity for sentience, there are compelling ethical reasons to extend ethical consideration and protection to them.

In contrast, restrictions on the reproductive use of SCBEMs are primarily driven by safety and risk concerns. Although currently stringent, these prohibitions might not be permanent. Indeed, the Nuffield Council on Bioethics has suggested that the ethical permissibility of using SCBEMs for reproductive purposes might become open to inclusive public deliberation if associated safety concerns are adequately resolved in the future. Nevertheless, before we can ethically or legally reconsider reproductive use, SCBEMs will need to both (i) demonstrate the capacity for implantation and development *in vivo* and (ii) undergo large-scale clinical trials, which are inherently risky [12, p. 33]. Thus, for now, maintaining a legal prohibition on the implantation of SCBEMs into human reproductive tracts might serve as an important safeguard, one that preserves public trust and delineates the current limits of acceptable SCBEM research [12, p. 77]. As before, this observation reminds us of the important distinction between urgent ethical questions—what should be (im)permissible in today's society?—and anticipatory, future-oriented questions—what might be (im)permissible in the future?

For the sake of regulatory coherence, ethical evaluations of SCBEMs' reproductive use should also be guided by norms that are established in associated fields, such as those related to somatic cell nuclear transfer and human cloning. Notably, the United Nations Declaration on Human Cloning prohibits reproductive human cloning based on respect for human dignity [31]. Unless there is a compelling ethical justification for treating SCBEMs differently, or even for reconsidering such a ban altogether, the existing normative consensus should apply provisionally to SCBEMs derived from iPSCs as well.

2.5 Anticipating Future Ethical Challenges: Techno-Moral Change

Anticipating SCBEM technology's far-future possibilities is, arguably, not feasible in light of the current state of the relevant science. Nonetheless, doing so can highlight some potential ethical challenges associated with the pertinent reproductive applications. Such challenges are, no doubt, worth consideration over the course of responsible innovation. For instance, hypothetical future scenarios, such as solo reproduction or multiplex parenting enabled by advanced SCBEM technologies, could pose risks or disadvantages to individuals developing from SCBEMs. Specifically, such scenarios invoke important questions related to identity, stigmatization due to their biological origin, and long-term genetic health risks (Domain a3).

These issues are more or less analogous to those faced by people born through assisted reproductive technologies, including in vitro fertilization [32], gamete donation [33], and surrogacy [34]. Reproductive applications might also affect closely related parties, such as donors and would-be parents. Specifically, such applications of SCBEMs would implicate these stakeholders in ethically sensitive practices, raise intergenerational issues surrounding heritable genetic risks, and create complex ethical (and psychological) quandaries about whether, when, and how to inform SCBEM children of their biological origin. Again, these ethical challenges resemble those routinely faced by parents before and after using assisted reproductive technologies.

Such ethical issues can be anticipated given current scientific knowledge, extrapolation from past or present cases, and sheer imagination. But there is always a challenge involved in examining the long-term ethical implications of emerging technologies (Domains a3, b3, c3). Not only is innovation's precise trajectory difficult to predict, but we should also be wary of the social and ethical consequences that scientific and technological advances can bring about, especially when it comes to how they can reshape values, norms, and interpersonal relationships over time. This dynamic interplay between innovation and changing moral frameworks is often referred to as *techno-moral change* [35–38].

The co-emergence of technology and social norms deserves close attention for at least two reasons. First, social norms embraced when a technology is introduced often differ from those that prevail once the technology becomes widely embedded in society. A well-known example is the diffusion of contraceptive pills, which contributed to shifting norms surrounding premarital sex and reproductive autonomy. Second, such normative transitions do not always occur smoothly. New technologies can provoke emotional reactions, public anxiety, or value-based conflicts. As Swierstra and colleagues argue, it is thus crucial to proactively anticipate plausible techno-moral scenarios rather than merely respond to them [35]. This can be done through inclusive, participatory ethical deliberation.

In our case, this means envisioning multiple possible future scenarios vis-à-vis SCBEMs' clinical, reproductive, and/or commercial applications. These could range from broadly permissive environments that normalize clinical use to restrictive frameworks shaped by strong ethical resistance and hybrid models that allow certain applications but not others. Scenario-based foresight, in turn, enables researchers, policymakers, and the public to reflect on both what *could* and what *should* happen. This can embed normative guidance into governance planning.

It is also important to note that techno-moral change is not an impersonal force. It is shaped by the actions of key actors, including scientists, regulators, patient advocacy groups, religious institutions, and the media. These stakeholders influence how SCBEM technologies are framed, perceived, and ultimately governed. Acknowledging this agential and multifaceted process highlights the importance of transparent communication and inclusive dialogue during the early stages of SCBEM development processes.

As intimated, in a future in which SCBEMs are clinically or reproductively viable, public attitudes and ethical evaluations could differ markedly from those held today. It is consequently essential to (i) prepare for a wide range of potential techno-moral trajectories and (ii) maintain governance systems that are both ethically grounded and adaptable to evolving social contexts. Indeed, embedding forward-looking perspectives into governance not only helps to prevent public backlash and regulatory vacuums but also supports responsible and socially responsive innovation in the field [39].

3 The Future Regulation and Governance of SCBEM Research

In this section, we outline directions for governance that support the responsible development of SCBEM research, building on both the ethical issues previously discussed and current regulatory approaches in the field. Given the rapid pace of research, the future trajectory and social implications of SCBEMs remain highly uncertain. To address this uncertainty, governance frameworks should, ideally, be both flexible and adaptive (as intimated in Subheading 2.5).

A key challenge lies in the fact that, under many countries' existing legal frameworks, SCBEMs often fall into regulatory gray zones or might be left without explicit legal definitions altogether. Most national legislation is designed to govern human embryos, which leaves SCBEMs unregulated or, at best, ambiguously categorized. This regulatory gap creates significant barriers to effective oversight. Moreover, policy responses often fail to keep pace with scientific and technological developments, which increases the risk of regulatory vacuums on the one hand or overly restrictive regulations on the other.

At the domestic level, Japan has advanced regulatory deliberations on SCBEM research. In 2024, the Cabinet Office's Expert Panel on Bioethics issued an interim report recommending revisions to existing stem cell research guidelines rather than creating an entirely new framework, thereby helping to avoid unnecessary proliferation of regulatory instruments [40]. However, the current multi-law approach generates conceptual and procedural inconsistencies. These include varying requirements for ethical review and notification (depending on the cell source) and fragmented processes for studies involving multiple cell types. These issues highlight the need for a unified regulatory framework that sets non-negotiable baseline ethical standards for all forms of SCBEM research, while explicitly accommodating cultural, social, and other backgrounds influencing research practices in Japan.

The last point concerns the general issue of how to establish ethical guidelines that reconcile universal ethical principles, which should apply to all societies, with culture-sensitive considerations. In response to this challenge, several governance initiatives and SCBEM-specific codes of conduct have begun to emerge. A notable example is the *Code of Practice for the Generation and Use of Human Stem Cell-Based Embryo Models*, published by Cambridge Reproduction and Progress Educational Trust [30]. Although not legally binding, this code serves as a practical guide for researchers and funders (both inside and outside the UK). Developed by a working group of scientists and legal or ethical experts, it articulates three key principles: (i) compliance with relevant existing regulatory standards, (ii) establishing specialized review committees, and (iii) transparency in public communication.

The Nuffield Council on Bioethics has, in turn, proposed a comprehensive governance framework grounded in an expert review of the ethical and regulatory challenges associated with SCBEM research [12]. Its model seeks to balance the protection of ethical standards and public interest with the need to facilitate scientific progress. As before, this governance framework is structured into three phases:

1. The first phase involves voluntary oversight that incorporates the SCBEM Code developed by Cambridge Reproduction.
2. The second phase proposes revisions to the UK's 1990 Human Fertilisation and Embryology Act. It also introduces "regulatory sandboxes" (controlled environments that allow for experimenting with new regulatory approaches).
3. The third phase develops formal regulatory schemes for SCBEMs on a broader scale.

This phased approach aims to (i) incorporate diverse perspectives and the interests of relevant stakeholders, (ii) promote responsible investment and stable research development, (iii) manage potential risks, and (iv) remain responsive to the uncertainties that are inherent in scientific progress.

Although such regulatory progress is promising, there is also some room to further refine and strengthen SCBEM governance. In particular, there is a crucial need to address gaps between national regulatory frameworks. This can help to foster internationally harmonized regulations, which must take into account the legal traditions and cultural contexts of different countries, including non-Western ones. Indeed, a sensitivity to cultural diversity is especially relevant in this context because the ethical treatment of SCBEMs (similar to that of natural human embryos) might hinge on deeper ethical, social, and spiritual outlooks relating to what constitutes "a life," when life starts (and ends), and what kind of entities comprise our community in the broadest sense [41, p. 4]. Some researchers have emphasized the need to reconsider our universally accepted definition of "embryo" [42, 43].

Moreover, ensuring consistency between national legislation and international governance frameworks (including, but not limited to, guidelines) is essential for the responsible advancement of research. Incorporating public values into ethical and governance discussions surrounding SCBEM research is equally important. As both Cambridge Reproduction's Code and the Nuffield Council's report emphasize, this requires moving beyond expert-led ethical deliberations and incorporating empirical research on public awareness and attitudes. Such an evidence-based approach is necessary to develop governance schemes that more accurately reflect diverse voices and perspectives of citizens from various backgrounds, rather than the hypothetical "citizens" often imagined by experts.

Finally, drawing on the insights related to techno-moral changes discussed in Subheading 2.5, it is essential to anticipate future shifts in societal values and public discourse. Governance systems should be developed with multiple potential futures in mind—even though this is of course a complex task. In this way, they can incorporate anticipatory elements that help to prevent backlash or regulatory gaps. Such foresight should ensure the responsible advancement of SCBEM research while responding to both pertinent scientific goals and evolving societal expectations.

4 Conclusion

We have provided an overview of the ethical issues in SCBEM research. In so doing, we have focused on the different parties that are variously affected by ethical or unethical practices in that field. Although this field has garnered attention for its potential to deepen our understanding of early human development and contribute to medical applications, it also invokes a range of ethical concerns. Some of the key issues are (i) the moral status of SCBEMs, (ii) the interests and autonomy of donors of biological materials, (iii) the broader societal implications of the research, and (iv) the scope of ethically permissible applications. Each of these issues presents complex challenges that demand careful and ongoing examination.

When planning governance frameworks for SCBEM research, several practical challenges need to be addressed. First, regulatory harmonization is not merely a matter of agreeing on broad principles; rather, it requires careful mapping of existing legal systems, identifying points of disagreement, and developing mechanisms for meaningful negotiation. This is particularly important when considering *global* governance in diverse contexts (including non-Western ones) where legal traditions and societal priorities may differ sharply. Second, sound governance should be informed by more than just expert consensus. Public values, attitudes, and concerns must also be incorporated into a systematic way through empirical research (e.g., targeted surveys, deliberative forums, and longitudinal studies tracking opinion over time) to avoid policies that are technically sound but socially disconnected. Finally, effective governance should also be resilient enough to adapt to "techno-moral change," i.e., the shifts in ethical norms and social expectations that often follow major innovations. This calls for policy tools such as periodic review, scenario planning, and policy stress-testing. Their effectiveness should be carefully assessed by empirical findings [44] so that the governance framework evolves with society rather than lags behind it.

Through these adaptive and inclusive approaches, SCBEM research can be guided along a trajectory that supports scientific innovation while safeguarding public interests. This can help to ensure that the field develops in a way that is responsible, ethically sound, and worthy of public trust.

Acknowledgments

TS was funded in whole, or in part, by the Japan Society for the Promotion of Science (JSPS) KAKENHI [Grant Number JP24K00039, JP24H00813], the JST Research Institute of Science and Technology for Society (RISTEX) [Grant Number JPMJRS22J4], the Japan Agency for Medical Research and Development (AMED) [Grant Number JP24wm0425021], and the Uehiro Foundation on Ethics and Education [Grant Number UEHIRO2023-0128].

References

1. Van den Brink SC, Baillie-Johnson P, Balayo T et al (2014) Symmetry breaking, germ layer specification and axial organisation in aggregates of mouse embryonic stem cells. Development 141(22):4231–4242. https://doi.org/10.1242/dev.113001
2. Warmflash A, Sorre B, Etoc F et al (2014) A method to recapitulate early embryonic spatial patterning in human embryonic stem cells. Nat Methods 11(8):847–854. https://doi.org/10.1038/nmeth.3016
3. Baillie-Benson P, Moris N, Martinez Arias A (2020) Pluripotent stem cell models of early mammalian development. Curr Opin Cell Bio 66(1):89–96. https://doi.org/10.1016/j.ceb.2020.05.010
4. Matthews KRW, Wagner DS, Warmflash A (2021) Stem cell-based models of embryos: the need for improved naming conventions. Stem Cell Rep 16(5):1014–1020. https://doi.org/10.1016/j.stemcr.2021.02.018

5. International Society for Stem Cell Research (2021) Guidelines for stem cell research and clinical translation. International Society for Stem Cell Research (ISSCR). https://www.isscr.org/guidelines. Accessed 10 Aug 2025
6. Matthews KRW, Moralí D (2020) National human embryo and embryoid research policies: a survey of 22 top research-intensive countries. Regen Med 15(7):1905–1917. https://doi.org/10.2217/rme-2019-0138
7. Nicolas P, Etoc F, Brivanlou AH (2021) The ethics of human-embryoids model: a call for consistency. J Mol Med 99(4):569–579. https://doi.org/10.1007/s00109-021-02053-7
8. Brivanlou AH, Rivron N, Gleicher N (2021) How will our understanding of human development evolve over the next 10 years? Nat Commun 12(1):4614. https://doi.org/10.1038/s41467-021-24794-2
9. Clark AT, Brivanlou A, Fu J et al (2021) Human embryo research, stem cell-derived embryo models and in vitro gametogenesis: considerations leading to the revised ISSCR guidelines. Stem Cell Rep 16(6):1416–1424. https://doi.org/10.1016/j.stemcr.2021.05.008
10. Landecker HL, Clark AT (2023) Human embryo models made from pluripotent stem cells are not synthetic; they aren't embryos, either. Cell Stem Cell 30(10):1290–1293. https://doi.org/10.1016/j.stem.2023.09.006
11. Monasterio Astobiza A, Molina Pérez A (2022) Why iblastoids (embryo-like structures) do not raise significant ethical issues. Am J Bioeth 22(1):59–61. https://doi.org/10.1080/15265161.2021.2001103
12. Nuffield Council on Bioethics (2024) Stem cell-based embryo models: a review of ethical and governance questions. https://www.nuffieldbioethics.org/publication/human-stem-cell-based-embryo-models-a-review-of-ethical-and-governance-questions/. Accessed 10 Aug 2025
13. Barnhart AJ, Dierickx K (2022) The many moral matters of organoid models: a systematic review of reasons. Med Health Care Philos 25(3):545–560. https://doi.org/10.1007/s11019-022-10082-3
14. Warren MA (1997) Moral status: obligations to persons and other living things. Oxford University Press, Oxford
15. Iltis AS, Koster G, Reeves E et al (2023) Ethical, legal, regulatory, and policy issues concerning embryoids: a systematic review of the literature. Stem Cell Res Ther 14(1):209. https://doi.org/10.1186/s13287-023-03448-8
16. Aach J, Lunshof J, Iyer E et al (2017) Addressing the ethical issues raised by synthetic human entities with embryo-like features. eLife 6:e20674. https://doi.org/10.7554/eLife.20674
17. McCully S (2021) The time has come to extend the 14-day limit. J Med Ethics 47(12):e66–e66. https://doi.org/10.1136/medethics-2020-106406
18. Sawai T, Minakawa T, Pugh J et al (2020) The moral status of human embryo-like structures: potentiality matters? EMBO Rep 21(8):e50984. https://doi.org/10.15252/embr.202050984
19. Stier M, Schoene-Seifert B (2013) The argument from potentiality in the embryo protection debate: finally "depotentialized"? Am J Bioeth 13(1):19–27. https://doi.org/10.1080/15265161.2012.743619
20. Bortolotti L, Harris J (2006) Embryos and eagles: symbolic value in research and reproduction. Camb Q Healthc Ethics 15(1):22–34. https://doi.org/10.1017/S0963180106060038
21. Pereira Daoud AM, Popovic M et al (2020) Modelling human embryogenesis: embryo-like structures spark ethical and policy debate. Hum Reprod Update 26(6):779–798. https://doi.org/10.1093/humupd/dmaa027
22. Neaves W (2017) The status of the human embryo in various religions. Development 144(14):2541–2543. https://doi.org/10.1242/dev.151886
23. Munsie M, Hyun I, Sugarman J (2017) Ethical issues in human organoid and gastruloid research. Development 144(6):942–945. https://doi.org/10.1242/dev.140111
24. Mittleman B (2024) Rethinking general consent for stem cell-based embryo model research. Stan J CR & CL 20(1):181–218. https://doi.org/10.1234/sjcrcl.2024.20.1.181
25. Kataoka M, Gyngell C, Savulescu J et al (2024) The donation of human biological material for brain organoid research: the problems of consciousness and consent. Sci Eng Ethics 30(1):3. https://doi.org/10.1007/s11948-024-00471-7
26. Wei Y, Zhang C, Fan G et al (2021) Organoids as novel models for embryo implantation study. Reprod Sci 28(6):1637–1643. https://doi.org/10.1007/s43032-021-00501-w
27. De Groot H, van Daal M, Hofland RW et al (2025) The ethics and economics of organoid commercialization: potential donors'

perspectives. BMC Med Ethics 26(1):109. https://doi.org/10.1186/s12910-025-01269-3

28. Paşca SP, Arlotta P, Bateup HS et al (2022) A nomenclature consensus for nervous system organoids and assembloids. Nature 609 (7929):7929. https://doi.org/10.1038/s41586-022-05219-6
29. M'hamdi HI (2025) Language and labels from the lab: definitions in the stem cell-based embryo model debate. Stem Cell Rep 20(5): 102477. https://doi.org/10.1016/j.stemcr.2025.102477
30. Cambridge Reproduction (2024) Code of practice for the generation and use of human stem cell-based embryo models. Cambridge Reproduction. https://www.repro.cam.ac.uk/scbemcode. Accessed 10 Aug 2025
31. United Nations (2005) United Nations Declaration on Human Cloning, Pub. L. No. A/C.6/59/L.27/Add.1. https://digitallibrary.un.org/record/541409. Accessed 10 Aug 2025
32. Asplund K (2020) Use of in vitro fertilization: ethical issues. Upsala J Med Sci 125(2):192–199. https://doi.org/10.1080/03009734.2019.1684405
33. Alon I, Cassou M, Golan OC et al (2024) Mapping ethical, legal, and social implications (ELSI) of gamete donation. J Assist Reprod Genet 41(11):2855–2875. https://doi.org/10.1007/s10815-024-03229-z
34. Writing Group on behalf of the ESHRE Ethics Committee, Shenfield F, Tarlatzis B et al (2025) Ethical considerations on surrogacy. Hum Reprod 40(3):420–425. https://doi.org/10.1093/humrep/deaf006
35. Swierstra T, Stemerding D, Boenink M (2009) Exploring techno-moral change: the case of the obesity pill. In: Sollie P, Düwell M (eds) Evaluating new technologies: methodological problems for the ethical assessment of technology developments. Springer, Dordrecht. https://doi.org/10.1007/978-90-481-2229-5_9
36. Nickel PJ, Kudina O, van de Poel I (2022) Moral uncertainty in technomoral change: bridging the explanatory gap. Perspect Sci 30 (2):260–283. https://doi.org/10.1162/posc_a_00414
37. Danaher J, Sætra HS (2023) Mechanisms of techno-moral change: a taxonomy and overview. Ethical Theory Moral Pract 26(5):763–784. https://doi.org/10.1007/s10677-023-10397-x
38. Hopster JKG, Arora C, Blunden C et al (2022) Pistols, pills, pork and ploughs: the structure of technomoral revolutions. Inquiry 68(2):264–296. https://doi.org/10.1080/0020174X.2022.2090434
39. Stilgoe J, Owen R, Macnaghten P (2013) Developing a framework for responsible innovation. Res Policy 42(9):1568–1580. https://doi.org/10.1016/j.respol.2013.05.008
40. Sawai T, Ishida S, Kobayashi C et al (2025) Regulating stem cell–based embryo model research in Japan. EMBO Rep 26(7):1682–1687. https://doi.org/10.1038/s44319-025-00409-5
41. Akabayashi A (2020) Bioethics across the globe: rebirthing bioethics. Asian Bioeth Rev 12(2):99–113. https://doi.org/10.1007/s41649-020-00113-3
42. Rivron NC, Arias AM, Pera MF et al (2023) An ethical framework for human embryology with embryo models. Cell 186(17):3548–3557. https://doi.org/10.1016/j.cell.2023.07.028
43. De Miguel Beriain I, Rueda J, Villalba A (2024) Re-defining the human embryo. EMBO Rep 25(2):467–470. https://doi.org/10.1038/s44319-023-00034-0
44. Sharma-Wallace L, Velarde SJ, Wreford A (2018) Adaptive governance good practice: show me the evidence! J Environ Manag 222: 174–184. https://doi.org/10.1016/j.jenvman.2018.05.067

Methods in Molecular Biology (2026) 3048: 17–32
DOI 10.1007/7651_2025_647

Published online: 22 May 2025

From Single Stem Cells to an In Vitro Model of the Post-implantation Human Embryo: A Step-by-Step Guide

Seher Ipek Gassaloglu, Monique Pedroza, and Berna Sozen

Abstract

Recent advances in three-dimensional (3D) modeling of post-implantation human embryos using human pluripotent stem cells (hPSCs) have revolutionized our ability to investigate this crucial yet enigmatic stage of development. Here we detail the generation of *the human extra-embryoid* (hEE), a 3D stem cell-based embryo model that uniquely captures key spatiotemporal events of peri-gastrulation development through the formation and co-development of post-implantation embryonic and extra-embryonic lineages, with high efficiency and robustness across genetic backgrounds. This chapter provides a detailed protocol for generating hEEs in vitro, including guidance on hPSC maintenance, expected cell morphology, troubleshooting strategies, and key culture techniques.

Key words 3D culture, Differentiation, Embryo model, Human embryo, Human embryonic stem cells, Human extra-embryoid, Pluripotency, Post-implantation, Stem cell

1 Introduction

Investigating the origins and early differentiation events of our species remains one of the most challenging and obscure areas of research, primarily due to the technical and ethical barriers associated with human embryo studies. Specifically, the stages following human embryo implantation into the uterus (Carnegie stages [CS] 4-7) remains extremely poorly understood. This period is critical, as it encompasses major morphological and differentiation events that establish the body axes and lay foundation for subsequent development. The post-implantation embryo comprises embryonic lineages that will give rise to the embryo proper and extra-embryonic lineages that support patterning and nutrient supply necessary for continued development. Recent advances have

Seher Ipek Gassaloglu and Monique Pedroza contributed equally with all other contributors.

led to the development of various stem cell-based platforms designed to model aspects of primate post-implantation stages [1–16]. We demonstrated that human pluripotent stem cells (hPSCs) with a distinct pluripotency state could be triggered to self-organize into 3D models that closely resemble human embryos at 2- to 3-week old. This platform, termed ***human extra-embryoid (hEE)***, captures both embryonic and extra-embryonic cell states, including epiblast-, amnion-, mesoderm-, primitive streak-, and hypoblast-like lineages, and their co-development [6]. The hEE system is a tractable, high-efficiency platform, enabling perturbation experiments to understand fundamental questions at this otherwise inaccessible stage of human development. In addition, generation of hEEs does not rely on genetically inducible factors, further enhancing its simplicity and reproducibility.

Here, we elaborate on a robust and efficient protocol for the generation of hEEs from hPSCs in vitro. The intrinsic differentiation potential and adaptability of hPSCs make them ideal building blocks for stem cell-based modeling. For hEE generation, we primarily utilize hPSCs maintained under specific conditions including either human extended potential (EP) [17] or partial capacitation [18] or RseT [19] that support an intermediate pluripotency state between naïve and primed pluripotency. This intermediate state allows to facilitate differentiation into embryonic and extra-embryonic lineages [17]. These hPSCs are aggregated in 3D culture with an exposure to a sequential 2-step media composition. Initially, the cells are cultured in spontaneous differentiation media (SDM), a formulation we optimized that includes basic nutritional supplements without added growth factors, promoting spontaneous differentiation of the epiblast- and hypoblast-like lineages in 3D culture. After 72 h, the SDM is replaced with a modified in vitro culture medium 2 (mIVC2), previously demonstrated for culturing post-implantation human embryos [20] (Fig. 1). By 120 h, the hEE forms distinct compartments: an inner embryonic epiblast-like compartment with a central cavity surrounded by outer extra-embryonic multilayered hypoblast-like compartment. These structures are accompanied by the concurrent differentiation of key lineages, including the amnion-like and anterior hypoblast-like cell types, whose critical interactions with epiblast-like lineage orchestrate the complex process of embryonic symmetry breaking, closely mimicking key features of early post-implantation human development [6] (Fig. 2). The hEE system offers a readily accessible and scalable 3D model of the early post-implantation human embryo, allowing for precise genetic and chemical manipulations to investigate early pregnancy loss and developmental abnormalities.

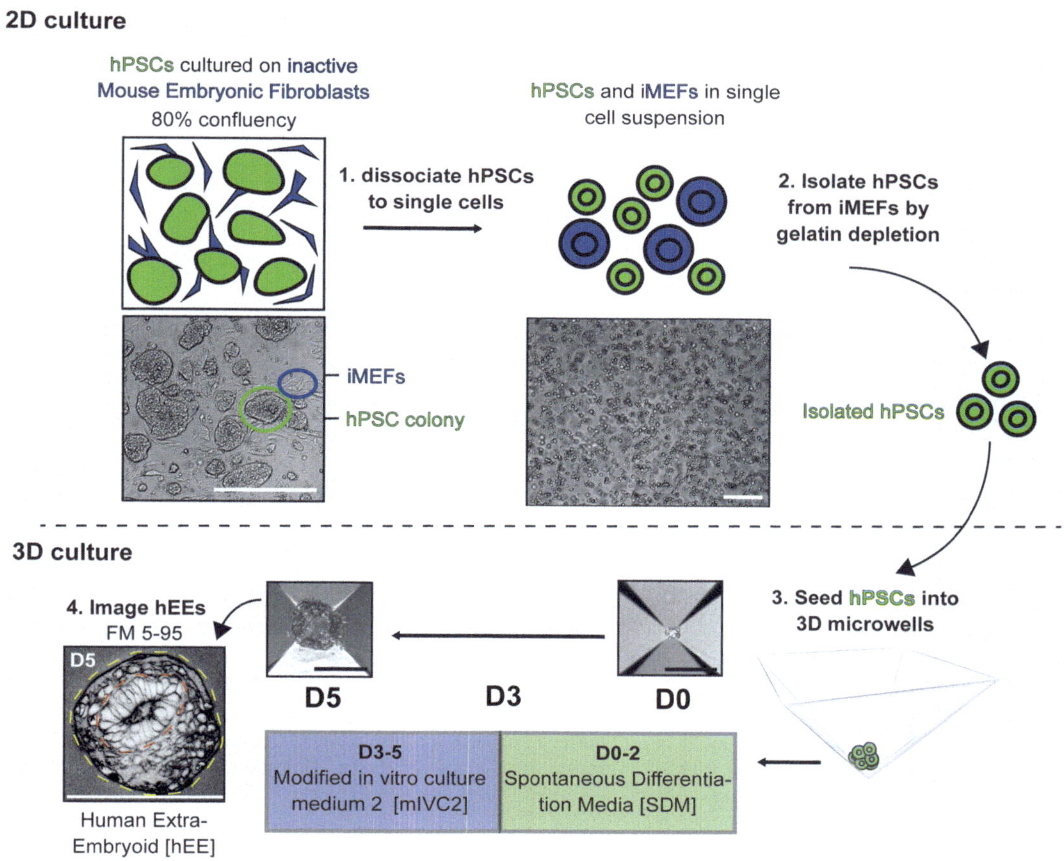

Fig. 1 Schematic of human extra-embryoid generation. *Top: 2D culture.* **Step 1**: Human pluripotent stem cells (hPSCs) are cultured on inactive mouse embryonic fibroblasts (iMEFs) until reaching approximately 80% confluency. **Step 2**: hPSCs are isolated from iMEFs by gelatin depletion. Green cells: hPSCs, blue cells: iMEFs. *Bottom: 3D culture.* **Step 3**: Depending on cell type (5–6 cells per microwell [6000–7200 cells total] for extended potential (EP) hPSCs; 6–10 cells per microwell [7200–12,000 cells total] for RSeT; and 6–10 cells per microwell [7200–12,000 cells total] for partially capacitated naïve hPSCs), single cells were seeded to AggreWell 400 format and cultured with spontaneous differentiation media (SDM) + ROCK inhibitor on Day 0, SDM without ROCK inhibitor on Days 1–2, and modified in vitro culture medium 2 (mIVC2) on Days 3–5 or 6. **Step 4**: On Day 5 or 6, hEEs are prepared for imaging. FM 5–95 membrane staining: inverted LUT in Fiji. Orange and yellow dashed lines indicate the epiblast-like and hypoblast-like compartments, respectively. Scale bars are 200 μM

2 Materials

2.1 Human Pluripotent Stem Cell Maintenance and Cryopreservation

1. 12-Well sterile, culture plate.
2. 0.1% Gelatin solution.
3. Human Extended Potential Stem Cell—base medium (EP-base) (*see alternatives in* Subheading 2.2): 1:1 mixture of DMEM/F12 and neurobasal supplemented with 0.5% N2 supplement, 1% B27 without vitamin A supplement, 1% 100X

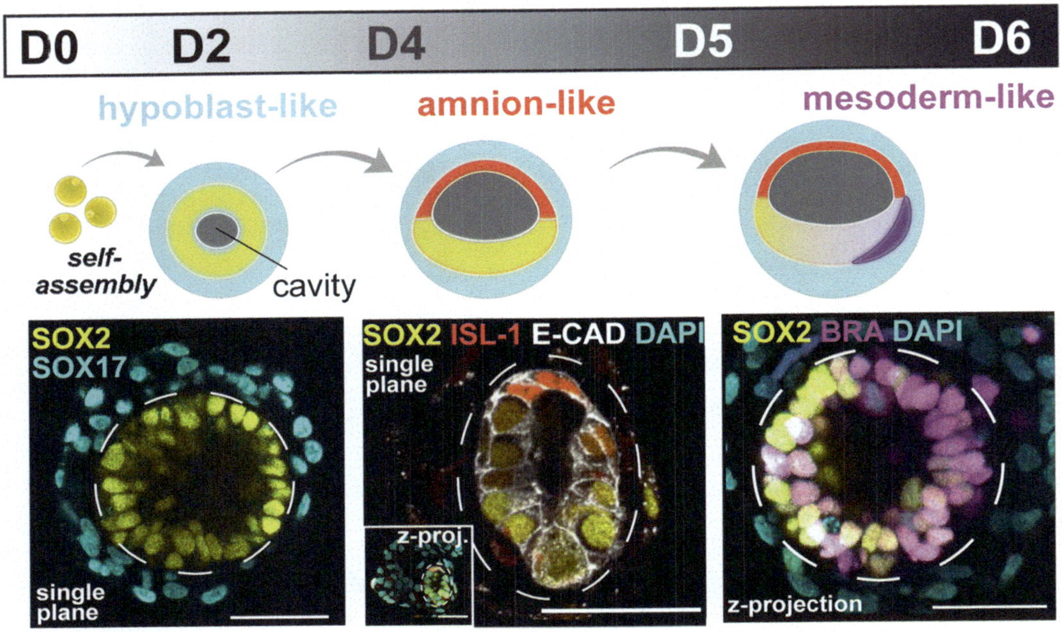

Fig. 2 Morphological features and key cell types present in the human extra-embryoid at a spatial-temporal resolution. Yellow: SOX2+ (epiblast-like), blue: SOX17+ (hypoblast-like), red: ISL1 (amnion-like), pink: BRA (mesoderm-like). *The inset middle panel*: maximum intensity projection (MIP) of z-stacks 18–23 out of 1–48. Z-proj., z-projection. *Right panel*: MIP of z-stacks 18–22 out of 1–34. Dotted line indicates the inner compartment. Scale bars are 50 μM

GlutaMAX, 1% 100X MEM nonessential amino acids, 0.2% 50 mM β-mercaptoethanol, 1% 10,000 U mL−1 penicillin–streptomycin, and 5% knockout serum replacement (KSR). Store at 4 °C.

4. Human Extended Potential—complete medium (EP-complete medium): EP-base medium supplemented with 10 ng/mL recombinant human LIF (hLIF), 1 μM CHIR 99021, 2 μM (S)-(+)-dimethindene maleate, 2 μM minocycline hydrochloride, 1 μM WNT inhibitor IWR-1, and 2 μM ROCK inhibitor (Y27632). Store at 4 °C. Prepare 10 mL fresh aliquots.
5. Gibco StemPro Accutase Cell Dissociation Reagent.
6. Thermo Scientific™ Mr. Frosty™ Freezing Container.
7. NutriFreez® D10 Cryopreservation Medium (freezing medium).
8. Feeder cell (FC) medium: DMEM (high glucose, pyruvate) supplemented with 18% inactivated FBS supplemented with 1.2% 10,000 U mL−1 penicillin-streptomycin, 1.2% 100X Glutamax, 1.2% MEM nonessential amino acids, 1.2% 100 mM sodium pyruvate, and 0.24% 50 mM 2-mercaptoethanol.

2.2 Alternative Starting hPSCs for hEE Generation: RSeT or Partial Capacitation of hPSCs

1. RSeT™ Medium (2-component; Basal and 5× supplement) medium is commercially available from STEMCELL Technologies. Prepare and store as per the manufacturer's instructions.
2. Geltrex™ LDEV-Free Reduced Growth Factor Basement Membrane Matrix (for partial capacitation of naïve hPSCs).
3. N2B27-based medium: 1:1 mixture of DMEM/F12 and neurobasal supplemented with 1% B27, 0.5% N2, 1% 100X Glutamax, 0.5% 10,000 U mL−1 penicillin–streptomycin and 0.1 mM β-mercaptoethanol (for partial capacitation of naïve hPSCs).
4. PXGL medium: N2B27-based medium supplemented with 10 ng/mL human LIF, 2 μM Gö6983, 1 μM PD0325901, and 2 μM XAV939 (for maintenance of naïve hPSCs).

2.3 Mouse Embryonic Fibroblast (MEF)-Conditioned Medium Preparation

1. Feeder cell (FC) medium: DMEM (high glucose, pyruvate) supplemented with 18% inactivated FBS supplemented with 1.2% 10,000 U mL−1 penicillin-streptomycin, 1.2% 100X Glutamax, 1.2% MEM nonessential amino acids, 1.2% 100 mM sodium pyruvate, and 0.24% 50 mM 2-mercaptoethanol.
2. Commercially available Gibco™ CF1 mitotically inactivated Mouse Embryonic Fibroblasts (iMEFs), irradiated.
3. 100 × 15 mm Petri dish.

2.4 Human Extra-Embryoid Generation

1. MEF-conditioned medium: two million mitotically inactivated CF-1 iMEFs were seeded onto a 100 × 15 mm Petri dish and cultured in FC medium. 7–10 mL of the 100% MEF-conditioned medium was collected every 3 days. A maximum of three batches were collected per dish. Store at −20 °C.
2. Spontaneous differentiation medium (SDM): 50% of hEP-base medium* and 50% iMEF-conditioned medium (*see* **Note 1**). *Depending on the alternative starting hPSCs used for hEE generation, the following SDM is prepared:
   (a) Alternative 1: 50% of RSeT™ Basal Medium and 50% iMEF-conditioned medium.
   (b) Alternative 2 (using partially capacitated naïve hPSCs): 50% of N2B27-based media and 50% iMEF-conditioned medium (*see* **Note 1**).
3. Modified In Vitro Culture Medium 2 (mIVC2) [20]: Advanced DMEM/F12 supplemented with 1× ITS-X, 2 mM l-glutamine, 30% (v/v) KSR, 1 mM sodium pyruvate, 8 nM β-estradiol, 25 μM N-acetyl-l-cysteine, 0.22% (v/v) sodium lactate, 200 ng mL^{-1} progesterone.
4. AggreWell 400 format 24-well plates.
5. Anti-Adherence Rinsing Solution.
6. Phosphate-Buffered Saline (PBS) 1×.

2.5 Immunostaining

1. 4% Paraformaldehyde.
2. PBST: PBS supplemented with 0.05% Tween-20. Store at room temperature.
3. Permeabilization buffer (PB): PBS supplemented with 0.1 M glycine and 0.3% Triton X-100. Prepare fresh.
4. Blocking buffer: PBS supplemented with 10% filtered FBS and 0.1% Tween-20. Prepare fresh and store at 4 °C.
5. PBS-BSA: PBS supplemented with 1% of 7.5% BSA. Prepare fresh and store at 4 °C.
6. Mineral oil.
7. Glass-bottom microwell dishes (MatTek Corporation).

3 Methods

3.1 Preparation of Inactive Mouse Embryonic Fibroblasts

1. Coat a 12-well sterile, culture plate with 500 μL/well of 0.1% gelatin solution.
2. Leave gelatin-coated plate in 37 °C, 5% CO_2, 20% O_2 incubator overnight (*see* **Note 2**).
3. Once gelatin-coated plates are ready for use, pre-warm FC medium in the 37 °C water bath for about 5 min or until warm. Prepare a 15 mL Falcon tube with 5 mL of FC medium.
4. Thaw a vial of two million mitotically inactive CF-1 iMEFs (for two 12-well plates; one million iMEFs per 12 well plate) at 37 °C in the water bath for about 1 min or until half the vial is thawed (*see* **Note 3**).
5. To thaw the rest of the vial, add 1 mL of pre-warmed FC medium directly into the vial and gently pipette up and down until the remaining vial is completely thawed. In an empty Falcon tube, collect all the media containing iMEF cells from the vial (*see* **Note 4**).
6. Centrifuge the cells at 1000 RPM (0.2 RCF) for 3 min.
7. Aspirate the supernatant from the 15 mL Falcon tube containing iMEF cells and resuspend the cell pellet to about 83,333 cells per mL per well (*see* **Note 5**).
8. Pipette up and down to make the media homogenous and add 1 mL of cell suspension to each gelatin-coated well in a 12-well plate (*see* **Note 6**).
9. Place the plate in 37 °C, 5% CO_2, 20% O_2 incubator and gently rock plate side to side to ensure cells are evenly distributed. Leave in incubator overnight.

3.2 Preparation of Inactive Mouse Embryonic Fibroblasts-Conditioned Medium (see Note 7)

1. Coat a sterile, 15 mm Petri dish with 4 mL of 0.1% gelatin solution.

   Same as **steps 2–6** *in* Subheading 3.1 *above*.
2. Aspirate the supernatant from the 15 mL Falcon tube containing two million mitotically inactivated CF-1 MEFs (Gibco) and resuspend the cell pellet in 10 mL of FC medium (200,000 cells per mL concentration). Two million iMEFs are seeded to a 15 mm dish (*see* **Note 8**).
3. Pipette up and down to make the medium homogenous. Aspirate the 0.1% gelatin from the Petri dish and add up all 10 mL of cell suspension to the 15 mm gelatin-coated dish (*see* **Note 9**).
4. Place the plate in 37 °C, 5% CO_2, 20% O_2 incubator and gently rock dish side to side to ensure cells are evenly distributed. Leave in incubator overnight.
5. Collect all medium (7–10 mL) after 3 days of incubation with iMEFs and either use immediately or place in −20 for later use (*see* **Note 10**).

3.3 Seeding Human Pluripotent Stem Cells

1. Prepare a 12-well sterile, culture plate with iMEFs at least one day before.
2. On the day of seeding human cells, prepare EP-complete medium (or RSeT or PXGL medium, see in Subheading 2.2) (*see* **Note 11**).
3. Pre-warm EP-complete medium (or RSeT or PXGL medium) and FC medium at 37 °C in the water bath for about 5 min or until warm. Prepare a 15 mL Falcon tube with 5 mL of FC medium.
4. Thaw a vial of EP hPSCs (or RSeT or naïve hPSCs for partial capacitation. See in Subheading 3.6) at 37 °C in the water bath for about 1 min or until half the vial is thawed.
5. To thaw the rest of the vial, add 1 mL of pre-warmed FC medium directly into the vial and gently pipette up and down until the remaining vial is completely thawed. In an empty Falcon tube, collect all the media containing EP hPSCs (or RSeT or naïve hPSCs for partial capacitation. See in Subheading 3.6) from the vial.
6. Centrifuge the cells at 1000 RPM (0.2 RCF) for 3 min.
7. Aspirate the supernatant from the 15 mL Falcon tube containing hPSCs and resuspend the cell pellet in 1 mL of EP-complete medium (or RSeT or PXGL medium).
8. Gently pipette up and down to make the medium homogeneous. Aspirate the FC medium in an iMEF coated well and

add 1 mL of cell suspension to one well in 12wp iMEF coated plate (*see* **Note 12**).

9. Place the plate with EP hPSCs in 37 °C, 5% CO_2, 20% O_2 incubator and gently rock the plate side to side to ensure cells are evenly distributed. RseT or naïve hPSCs are placed in hypoxic conditions: 37 °C, 5% CO_2, 5% O_2. Incubate overnight and check for cell attachment after 24 h.

3.4 Cryopreserving Human Pluripotent Stem Cells

1. Prepare freezing medium and pre-warm FC medium and EP-complete medium (or RSeT or N2B27-based medium. See in Subheading 3.6) to 37 °C in the water bath and Accutase at room temperature. Label cryovials with cell type, passage number, media, and date (*see* **Note 13**).
2. Aspirate the existing media and dissociate EP hPSCs (or RSeT or partially capacitated hPSCs, *see in* Subheading 3.6) with 500 μL of Accutase for at least 3 min at 37 °C in the incubator. Do not exceed 8 min total (*see* **Note 14**).
3. Once the majority of EP hPSC (or RSeT or partially capacitated) colonies are dislodged from the bottom of the plate, neutralize the reaction with 1 mL of FC medium into the well of EP hPSCs (or RSeT or partially capacitated hPSCs). Using the same medium within the well, pipette up and down to detach the colonies from the plate and transfer medium containing cells into an empty 15 mL Falcon tube (*see* **Note 15**).
4. Centrifuge the cells at 1000 RPM (0.2 RCF) for 3 min.
5. Aspirate the supernatant from the 15 mL Falcon tube containing EP hPSCs (or RSeT or partially capacitated hPSCs) and resuspend the cell pellet in 1 mL of freezing medium.
6. Gently pipette up and down and add into pre-labeled cryovial.
7. Place vial in Mr. Frosty in −80 overnight.
8. Transfer vial from −80 to liquid nitrogen storage.

3.5 Maintenance of Human Pluripotent Stem Cells (see Note 16)

1. If cells are cultured in an iMEF-coated 12-well plate, aspirate the used media and replace it with 1 mL of pre-warmed EP-complete medium (or RseT medium) per well. *See* Subheading 3.6 for the maintenance of naïve hPSCs for partial capacitation.
2. Observe the growth of the colonies under the microscope. Once dome-shaped colonies reach approximately 80% confluency, passage the cells to avoid overcrowding (Fig. 3) (*see* **Note 17**).
3. To passage the cells, follow **steps 2–4** of the Subheading 3.4 to dissociate them.

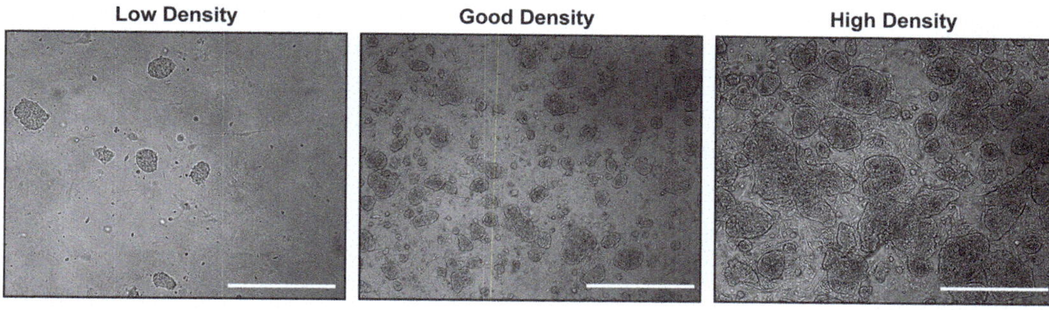

Fig. 3 Optimal human pluripotent stem cell density for generating human extra-embryoids (hEEs). Low and high cell densities are suboptimal. Ensure colony sizes are uniform and an optimal density (middle panel, ~80 efficiency) before starting the protocol for successful hEE formation. Scale bars are 400 μM

4. After centrifugation, aspirate the supernatant from the 15 mL Falcon tube containing EP hPSCs (or RseT) and resuspend the cell pellet in 1 mL EP-complete medium (or RseT medium).
5. Passaging ratio: Adjust the ratio of passaging based on the confluency of hPSCs:
   (a) *>80% confluency: 1:20 ratio.*
   (b) *~80% confluency: 1:10 ratio.*
   (c) *<80% confluency: 1:5 or 1:8 ratio.*
6. Calculate the cell concentration required to make up 1 mL of EP-complete medium (or RseT medium) per well.
7. Aspirate the FC medium from a new well and add the EP-complete medium (or RseT medium) containing the cells into the MEF-coated well.
8. Gently shake the plate to ensure even distribution.
9. Incubate the plate at 37 °C, 5% CO_2, 20% O_2 for EP condition and 37 °C, 5% CO_2, 5% O_2 for RseT condition.

3.6 Alternative: Partial Capacitation of Naïve hPSCs

Partial capacitation of naïve hPSCs and the required medium conditions were performed following a previously published protocol with slight modifications [18].

1. To seed naïve hPSCs, follow **steps 1–8** of the Subheading 3.3.
2. Place the plate in 37 °C, 5% CO_2, 5% O_2 incubator and gently rock it side to side to ensure even cell distribution.
3. On Day 0, supplement the PXGL medium with 10 μM ROCK inhibitor. The following morning, replace the medium with fresh PXGL medium lacking the ROCK inhibitor. Change the medium daily.
4. Once the cells reach approximately 80% confluency, proceed to the next steps for capacitation preparation.

5. Prepare a gelatin-coated plate to deplete the iMEFs by adding 500 μL of 0.1% gelatin solution to a well of a 12-well plate and incubate at 37 °C for 20 min.
6. After 20 min, passage the cells by following **steps 2–4** of the Subheading 3.4 and resuspend the cells in 1 mL of PXGL medium.
7. Aspirate the gelatin solution from the coated plate and seed the naïve hPSCs in PXGL medium onto the gelatin-coated plate.
8. Incubate the plate for 40 min at 37 °C, 5% CO_2, 5% O_2 to allow iMEFs to adhere.
9. Prepare Geltrex-coated plates by coating one well of a 12-well plate with 25 μL Geltrex diluted in 750 μL of DMEM (1:30 dilution).
10. Incubate the plate for 30 min at room temperature followed by 1 h at 37 °C.
11. After the gelatin depletion step, collect the naïve hPSCs and resuspended in PXGL medium supplemented with 10 μM ROCK inhibitor (Y27632).
12. Seed the cells onto Geltrex-coated plates at a density of 1.6×10^4 cells/cm^2.
13. After 24 h, aspirate the medium and replace it with fresh PXGL medium without ROCK inhibitor.
14. The following day, wash the naïve hPSCs cells once with DMEM/F12 supplemented with 0.1% BSA.
15. At this step, the capacitation begins by replacing the PXGL medium with N2B27-based medium supplemented with the WNT inhibitor 2 μM XAV939. Change the medium daily.
16. Collect the cells on day 5 when they exhibit characteristics of being partially pluripotent and partially differentiated, suitable for generating human extra–embryoids (*see* **Note 18**).

3.7 Generation of Human Extra-Embryoids (hEE) in 3D (see Note 19)

3.7.1 Preparing the AggreWell Plate

1. Add 500 μL of anti-adherence solution to the experimental wells of the AggreWell plate. Check under the microscope for bubble formation (*see* **Note 20**).
2. Place the AggreWell plate and a balance plate of equal weight in the centrifuge. Spin down at 3600 RPM (2000 RCF) for 5 min.
3. Recheck the wells under the microscope to ensure no bubbles remain.
4. Incubate the experiment plate at room temperature for 30 min.
5. Aspirate the anti-adherence solution and gently wash the wells with 600 μL of PBS, pipetting along the side of the well to avoid bubble formation.

6. Aspirate the PBS, then fill any unused wells with PBS (*see* **Note 21**).
7. Prepare and add 500 μL of SDM supplemented with ROCK inhibitor (*see* **Note 1**) to the experimental wells. Check under the microscope to ensure no bubbles are present.
8. Incubate the plate at 37 °C, 5% CO_2, 5% O_2 until the cells are ready to be added. This step equilibrates the medium to the incubator's temperature and pH.

3.7.2 Isolation of hPSCs from iMEFs by Gelatin Depletion

1. While the anti-adherence solution incubates in the AggreWell plate, prepare for iMEF depletion.
2. Add 500 μL of 0.1% gelatin solution to a 12-well plate and incubate at 37 °C for 20 min (*see* **Note 22**).
3. Follow **steps 2–4** of Subheading 3.4 to dissociate the cells.
4. After centrifugation, aspirate the supernatant from the 15 mL Falcon tube containing EP hPSCs (or RseT or partially capacitated hPSCs) and resuspend the cell pellet in EP-complete media (or RseT or N2B27-based medium supplemented with the WNT inhibitor 2 μM XAV939; 1 mL per coated well).
5. Aspirate the gelatin solution from the coated plate and add 1 mL of EP hPSCs in EP-complete medium (or RseT or N2B27-based medium supplemented with the WNT inhibitor 2 μM XAV939) per well.
6. Incubate for 30 min at 37 °C.
7. After 30 min, check under the microscope to confirm that iMEFs have adhered. If attachment is incomplete, incubate for an additional 10 min (*see* **Note 23**).

3.7.3 AggreWell Plating (Day 0)

1. Collect EP hPSCs (or RseT or partially capacitated hPSCs) from the gelatin-coated wells and transfer to 15 mL Falcon tubes.
2. Centrifuge the cells at 1000 RPM (0.2 RCF) for 3 min.
3. Aspirate the supernatant and resuspend the cell pellet in 1 mL of SDM supplemented with 5 μM ROCK inhibitor (*see* **Note 1**) by gently pipetting.
4. Count the cells using a cell counter or hemocytometer, ensuring 6000–7200 cells total [5–6 cells per microwell] for hPSCs in extended potential condition; 7200–12000 cells total [6–10 cells per microwell] for RseT; and 7200–12000 cells total [6–10 cells per microwell] for partially capacitated hPSCs (*see* **Notes 24** and **25**).
5. Add 500 μL of SDM-containing cells to each well of the pre-equilibrated AggreWell plate. Dispense the 500 μL gradually, drop by drop, around the well to ensure even distribution.

6. Check under the microscope to confirm only a few cells are present in each microwell.
7. Incubate until the next day at 37 °C in 5% O2 and 5% CO2 (*see* **Note 26**).

3.7.4 Experimental Time Course

1. From day 0 to day 1, SDM with 5 μM ROCK inhibitor is used. On day 1, check the AggreWell plate under the microscope for the formation of small aggregates (Fig. 1).
2. From day 1 to day 6, change the medium daily (*see* **Note 27**):
   (a) Day 1–2: SDM without ROCK inhibitor.
   (b) Day 3–6 mIVC2 medium.
3. To change the medium, aspirate 800 μL of used medium from the side of the well, starting at the surface and slowly moving downward. Avoid disturbing the microwell bottom.
4. Add 1 mL of pre-warmed, freshly prepared medium (SDM or mIVC2) along the side of the well without touching the microwells.
5. Collect the structures on the desired day for immunostaining or other assays.

3.8 Collection of Human Extra-Embryoids for Immunostaining

1. To collect the hEEs, set a P1000 pipette to 1000 μL and gently pipette up and down to float the structures into the medium.
2. Transfer the medium containing the structures to a 15 mL Falcon tube.
3. Check the well under the microscope to ensure all the structures have been collected. If not, wash the well with 1 mL of SDM or mIVC2 media (depending on day of collection) and transfer the wash to the same Falcon tube.
4. Loosen the lid of the Falcon tube and incubate the structures at 37 °C for 5 min before proceeding with downstream assays.
5. For immunostaining, gently aspirate the supernatant with a P1000 pipette and add 1 mL of 4% paraformaldehyde (PFA) to the Falcon tube.
6. Incubate at room temperature for 20 min.
7. After incubation, centrifuge the Falcon tube at 1000 RPM (0.2 RCF) for 1 min to pellet the structures.
8. Carefully aspirate the supernatant without disturbing the pellet.
9. Wash the pellet with 1 mL of PBST and repeat **steps 7–8**.
10. Add 1 mL of permeabilization buffer (PB) and incubate at room temperature for 20 min.
11. Repeat **steps 7–9**.

12. Add 100 μL of primary antibodies prepared in blocking solution and incubate at 4 °C overnight.
13. The next day, after repeating **steps 7–9**, add 100 μL of secondary antibodies prepared in blocking solution:
    (a) For same-day imaging: Incubate at room temperature for 2 h.
    (b) For next-day imaging: Incubate at 4 °C overnight.
14. After incubation, repeat **steps 7–9**, add 200 μL of PBS-BSA, and centrifuge at 1000 RPM (0.2 RCF) for 1 min to pellet the structures.

3.9 Imaging Plate Preparation

1. Transfer the hEEs to a 35 × 10 mm Petri dish by pipetting a 200 μL droplet into the center of the plate (Fig. 4, **step 1**).
2. Using a P10 pipette set to 10 μL, carefully collect hEEs from the 200 μL droplet and dispense a single droplet containing hEEs on a MATTEK glass-bottom dish (Fig. 4, **steps 2** and **3**).

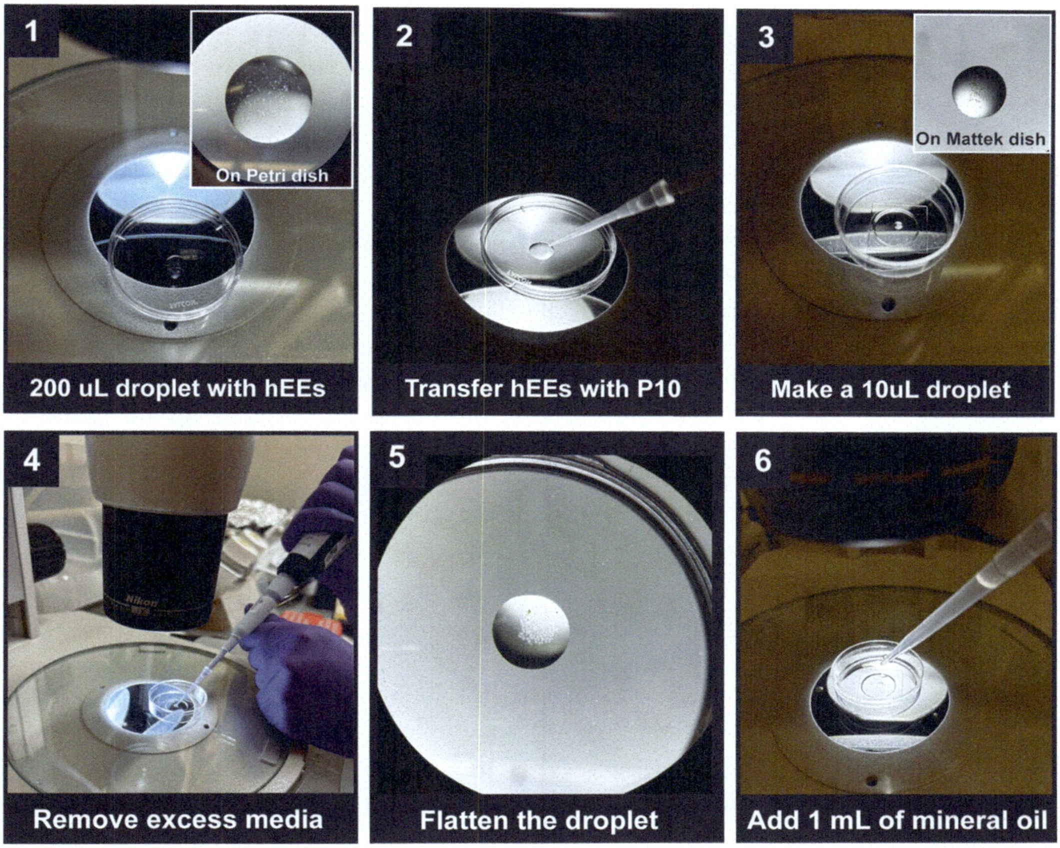

Fig. 4 Imaging Plate Preparation. **Step 1**: Transfer hEES to a Petri dish in a 200 μL droplet. **Steps 2–3**: Collect hEEs with P10 pipette and dispense them into a single 10 uL droplet on a MATTEK glass-bottom dish. **Steps 4–5**: Flatten the droplet by gently removing excess media. **Step 6**: Cover the droplet with 1 mL of mineral oil

3. Gently remove excess media from the droplet to flatten it. Use the P10 pipette to slowly aspirate media from the top of the droplet without disturbing the hEEs. If additional hEEs remain in the 200 μL droplet, repeat **steps 2** and **3** to add more hEEs to the same 10 μL on the MATTEK glass-bottom dish (Fig. 3, **steps 4** and **5**) (*see* **Note 28**).
4. Add 1 mL of mineral oil to the dish, ensuring the droplet is completely covered (Fig. 4, **step 6**).
5. Place the dish in a dark container and proceed with imaging.

4 Notes

1. On day 0 of extra-embryoid generation, SDM medium is supplemented with 5 μM of ROCK inhibitor (Y27632) for the first 24 h of aggregation. If the cell line used is sensitive, the concentration can be increased to 10 μM during this period.
2. Gelatin-coated plates must be prepared one day before seeding iMEFs. Do not store gelatin-coated plates for later use.
3. Hold vial and partially submerge vial in water bath and shake until about half the vial is thawed.
4. Rinse vial with additional 1 mL of pre-warmed FC medium to collect any remaining cells in vial and add to Falcon tube containing all the iMEF cells.
5. Before freezing, viable iMEF cells are counted and labeled on vial. Use the number of iMEFs in vial to calculate concentration of iMEFs for cell pellet resuspension. About one million iMEF cells are plated per 12 well plate for hPSCs (about 83 k iMEFs per well in a 12wp).
6. Before seeding cells onto each well, pipette up and down to ensure cell suspension is homogeneous. Once cells are added to wells, check under the microscope to ensure that each well receives about the same amount of iMEFs.
7. Similar to Subheading 3.1 above except iMEFs were seeded onto a gelatin-coated 100 × 15 mm Petri dish.
8. Gibco CF-1 iMEFs (*see* Subheading 2) is recommended to be used for preparation of MEF-conditioned FC medium. In our hands, we obtained the highest efficiency of hEE generation with MEF-conditioned FC medium from Gibco CF-1 iMEFs.
9. 7–10 mL of FC medium can be used to resuspend two million iMEFs and added to Petri dish. This is considered one batch and the 7–10 mL of medium is what will be conditioned.
10. A maximum of three batches of iMEF-conditioned FC medium were collected per dish.

11. hPSCs in EP conditions are recommended as starting cell population for highest efficiency of hEE generation. Alternative options are RseT or partially capacitated conditions.
12. Ensure that the attached iMEFs in the well of the 12wp are sufficiently confluent, as inadequate density can negatively impact the growth of the hPSC colonies.
13. Once the freezing medium is prepared, keep at 4 °C until **step 5**.
14. While looking under the microscope, shake the plate with hPSCs to check if the majority of the colonies have detached from the plate.
15. Tilt the plate at approximately a 45° angle while pipetting up and down along the rim of the well to ensure all cells have detached from the plate.
16. Depending on the genetic background of the hPSCs, they can be maintained for up to passage 70. During maintenance, replace the hPSC medium daily.
17. hPSCs typically reach confluency within 4–5 days, but this duration may vary depending on the cell line's genetic background.
18. The highest potential for hEE generation from partially capacitated hPSCs was observed on day 5 of the naive-to-primed (intermediate) pluripotency transition. Beyond day 5, this ability rapidly declined, similar to our observation with conventional primed pluripotent hPSCs.
19. Ensure that hPSCs have reached approximately 80% confluency before starting the aggregation protocol.
20. Anti-adherence solution prevents cell attachment to the wells, promoting proper aggregate formation in suspension.
21. Ensure no well is left empty. Filling unused wells maintains consistent humidity across the plate.
22. The number of wells to coat depends on the confluency of hPSCs (EP hPSCs are recommended for highest efficiency of hEE generation. Alternative options: RseT or partially capacitated hPSCs) (Fig. 3):
    (a) 60–80% confluency: 1 gelatin-coated well per starting well.
    (b) ~80% confluency: 2 gelatin-coated wells per starting well.
    (c) >80% confluency: recommended not to use if cells are over-crowded.
23. Total incubation time for gelatin-depletion step should not exceed 40 min to prevent the loss and attachment of hPSCs along with iMEFs as iMEFS adhere faster than hPSCs.
24. AggreWell 400 format contains 1200 microwells per well. 5 cells × 1200 microwells = 6000 cells in total.

25. The total volume in the AggreWell should add up to 1 mL. Since 500 μL was previously added for temperature equilibration, calculations must be done by ~6000 cells/500 μL per well.
26. Total experimental time for hPSC seeding into Aggrewells should not exceed 1.5–2 h. Optimum results are achieved by minimizing the time hPSCs spend outside the incubator.
27. Daily media changes should not take longer than 5 min, as exposure to varying temperature and oxygen levels can affect the cell aggregation. To minimize this risk, divide your experiments across separate plates and avoid using all the wells in the Aggrewell plate at a time. Change media every 24 hours.
28. Ensure the droplets are not overly domed, as this can cause the hEEs to float away after the addition of the mineral oil.

References

1. Ai Z et al (2023) Dissecting peri-implantation development using cultured human embryos and embryo-like assembloids. Cell Res 33:661–678
2. Oldak B et al (2023) Complete human day 14 post-implantation embryo models from naive ES cells. Nature. https://doi.org/10.1038/s41586-023-06604-5
3. Hislop J et al (2024) Modelling post-implantation human development to yolk sac blood emergence. Nature 626:367–376
4. Zheng Y et al (2019) Controlled modelling of human epiblast and amnion development using stem cells. Nature 573:421–425
5. Karvas RM et al (2023) 3D-cultured blastoids model human embryogenesis from pre-implantation to early gastrulation stages. Cell Stem Cell 30:1148–1165.e7
6. Pedroza M et al (2023) Self-patterning of human stem cells into post-implantation lineages. Nature 622:574–583
7. Liu L et al (2023) Modeling post-implantation stages of human development into early organogenesis with stem-cell-derived peri-gastruloids. Cell 186:3776–3792.e16
8. Weatherbee BAT et al (2023) Pluripotent stem cell-derived model of the post-implantation human embryo. Nature 622:584–593
9. Moris N et al (2020) An in vitro model of early anteroposterior organization during human development. Nature 582:410–415
10. Yamanaka Y et al (2023) Reconstituting human somitogenesis in vitro. Nature 614:509–520
11. Okubo T et al (2024) Hypoblast from human pluripotent stem cells regulates epiblast development. Nature 626:357–366
12. Miao Y et al (2023) Reconstruction and deconstruction of human somitogenesis in vitro. Nature 614:500–508
13. Karzbrun E et al (2021) Human neural tube morphogenesis in vitro by geometric constraints. Nature 599:268–272
14. Sanaki-Matsumiya M et al (2022) Periodic formation of epithelial somites from human pluripotent stem cells. Nat Commun 13:2325
15. Li J et al (2023) Cynomolgus monkey embryo model captures gastrulation and early pregnancy. Cell Stem Cell 30(4):362–377.e7. https://doi.org/10.1016/j.stem.2023.03.009
16. Simunovic M et al (2019) A 3D model of a human epiblast reveals BMP4-driven symmetry breaking. Nat Cell Biol 21(7):900–910. https://doi.org/10.1038/s41556-019-0349-7
17. Yang Y et al (2017) Derivation of pluripotent stem cells with in vivo embryonic and extraembryonic potency. Cell 169:243–257.e25
18. Rostovskaya M, Stirparo GG, Smith A (2019) Capacitation of human naïve pluripotent stem cells for multi-lineage differentiation. Development 146:dev172916
19. Gafni O et al (2013) Derivation of novel human ground state naive pluripotent stem cells. Nature 504:282–286
20. Xiang L et al (2020) A developmental landscape of 3D-cultured human pre-gastrulation embryos. Nature 577:537–542

Methods in Molecular Biology (2026) 3048: 33–45
DOI 10.1007/7651_2025_649

Published online: 03 July 2025

Generating Human Blastoids Following Transient Naïve Resetting of Primed Pluripotent Stem Cells

Maria Carolina Zimara, Carlos A. Pinzon-Arteaga, Kun Liu, Toshihiko Ezashi, Seiya Oura, Shijian Lyu, Menaka Sanghvi, Ye Yuan, and Jun Wu

Abstract

Human pluripotent stem cells (hPSCs) exist in at least two distinct states of pluripotency: naïve and primed. While naïve hPSCs possess the unique ability to generate blastocyst-like structures, they are often genetically and epigenetically unstable, which compromises the quality and developmental potential of naïve hPSC-derived blastoids. This protocol presents an optimized human blastoid protocol through a transient resetting method that converts primed hPSCs into a naïve-like state, addressing the stability issues associated with long-term naïve hPSC maintenance. The approach is compatible with both feeder-free and feeder-based culture systems and demonstrates high efficiency in generating human blastoids directly from primed hPSCs. This advancement provides a more robust and reliable strategy for blastoid formation, circumventing the limitations of suboptimal naïve hPSC cultures.

Key words Embryonic stem cells, Human blastoids, Naïve pluripotent stem cells, Primed pluripotent stem cells

1 Introduction

When working with pluripotent stem cells (PSCs), the scientific consensus broadly recognizes two distinct states at opposite ends of the pluripotency spectrum: the naïve and primed states [16, 24]. This distinction in pluripotency potentials is closely linked to the developmental stage of the epiblast cells. Naïve PSCs closely resemble the pre-implantation epiblast, while primed PSCs more accurately reflect the post-implantation epiblast [6].

In the human naïve pluripotency state, cells retain the plasticity characteristic of the inner cell mass, enabling differentiation into both embryonic and extraembryonic lineages, including the trophoblast and hypoblast [5, 8]. This multilineage differentiation capacity is attributed to the unique transcriptional and epigenetic landscape of naïve pluripotent stem cells (PSCs), which maintains a chromatin state conducive to rapid fate transitions. A defining feature of this state is the presence of bivalent chromatin domains, where activating (H3K4me3) and repressive (H3K27me3) histone modifications coexist, keeping key developmental genes in a transcriptionally poised but repressed state [1, 20]. This configuration ensures a rapid transcriptional response upon exposure to differentiation signals, allowing naïve PSCs to efficiently engage alternative lineage trajectories.

Additionally, the epigenome of naïve PSCs is more permissive than that of primed PSCs, as evidenced by globally reduced DNA methylation levels and the retention of two active X chromosomes in female cells [15]. The global hypomethylation pattern, a hallmark of naïve pluripotency, facilitates enhancer accessibility and transcriptional flexibility, supporting broad developmental potential [6, 13]. However, certain imprinted loci remain stably methylated, preserving essential epigenetic marks required for proper lineage commitment. This highly dynamic chromatin and epigenetic state distinguishes naïve PSCs from primed PSCs, which exhibit increased DNA methylation, loss of bivalent domains, and a more restricted differentiation potential.

Studies also suggest the presence of intermediate, or formative, pluripotency states in humans [3, 12, 19, 22, 27, 28]. However, the stabilization of these intermediate PSCs has been achieved using diverse culture conditions, and a consensus on defining characteristics of human intermediate/formative pluripotency has yet to be established [6]. Therefore, this topic will not be discussed further here.

Naïve human PSCs (hPSCs) can be obtained through various approaches. The most direct method involves capturing them from a pre-implantation human blastocyst [10]. However, access to human embryos is often restricted and varies between jurisdictions and laboratories. Alternatively, naïve hPSCs can be generated by reprogramming somatic cells or by resetting primed hPSCs through transient expression of naïve pluripotency-related transcription factors such as KLF2 and NANOG, in combination with specific cytokines and small-molecule inhibitors.

A notable challenge that scientists face when culturing naïve hPSCs long term is the loss of epigenetic stability and the accumulation of karyotypic aberrations resulting from repeated passaging under naive conditions [14, 17]. However, this issue appears to be significantly less pronounced when working with hPSCs in a primed state. Notably, the ability to transiently reset primed cells to a naive state as needed presents a clear advantage. Several protocols have been developed to reset primed hPSCs into various naïve hPSC cultures. Among these, two culture conditions, 5iLA [23, 27, 28] and PXGL [26], have garnered the most attention, both of which have been successfully used to generate human blastocyst-like structures (blastoids) [11, 23, 26–28]. The resetting of primed hPSCs into the 5iLA condition has been achieved through at least two different approaches, both of which emphasize the maintenance of endogenous *POU5F1* (aka *OCT4*) distal enhancer activity, a widely recognized hallmark of naïve pluripotency. The first method involves directly culturing primed hPSCs in the 5iL/A medium, which consists of five kinase inhibitors (MEK, GSK3, BRAF, SRC, and ROCK inhibitors) and two growth factors (LIF and Activin A) [7, 21]. The second approach begins with transient histone deacetylase (HDAC) inhibition, as HDAC2 acts as a as a major reprogramming impediment acting through the CoREST complex, and its inhibition promotes a more permissive chromatin state before transitioning the cells into the 5iL/A medium [18, 29].

In this chapter, we validate an optimized version of the Rugg-Gunn [18] protocol for the generation of naïve human embryonic stem cells (hESCs) from the primed pluripotency state. While the Rugg-Gunn method is favored for its ability to minimize unwanted differentiation [9], its practical application is hindered by the complexity of media preparation. To address this limitation, we streamlined the protocol by reducing the number of required media from two to a single formulation, while aiming to retain its beneficial effects on naïve pluripotency. Additionally, we refined the 5iL/A medium by incorporating Gö6983 (a pan-protein kinase C inhibitor), XAV939 (a tankyrase 1/2 inhibitor to prevent trophectoderm (TE) formation by suppressing YAP activity independent of β-catenin) [4, 9, 10], and a halved concentration of the GSK-3β inhibitor IM-12. This improved medium, designated 4.5iL/A/G/X, synergizes key elements from previous protocols while simplifying experimental workflows (Fig. 1).

Using a WIBR3 hESC line, which carries a GFP reporter under the control of the *POU5F1* distal enhancer (*POU5F1-ΔPE-GFP*), we were able to visually monitor the process and efficiency of naïve resetting in real time (Fig. 1). This fluorescent marker, combined with routine morphological assessments, provided a reliable and dynamic means of evaluating the transition to naïve pluripotency throughout the entire experiment. Further functional validation was conducted by subjecting the naïve cells to controlled trilineage differentiation, demonstrating their ability to generate epiblast-, trophoblast- and hypoblast-like cells, and human blastoids (Fig. 2).

2 Materials

2.1 Preparation of Matrigel Plates

1. Matrigel (Corning) solution: 0.11–0.12 mg/mL in DMEM/F12.

2.2 Preparation of iMEF Plates

1. Mitotically inactivated mouse embryonic fibroblasts (iMEFs).
2. Gelatin solution: Autoclave for 30 min 0.1% gelatin dissolved in MilliQ water (*see* **Note 1**).

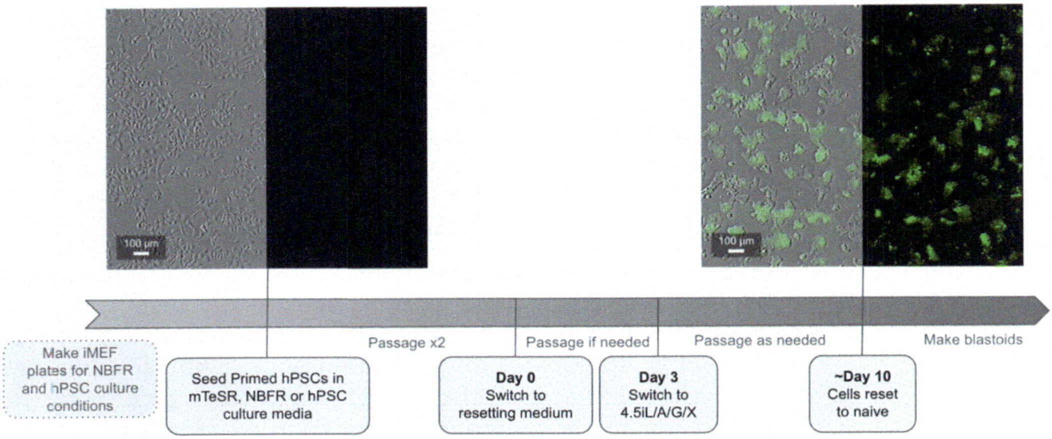

Fig. 1 Schematic depicting the timeline for resetting primed embryonic stem cells to their ground state. The diagram includes pre- and post-resetting pictures of WIBR3 cells cultured in feeder-free conditions using mTeSR complete medium and reset using resetting medium. The right half of each picture has been taken using GFP illumination only, whereas the left half merges GFP signal and brightfield acquisition. The experiment has also been replicated using WIBR3 cells cultured on iMEF feeder cells under NBFR and hPSC culture media

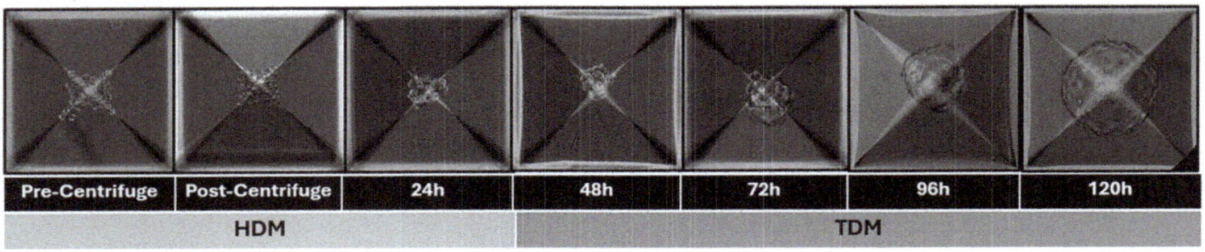

Fig. 2 Visual portraying the chronological progression of blastoids generation

3. MEF medium: Advanced DMEM supplemented with 5% FBS or FetalGro EX, 1% GlutaMax. Filter-sterilize the solution using a 0.22 μm filter and store at 4 °C for up to 4 weeks.
4. Washing solution: DMEM/F12 supplemented with 0.1% BSA. Filter-sterilize the solution using a 0.2 μm filter and store at 4 °C for up to 4 weeks.

2.3 Primed hPSCs Culture

1. WIBR3 cell line with integrated *POU5F1-ΔPE-GFP* naïve reporter system [21].
2. Basal N2B27 Medium: 1:1 solution of DMEM/F12 and Neurobasal Medium, supplemented with 1% N2, 2% B27, 1% GlutaMAX, 1% Non-Essential Amino Acids (NEAA), and β-mercaptoethanol (0.1 mM). Filter-sterilize the solution using a 0.2 μm filter and store at 4 °C for up to 4 weeks.

3. Complete mTeSR Plus Medium: mTeSR Plus Basal Medium supplemented with 20% mTeSR Plus 5X Supplement (*see* **Note 2**).
4. NBFR: N2B27 Basal Medium supplemented with FGF2 (20 ng/mL), Activin A (50 ng/mL), human Leukemia Inhibitory Factor (hLIF, 20 ng/mL), and IWR1 (2.5 μM) [25]. Filter-sterilize the solution using a 0.22 μm filter and store at 4 °C for up to 1 week.
5. hPSCs Culture Medium: 80% DMEM/F12 and 20% Knockout Serum Replacement (KOSR), supplemented with GlutaMax (1 mM), NEAA (0.1 mM), β-mercaptoethanol (0.1 mM), and FGF2 (4 ng/mL). Filter-sterilize the solution using a 0.22 μm filter and store at 4 °C for up to 1 week.
6. CEPT cocktail [2]: Chroman 1 (50 nM), Emricasan (5 μM), trans-ISRIB (0.7 μM). Store aliquots at −20 °C. Freshly add polyamine diluted at 1:1000.

2.4 Preparation of Geltrex Plates

1. Geltrex (Gibco) solution: 0.12–0.18 mg/mL in DMEM/F12.

2.5 Naïve hPSCs Resetting and Culture

1. Chemical Reset Medium: Basal N2B27 Medium, PD0325901 (1 μM), human Leukemia Inhibitory Factor (10 ng/mL), Activin A (50 ng/mL). Supplement freshly with BRD4884 (25 μM) and CEPT cocktail. Filter-sterilize the solution using a 0.22 μm filter and store at 4 °C for up to 1 week.
2. 4.5iL/A/G/X Medium: Basal N2B27 Medium supplemented with 0.5% KOSR, hLIF (20 ng/mL), PD0325901 (1 μM), IM-12 (0.5 μM), SB590885 (0.5 μM), WH-4-023 (1 μM), Activin A (20 ng/mL), Y-27632 (5 μM), 7.5% BSA (5%, 50 μg/mL), Gö6983 (2 μM), XAV939 (2 μM). Filter-sterilize the solution using a 0.22 μm filter and store at 4 °C for up to 1 week.

2.6 Blastoids Generation

1. Hypoblast Differentiation Medium (HDM): Basal N2B27 medium supplemented with FGF2 (20 ng/mL), Activin A (20 ng/mL), hLIF (20 ng/mL), and CHIR99021 (3 μM). Filter-sterilize the solution using a 0.22 μm filter and store at 4 °C for up to 1 week (*see* **Note 3**).
2. Trophoblast Differentiation Medium (TDM): Basal N2B27 medium supplemented with 0.5% ITS-X, PD0325901 (1 μM), SB590885 (0.5 μM), WH-4-023 (1 μM), hLIF (10 ng/mL). Filter-sterilize the solution using a 0.22 μm filter and store at 4 °C for up to 1 week. A77 (2 μM) and LPA (0.5 μM) to be supplemented freshly before medium employment (*see* **Note 3**).

3 Methods

3.1 Preparation of Matrigel Plates

1. Thaw Matrigel aliquots on ice to avoid gelatinization.
2. Chill washing solution on ice as well.
3. Fill a 15 mL conical tube with 6 mL of washing solution before adding the Matrigel aliquot.
4. Dispense 1 mL of Matrigel solution to each well.
5. Incubate at room temperature for at least 1 h.
6. After incubation, Matrigel solution can be transferred to a new six-well plate for the next use.
7. Store the new six-well plate in 4 °C.

3.2 Preparation of iMEF Plates

1. Warm up MEF medium in a 37 °C water bath.
2. Prepare gelatin-coated plate by adding 1 mL of 0.1% gelatin solution into each well and incubate at room temperature for at least 15 min.
3. Aspirate gelatin solution from wells and add 2 mL of MEF medium to each well.
4. Return MEF medium to water bath.
5. Thaw the iMEFs in the water bath.
6. Wipe the cryovial with 70% ethanol to sterilize before bringing it to the biohood.
7. Transfer iMEFs to a 15 mL conical tube. Slowly add 1 mL of washing solution to the conical tube.
8. Add another 1 mL of washing solution to the cryovial and transfer to the conical tube.
9. Keep adding washing solution to the conical tube, gently mixing as the medium is added. The ratio of the cell content to washing solution should be 1:10.
10. Centrifuge at 200× g for 5 min.
11. Add 1 mL of MEF medium to resuspend the cell pellet.
12. Dispense half a million cells to each well. Pipette every time before each dispense.
13. Distribute iMEFs evenly by moving the plate in back-and-forth and side-to-side motions.
14. Incubate at 37 °C under normoxic conditions (20% O_2, 5% CO_2) overnight (*see* **Note 4**).

3.3 Feeder-Free Culture of Primed hPSCs

1. Prepare a Matrigel-coated six-well plate at least 1 h prior to use.
2. Warm up Complete mTeSR Plus Medium to room temperature.

3. Aspirate Matrigel and add 2 mL of Complete mTeSR Plus Medium supplemented with CEPT to each well.
4. Thaw cryopreserved hPSCs in a 37 °C water bath.
5. Wipe the cryovial with 70% ethanol to sterilize before bringing it to the biohood.
6. Transfer hPSCs to a 15 mL conical tube.
7. Add 1 mL of washing solution dropwise to the conical tube.
8. Slowly add another 1 mL of washing solution to the cryovial and transfer to the conical tube.
9. Slowly add extra washing solution to the conical tube to dilute DMSO gradually, gently mixing as the medium is added. The ratio of the cell content to washing solution should be 1:10.
10. Centrifuge at 200× *g* for 3 min.
11. Resuspend cells gently in 1 mL of washing solution supplemented with CEPT. Transfer cells to Matrigel-coated wells.
12. Distribute hPSCs evenly by moving the plate in back-and-forth and side-to-side motions. Incubate at 37 °C under hypoxic conditions (5% O_2, 5% CO_2).
13. Refresh medium daily. Colonies should appear flat and well defined.

    Passage primed hPSCs when colonies begin to touch or show signs of differentiation.

3.4 Feeder-Based Culture of Primed hPSCs

1. Prepare an iMEF-coated six-well plate 48 h prior to use.
2. Warm up hPSCs Culture Medium and NBFR to room temperature.
3. Aspirate MEF medium from wells and wash twice with 1 mL of 1× DPBS.
4. Add 2 mL of hPSCs Culture Medium or NBFR supplemented with CEPT to each well.
5. Thaw cryopreserved hPSCs in a 37 °C water bath.
6. Wipe the cryovial with 70% ethanol to sterilize before bringing it to the biohood.
7. Transfer hPSCs to a 15 mL conical tube.
8. Add 1 mL of washing solution dropwise to the conical tube.
9. Slowly add another 1 mL of washing solution to the cryovial and transfer to the conical tube.
10. Slowly add extra washing solution to the conical tube to dilute DMSO gradually, gently mixing as the medium is added. The ratio of the cell content to washing solution should be 1:10.
11. Centrifuge at 200× *g* for 3 min.

12. Resuspend cells gently in 1 mL of washing solution supplemented with CEPT. Transfer cells to iMEF-coated wells.
13. Distribute hPSCs evenly by moving the plate in back-and-forth and side-to-side motions. Incubate at 37 °C under hypoxic conditions (5% O_2, 5% CO_2).
14. Refresh media daily. Colonies should appear flat and well defined.

    Passage primed hPSCs when colonies begin to touch or show signs of differentiation.

3.5 Passaging hPSCs

1. Prepare an iMEF-coated six-well plate 48 h prior to use.
2. Prepare a Matrigel-coated six-well plate at least 1 h prior to use.
3. Warm up complete mTeSR Plus medium, hPSCs Culture Medium, or NBFR to room temperature. Supplement media with CEPT.
4. Aspirate MEF medium and Matrigel from wells and wash twice with 1 mL of DPBS.
5. Add 2 mL of complete mTeSR Plus medium supplemented with CEPT to each Matrigel-coated well, 2 mL of hPSCs Culture Medium or NBFR to each iMEF-coated well.
6. Aspirate medium from primed hPSCs and add 0.5 mL of TrypLE to the cells. Incubate at 37 °C for 3 min.
7. Add 1:1 washing solution to quench the reaction and transfer to a 15 mL conical tube.
8. Centrifuge at 200× *g* for 3 min.
9. Resuspend cells gently in 1 mL of washing solution supplemented with CEPT.
10. Perform cell counting using a hemocytometer.
11. Seed 2×10^5 primed hPSCs to each well.
12. Distribute hPSCs evenly by moving the plate in back-and-forth and side-to-side motions.
13. Incubate at 37 °C under hypoxic conditions (5% O_2, 5% CO_2).
14. Perform daily medium change and passage cells when confluency reaches 80%.

3.6 Passaging hPSCs to Resetting Conditions (Day 0)

1. Prepare a Geltrex-coated six-well plate at least 1 h prior to use.
2. Warm up Reset Medium to room temperature. Supplement medium with CEPT.
3. Aspirate MEF medium and Geltrex from wells and wash twice with 1 mL of DPBS.
4. Add 4 mL of Reset Medium supplemented with CEPT to both Geltrex-coated and iMEF-coated wells.

5. Aspirate medium from primed hPSCs and add 0.5 mL of TrypLE to the cells. Incubate at 37 °C for 3 min.
6. Add 1:1 washing solution to quench the reaction and transfer to a 15 mL conical tube.
7. Centrifuge at 200× *g* for 3 min.
8. Resuspend cells gently in 1 mL of washing solution supplemented with CEPT.
9. Add DNase (10 μL/mL) to the cell suspension to release live primed hPSCs tangled by DNA mesh generated from dead cells. Leave 10 min at room temperature.
10. Rinse a 70 μm cell strainer and a 20 μm cell strainer with washing medium and pass the cell suspension through the filters in that order.
11. In order to remove death cells and cell debris, slowly layer the eluted cells onto 10 mL of washing medium, holding the tube at an angle.
12. Centrifuge the cells at 100× *g* for 10 min.
13. Aspirate the supernatant and resuspend the cell pellet in 1 mL of pre-warmed Reset Medium supplemented with CEPT.
14. Perform cell counting using a hemocytometer.
15. Seed 2×10^5 primed hPSCs to each well.
16. Distribute hPSCs evenly by moving the plate in back-and-forth and side-to-side motions.
17. Incubate at 37 °C under low oxygen conditions (5% O_2, 5% CO_2).

3.7 Maintenance and Expansion of Naïve hPSCs

1. After 48 h in resetting conditions, warm 4.5iL/A/G/X to room temperature. Supplement medium with CEPT.
2. Aspirate the Reset Medium and substitute it with pre-warmed 4.5iL/A/G/X.
3. Incubate at 37 °C under low oxygen conditions (5% O_2, 5% CO_2).
4. Perform daily medium change and passage cells when confluency reaches 80%.

3.8 Preparation of Human Blastoids

1. Pre-warm to room temperature sufficient 4.5i/L/A/G/X medium supplemented with CEPT cocktail, washing medium, and HDM medium equilibrated for at least 2 h at 37 °C under hypoxic conditions (5% O_2, 5% CO_2).
2. Wipe the sleeve containing an AggreWell plate with 70% ethanol to sterilize before bringing it into the biological safety cabinet.

3. Add 500 μL Anti-Adherence Rinsing Solution per well (*see* **Note 6**) and seal the plate to maintain the inside plate sterility during centrifugation.
4. Centrifuge the plate at 1500× *g* for 5 min.
5. Keep the plate treated with Anti-Adherence Rinsing Solution at room temperature until use.
6. Aspirate the medium from naïve hPSCs and wash twice with DPBS.
7. Dislodge the naive hPSCs from the well using 500 μL TrypLE per well at 37 °C for 3 min.
8. Use a P1000 pipette to dissociate the colonies into single cells and add 1 mL of washing medium to quench dissociation.
9. Centrifuge the cells at 200× *g* for 5 min.
10. Aspirate the supernatant and resuspend the cell pellet in 1 mL of 4.5i/L/A/G/X medium supplemented with CEPT.
11. Add DNase (10 μL/mL) to the cell suspension to release live naïve hPSCs tangled by DNA mesh generated from dead cells. Leave for 10 min at room temperature.
12. Rinse a 70 μm cell strainer and a 20 μm cell strainer with washing medium and pass the cell suspension through the filters in that order.
13. In order to remove death cells and cell debris, slowly layer the eluted cells onto 10 mL of washing medium, holding the tube at an angle.
14. Centrifuge the cells at 100× *g* for 10 min.
15. Aspirate the supernatant and resuspend the cell pellet in 1 mL of equilibrated HDM.
16. Gently pipette up and down 2–3 times to mix the cell content thoroughly.
17. Count the cells using a hemacytometer (*see* **Note 5**) and adjust the concentration to 0.384×10^5 cells/mL.
18. Aspirate Anti-Adherence Solution from the AggreWell plate.
19. Rinse the wells twice with DPBS.
20. Add 1 mL of cell suspension with the right concentration in HDM to each well.
21. Mix the cell suspension evenly by gently pipetting in zig-zag motions 2–3 times using a P200 pipette.
22. Leave standing still inside the hood for 15 min.
23. Balance the plate and centrifuge it at 1200× *g* for 2 min.
24. Incubate the AggreWell plate at 37 °C under hypoxic conditions in a humidified environment. The day of cell plating is designated as day 0.

25. The following day, carefully aspirate the media and completely replace it with TDM (1 mL per well). Refresh half of the medium with TDM every other day until blastoids are fully structured and ready to be used.

4 Notes

1. Gelatin solutions can be stored indefinitely at 4 °C. Gelatin-coated plates can be stored at 4 °C for up to 4 weeks. Seal the gelatin-coated plates well to prevent contamination and/or over-drying of gelatin. Do not use the product if discoloration or cracks appear on the surface of the coated material.
2. Aliquot complete mTeSR Plus medium into 40 mL and store at −20 °C until expiry date on label. Thaw the medium at room temperature or overnight at 4 °C. Do not thaw in a 37 °C water bath. Store the thawed complete medium at 4 °C for up to 2 weeks. Do not re-freeze.

   If cloudiness is observed in mTeSR Plus 5X Supplement after thawing, keep the supplement at room temperature or in a 37 °C water bath for approximately 5 min. Supplement must be clear before adding to basal medium.
3. Enhancement of cell recovery with both TDM and HDM can be attained by fresh addition of CEPT cocktail.
4. An iMEF confluency of approximately 85–90% is considered as the optimal feeder layer density. Ideally the iMEFs should be incubated for 48 h before use. iMEF plates may be used up to 4 days post seeding.
5. 28 ± 4 naïve WIBR3 hESCs should be seeded per microwell. There are 1200 microwells in each well.
6. Dispense Anti-Adherence Rinsing Solution slowly without introducing air bubbles to wells.

References

1. Bi Y, Tu Z, Zhou J, Zhu X, Wang H, Gao S, Wang Y (2022) Cell fate roadmap of human primed-to-naive transition reveals preimplantation cell lineage signatures. Nat Commun 13(1):3147
2. Chen Y, Tristan CA, Chen L, Jovanovic VM, Malley C, Chu PH, Ryu S, Deng T, Ormanoglu P, Tao D, Fang Y (2021) A versatile polypharmacology platform promotes cytoprotection and viability of human pluripotent and differentiated cells. Nat Methods 18(5):528–541
3. Cornacchia D, Zhang C, Zimmer B, Chung SY, Fan Y, Soliman MA, Tchieu J, Chambers SM, Shah H, Paull D, Konrad C (2019) Lipid deprivation induces a stable, naive-to-primed intermediate state of pluripotency in human PSCs. Cell Stem Cell 25(1):120–136
4. Dattani A, Huang T, Liddle C, Smith A, Guo G (2022) Suppression of YAP safeguards human naïve pluripotency. Development 149(24): dev200988
5. Dong C, Beltcheva M, Gontarz P, Zhang B, Popli P, Fischer LA, Khan SA, Park KM, Yoon EJ, Xing X, Kommagani R (2020) Derivation

of trophoblast stem cells from naïve human pluripotent stem cells. elife 9:e52504

6. Du P, Wu J (2024) Hallmarks of totipotent and pluripotent stem cell states. Cell Stem Cell 31(3):312–333
7. Fischer LA, Khan SA, Theunissen TW (2022) Induction of human naive pluripotency using 5i/L/A medium. In: Theunissen TW (ed) Human naïve pluripotent stem cells, pp 13–28
8. Guo G, Stirparo GG, Strawbridge SE, Spindlow D, Yang J, Clarke J, Dattani A, Yanagida A, Li MA, Myers S, Özel BN (2021) Human naive epiblast cells possess unrestricted lineage potential. Cell Stem Cell 28(6): 1040–1056
9. Guo G, von Meyenn F, Rostovskaya M, Clarke J, Dietmann S, Baker D, Smith A (2017) Epigenetic resetting of human pluripotency. Development 144(15):2748–2763
10. Guo G, von Meyenn F, Santos F, Chen Y, Reik W, Bertone P, Nichols J (2016) Naive pluripotent stem cells derived directly from isolated cells of the human inner cell mass. Stem cell Rep 6(4):437–446
11. Kagawa H, Javali A, Khoei HH, Sommer TM, Sestini G, Novatchkova M, Scholte op Reimer Y, Castel G, Bruneau A, Maenhoudt N, Lammers J (2022) Human blastoids model blastocyst development and implantation. Nature 601(7894):600–605
12. Kinoshita M, Barber M, Mansfield W, Cui Y, Spindlow D, Stirparo GG, Dietmann S, Nichols J, Smith A (2020) Capture of mouse and human stem cells with features of formative pluripotency. Cell Stem Cell 28(3): 453–471.e8
13. Leitch HG, McEwen KR, Turp A, Encheva V, Carroll T, Grabole N, Mansfield W, Nashun B, Knezovich JG, Smith A, Surani MA (2013) Naïve pluripotency is associated with global DNA hypomethylation. Nat Struct Mol Biol 20(3):311–316
14. Liu X, Nefzger CM, Rossello FJ, Chen J, Knaupp AS, Firas J, Ford E, Pflueger J, Paynter JM, Chy HS et al (2017) Comprehensive characterization of distinct states of human naive pluripotency generated by reprogramming. Nat Methods 14:1055–1062
15. Martin GR, Epstein CJ, Travis B, Tucker G, Yatziv S, Martin DW Jr, Clift S, Cohen S (1978) X-chromosome inactivation during differentiation of female teratocarcinoma stem cells in vitro. Nature 271(5643):329–333
16. Nichols J, Smith A (2009) Naive and primed pluripotent states. Cell Stem Cell 4(6): 487–492
17. Pastor WA, Chen D, Liu W, Kim R, Sahakyan A, Lukianchikov A, Plath K, Jacobsen SE, Clark AT (2016) Naive human pluripotent cells feature a methylation landscape devoid of blastocyst or germline memory. Cell Stem Cell 18:323–329
18. Rugg-Gunn PJ (2022) Induction of human naïve pluripotency using chemical resetting. In: Human naïve pluripotent stem cells. Springer, pp 29–37
19. Smith A (2017) Formative pluripotency: the executive phase in a developmental continuum. Development 144(3):365–373
20. de Sousa JA, Wong CW, Dunkel I, Owens T, Voigt P, Hodgson A, Baker D, Schulz EG, Reik W, Smith A, Rostovskaya M (2023) Epigenetic dynamics during capacitation of naïve human pluripotent stem cells. Science. Advances 9(39):eadg1936
21. Theunissen TW, Powell BE, Wang H, Mitalipova M, Faddah DA, Reddy J, Fan ZP, Maetzel D, Ganz K, Shi L, Lungjangwa T, Imsoonthornruksa S, Stelzer Y, Rangarajan S, D'Alessio A, Zhang J, Gao Q, Dawlaty MM, Young RA, Gray NS, Jaenisch R (2014) Systematic identification of culture conditions for induction and maintenance of naive human pluripotency. Cell Stem Cell 15(4):471–487
22. Wang X, Xiang Y, Yu Y, Wang R, Zhang Y, Xu Q, Sun H, Zhao ZA, Jiang X, Wang X, Lu X (2021) Formative pluripotent stem cells show features of epiblast cells poised for gastrulation. Cell Res 31(5):526–541
23. Wei Y, Liu K, Pinzon-Arteaga CA, Logsdon D, Yu L, Yuan Y, Wu J (2023) Generation of human blastoids from naive pluripotent stem cells. In: Wu J (ed) Embryo models in vitro: methods and protocols. Springer US, New York, pp 1–18
24. Weinberger L, Ayyash M, Novershtern N, Hanna JH (2016) Dynamic stem cell states: naive to primed pluripotency in rodents and humans. Nat Rev Mol Cell Biol 17(3): 155–169
25. Wu J, Okamura D, Li M, Suzuki K, Luo C, Ma L, He Y, Li Z, Benner C, Tamura I, Krause MN (2015) An alternative pluripotent state confers interspecies chimaeric competency. Nature 521(7552):316–321
26. Yanagida A, Spindlow D, Nichols J, Dattani A, Smith A, Guo G (2021) Naive stem cell blastocyst model captures human embryo lineage segregation. Cell Stem Cell 28(6):1016–1022
27. Yu L, Wei Y, Duan J, Schmitz DA, Sakurai M, Wang L, Wang K, Zhao S, Hon GC, Wu J (2021a) Blastocyst-like structures generated

from human pluripotent stem cells. Nature 591(7851):620–626

28. Yu L, Logsdon D, Pinzon-Arteaga CA, Duan J, Ezashi T, Wei Y, Ribeiro Orsi AE, Oura S, Liu L, Wang L, Liu K, Ding X, Zhan L, Zhang J, Nahar A, Stobbe C, Katz-Jaffe M, Schoolcraft WB, Tan T, Hon GC, Yuan Y, Wu J (2023) Large-scale production of human blastoids amenable to modeling blastocyst development and maternal-fetal cross talk. Cell Stem Cell. Sep 7;30(9):1246–1261.e9

29. Collier AJ, Bendall A, Fabian C, Malcolm AA, Tilgner K, Semprich CI, Wojdyla K, Nisi PS, Kishore K, Roamio Franklin VN, Mirshekar-Syahkal B, D'Santos C, Plath K, Yusa K, Rugg-Gunn PJ Genome-wide screening identifies Polycomb repressive complex 1.3 as an essential regulator of human naïve pluripotent cell reprogramming. Science Advances 8(12): eabk0013.

Methods in Molecular Biology (2026) 3048: 47–60
DOI 10.1007/7651_2025_642

Published online: 04 June 2025

Generation of Blastoids from Human Naïve Pluripotent Stem Cells

Hiroo Sasaki, Iori Sasaki, Jennifer Nichols, and Ayaka Yanagid

Abstract

Human embryo research is essential for understanding the development of human embryos and their unique features that cannot be investigated in mouse embryos. Natural development of human embryos remains challenging to study both in vivo and in vitro owing to ethical concerns, technical difficulties, and limited availability of embryos for research purposes. Until recently, access to human embryo research, especially the implantation period, was limited. However, optimization of a stem cell-based model known as a blastoid has opened a new era for human embryo research. In contrast with mouse embryonic stem cells, human naïve pluripotent stem cells retain extended cell lineage plasticity. They can differentiate into hypoblast and trophectoderm while retaining characteristic resembling the naïve epiblast in the inner cell mass region. Taking this unique differentiation potential and inherent propagation capacity as pluripotent stem cells, human blastoids are generated by the self-organization of naïve pluripotent stem cells. They resemble human blastocyst structures, consisting of the three founder cell lineages that closely resemble the gene expression profile of human blastocysts. This protocol for generating blastoids solely from naïve human pluripotent stem cells utilizing simple, efficient, and scalable procedures is a robust tool for advancing aspects of human embryo research.

Key words Blastocyst, Human blastoid, Human naïve embryonic stem cells, Naïve pluripotency, Stem cell-based embryo model

1 Introduction

How does a single cell give rise to an entire complex human body? Understanding human embryo development is essential to address this fundamental question. A fertilized egg undergoes sequential divisions and develops through 2-cell, 4-cell, and 8-cell stages to morula, which subsequently segregates into three lineages to form a blastocyst. The early blastocyst consists of two types of cells: trophectoderm, which gives rise to the placenta, and inner cell mass (ICM), which then differentiates into epiblast, founder of the fetus, and hypoblast, which will form the yolk sac. Epiblast and hypoblast cells comprise the ICM, arranged in a "salt and pepper" manner. By the late blastocyst stage, these cell segregate into two distinct populations [1, 2]. The mature blastocyst implants in the uterus and undergoes drastic morphological changes to further develop into a fetus, surrounded by extra-embryonic tissues. Blastocyst formation and implantation are hallmarks of mammalian development. However, as these events progress inside the uterus, studying human embryo development requires scalable and representative in vitro human embryo models.

A stem cell-based blastocyte model, the "blastoid," was initially devised for mice by mixing mouse embryonic stem cells (mESCs) and trophectoderm stem cells [3]. Subsequently, blastoids have commanded attention as a valuable in vitro model for mimicking embryonic development [4, 5]. Taking advantage of self-renewal and scalability of human pluripotent stem cells (hPSCs), blastoids have the potential to generate numerous 3D human blastocyst models [4], which enable genetic manipulation and overcome the limitation of available embryos for research. However, to investigate human embryo development, it is important to ensure that the blastoid model closely reflects the morphology and gene expression profile of human blastocysts. To apply this platform for further research, high reproducibility with simple techniques and minimum additional components in the medium is advantageous.

Naïve hPSCs are derived through various methods, including resetting from primed hPSCs [6], reprogramming from somatic cells [7], and derivation directly from embryos [8]. Similar to mESCs, naïve hPSCs exhibit the core hallmarks of pluripotency and form dome-shaped colonies with high clonogenicity. More importantly, they closely resemble the human epiblast of blastocysts in terms of gene expression profiles and differentiation capacities. Human naïve PSCs and epiblast retain cell lineage plasticity and can differentiate into trophectoderm or hypoblast when exposed directly to culture medium containing inhibitors of MEK/ERK and TGFβ signaling [9]. In contrast, mouse ICMs lose the capacity to regenerate trophectoderm at around the mid blastocyst stage [10–12] and mouse blastoids formation requires the addition of trophoblast stem cells to mESCs [3]. Utilizing the unique differentiation potential of naïve hPSCs to form all three cell lineages of the human blastocyst, blastoids can be generated solely from naïve human PSCs by optimizing cell seeding numbers and the timing of differentiation induction.

This chapter provides a platform for generating blastoids, which are similar in scale to and closely resemble human embryos. These blastoids enable the generation of genetically modified human embryo models for investigating particular gene functions during human embryo development as well as tracking particular cell lineages. Furthermore, they provide a powerful tool for investigating human embryo implantation and post-implantation development and addressing challenges in human fertility.

2 Materials

2.1 Equipment

1. 37 °C humidified incubator with 5% CO_2 and 5% O_2 (*see* **Note 1**).
2. Horizontal laminar flow hood.
3. Phase contrast microscope.
4. Stereo microscope.
5. Micropipettes P2, P20, P200, and P1000, 20-200 μL multichannel and sterile tips.
6. 6- and 12-well cell culture plates
7. 37 °C water bath.
8. Conical sterile polypropylene centrifuge tubes, 15 and 50 mL.
9. Sterile 1.5 or 2 mL screw tubes.
10. Centrifuge for microplates and conical tubes.
11. Vacuum pump.
12. Aspirator pipette.
13. Pipette controller.
14. Sterile graduated pipettes, 5 and 10 mL.
15. Hemocytometer or equivalent cell-counting device.
16. Reservoir.
17. 96-well clear round bottom ultra-low attachment microplate or 24-well plate with microwells (*see* **Notes 2** and **3**).
18. 96-well suspension culture microplate (*see* **Notes 3** and **4**).
19. Sterile-pulled glass Pasteur pipettes (*see* **Note 5**).
20. Mouth-controlled pipette with plastic tube and mouthpiece.

2.2 Reagents

1. Human naïve PSCs validated for trophectoderm differentiation potential.
2. 0.1% gelatin.
3. PBS.
4. Dulbecco's Modified Eagle's Medium—high glucose.
5. Fetal bovine serum (FBS).
6. Penicillin-Streptomycin-L-glutamine (×100).
7. Mitotically inactivated (irradiated or Mitomycin C treated) murine embryonic fibroblasts (iMEF) prepared in-house from E13.5 mouse fetuses (*see* **Note 6**).
8. Bovine serum Albumin fraction V.
9. DMEM/F-12 (*see* **Note 7**).
10. Apo-transferrin.

11. Insulin.
12. Sodium selenite.
13. Putrescine.
14. Progesterone.
15. L-glutamine (200 mM).
16. 2-mercaptoethanol (100 mM).
17. Neurobasal.
18. B27.
19. PD0325901 (MEK inhibitor).
20. XAV939.
21. Gö6983 (aPKC inhibitor).
22. Human leukemia inhibitory factor (LIF), prepared in-house or purchased commercially.
23. TrypLE Express.
24. Y-27632 (Y; ROCKi).
25. Geltrex (*see* **Note 8**).
26. A83-01 (Activin/Nodal/TGF-beta pathway inhibitor).

2.3 Culture Media

1. *iMEF medium*: 500 mL Dulbecco's Modified Eagle's Medium—high glucose, 50 mL FBS, 5.5 mL Penicillin-Streptomycin-L-Glutamine solution (×100).
2. *N2 (100 mL)*: DMEM/F-12, 1 g apo-transferrin, 250 mg insulin, 3 μM sodium selenite, 160 mg putrescine, 200 μg progesterone, 750 mg BSA.
3. *N2B27 medium*: 500 mL DMEM/F-12, 500 mL Neurobasal, 10 mL B27, 5 mL N2, 5 mL L-glutamine (final concentration: 2 mM), 1 mL 2-mercaptoethanol (*see* **Note 7**).
4. *Wash medium*: 0.1% BSA in DMEM/F-12.
5. *Human naïve pluripotent stem cell culture medium (PXGL)*: 1 μM PD0325901, 2 μM XAV939, 2 μM Gö6983, 10 ng/mL human LIF, 10 μM Y-27632 in N2B27.
6. *TE induction medium (PAY)*: 1.5 μM PD0325901, 1 μM A83-01, 10 μM Y-27632 in N2B27.
7. *Blastoid maturation medium*: 1 μM A83-01 in N2B27.

2.4 Materials for Examination of Trophectoderm Differentiation Potential of Human Naïve PSCs

1. Horizontal Laminar flow hood.
2. Phase contrast microscope.
3. Hemocytometer or equivalent cell-counting device.
4. Conical sterile polypropylene 15 mL centrifuge tubes.
5. Centrifuge for conical tubes.

6. Vacuum pump.
7. Aspirator pipet.
8. Pipette controller.
9. Sterile graduated pipettes, 5 and 10 mL.
10. Micropipettes P2, P20, P200, and P1000 and tips.
11. 48-well plates.
12. Human naïve PSCs.
13. Geltrex (*see* **Note 8**).
14. TrypLE Express.
15. Rho-associated kinase inhibitor (ROCKi, Y-27632).
16. Wash medium.
17. Human naïve pluripotent stem cell culture medium (PXGL).
18. TE induction medium (PAY).

3 Methods

All procedures should be carried out using aseptic techniques. The cells should be regularly checked for mycoplasma contamination.

3.1 Propagation of Human Naïve PSCs

This section is also described in the "Maintenance of Human Naïve Pluripotent Stem Cells" chapter [13]. Bring media and TrypLE at room temperature for at least 30 min before use.

3.1.1 Feeder Preparation

At least two days before use, prepare 6-well culture plates by thawing and plating batch-tested iMEF ($1.5–1.75 \times 10^4$ cells/cm^2) and allow them to settle in the incubator in iMEF medium (2 mL per well; *see* **Note 6**).

1. Add 1 mL of 0.1% gelatin in PBS to each well of the culture plate and incubate for at least 5 min at room temperature.
2. Add 9.5 mL of pre-warmed wash medium to a 50 mL centrifuge tube.
3. Quickly thaw an iMEF stock vial in a 37 °C water bath.
4. Gently mix the iMEF cell suspension by adding 500 μL–1 mL pre-warmed wash medium to the vial using a P1000 micropipette.
5. Transfer the entire cell suspension to the 50 mL centrifuge tube containing the wash medium.
6. Centrifuge the tube at 300× *g* for 3 min.
7. Aspirate the supernatant in the 50 mL centrifuge tube using a vacuum pump.
8. Resuspend the cell pellet with 1 mL of iMEF medium.

9. Add 9 mL of iMEF medium and gently mix with a 10 mL pipette.
10. Count the total cell number with a hemocytometer or an equivalent cell-counting device.
11. Adjust the cell suspension to 7.5–8.3 × 10^4 cells/mL using iMEF medium.
12. Aspirate the gelatin solution from the plates.
13. Seed the cell suspension onto the 0.1% gelatin-coated plates (2 mL/well of 6-well plates with a seeding density of 1.5–1.75 × 10^4 cells/cm^2).
14. Culture the iMEF in a humidified incubator at 37 °C with 5% CO_2.
15. Allow feeders to settle for two days before use.

3.1.2 Naïve Human PSCs Thawing and Propagation

At least one week prior to human blastoid generation, thaw human naïve PSCs and propagate them on batch-tested iMEF.

1. Add 9.5 mL of wash medium to a 15 mL centrifuge tube and warm it in a 37 °C water bath.
2. Quickly thaw a vial of human naïve PSCs (e.g., HNES1) in a 37 °C water bath.
3. Add 500 μL–1 mL of pre-warmed wash medium to the vial and mix gently using a P1000 micropipette.
4. Transfer the cell suspension into the 15 mL centrifuge tube containing warm wash medium.
5. Centrifuge the tube at 300× *g* for 3 min.
6. Aspirate the supernatant using a vacuum pump.
7. Flick the tube gently to loosen the cell pellet.
8. Resuspend the pellet with 2 mL of PXGL medium (*see* **Notes 9** and **10**).
9. Aspirate iMEF medium from one well of a pre-prepared feeder plate.
10. Seed the cell suspension into the iMEF-coated plates (2 mL per well of a 6-well plate).
11. Add 10 μM Y and diluted Geltrex (1:4 dilution; 2 μL/cm^2, typically 20 μL per well of a 6-well plate; *see* **Notes 8–10**).
12. Gently shake the plate to distribute the cells evenly.
13. Incubate the cells in a humidified incubator at 37 °C with 5% O_2 and 5% CO_2. The day of seeding human naïve PSCs is considered Day 0.
14. On Day1, add 1 mL of PXGL medium per well (half-volume feed; *see* **Note 11**).

15. On Day 2 and onwards, remove half of the medium and replace it with an equal volume of fresh PXGL medium daily until passaging (*see* **Note 12**).
16. Passage cells when they reach semi-confluency, usually on day 3–5.
17. Aspirate the medium from the well using a vacuum pump.
18. Add 1 mL of TrypLE to the well and incubate for 3–4 min in a 37 °C incubator.
19. Gently pipette with a P1000 micropipette to dissociate colonies into single cells.
20. Add 2 mL of wash medium to stop the enzyme reaction and gently mix using a P1000 micropipette or 5 mL plastic pipette.
21. Transfer 1 mL (or corresponding volume) of the cell suspension into a 15 mL centrifuge tube. Normally, a 1:3–1:5 passage is recommended.
22. Centrifuge the tube at 300× *g* for 3 min.
23. Aspirate the supernatant using a vacuum pump.
24. Flick the cell pellet several times to loosen it.
25. Resuspend the cells with 2 mL of PXGL medium.
26. Add 2 μL of Y and 20 μL of diluted Geltrex (1:4 Geltrex) to the cell suspension.
27. Aspirate iMEF medium from a well using a vacuum pump.
28. Gently seed the cell suspension onto the iMEF-coated plates (2 mL per well of a 6-well plate).
29. Culture the human naïve PSCs in a humidified incubator at 37 °C with 5% O_2 and 5% CO_2.
30. Propagate the cells until use by following **steps 14–29**.

3.1.3 Examination of the Trophectoderm Differentiation Potential of Human Naïve PSCs

1. Prepare Geltrex-coated plates in advance by adding diluted Geltrex in PBS (1:100 Geltrex) to a 48-well plate (100 μL/cm^2; 1 μL/cm^2 original Geltrex concentration) and incubate overnight at 4 °C. Alternatively, incubate Geltrex-added plates for a couple of hours at room temperature followed by 30 min at 37 °C. Equilibrate the plates at room temperature before use.
2. Dissociate the naïve hPSCs as described in Subheading 3.1.2, **steps 17–24**.
3. Resuspend the cells in PXGL + Y medium, aspirate the Geltrex solution from the Geltrex-coated plate and seed the cells onto the plate at a density of 1–4 × 10^4 cells/cm^2.
4. The following day, aspirate the media and gently wash the wells three times with PBS to remove excess Geltrex.

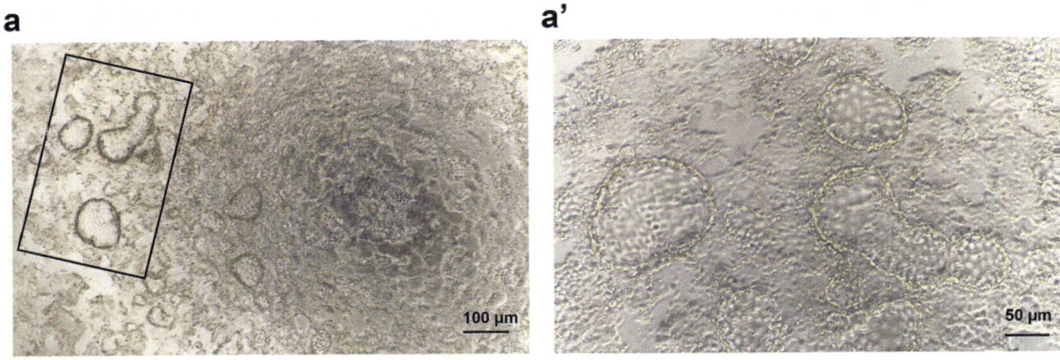

Fig. 1 Differentiation to trophectoderm lineage cells from human naïve PSCs. (**a**) Representative image of semi-spherical colonies after four days of TE induction and a higher-magnification image **(a')**

5. Aspirate PBS and replace with the TE induction medium (PAY).
6. Feed by medium changes every other day.
7. The cells form flat structures after 1–2 days of switching the medium.
8. Some colonies form semi-spheres after 4–5 days of switching the medium (*see* Fig. 1).

3.2 Blastoids

3.2.1 Blastoids Formation

It is vitally important to use human naïve PSCs validated for trophectoderm differentiation potential. (*See* Subheading 3.1.3 for the trophectoderm differentiation test).

1. Harvest naïve hPSC on one or two wells of a 6-well plate in PXGL medium, once they have reached semi- or full confluency.
2. Aspirate the culture medium from the wells (*see* **Note 13**).
3. Add 1 mL of TyrpLE or relevant dissociation reagent to the wells.
4. Incubate the plates for 3 min in a humidified CO_2 incubator at 37 °C.
5. Meanwhile, prepare PAY medium in a 50 mL centrifuge tube. We recommend preparing twice the estimated volume needed.
6. Tap the bottom of the plates to make the attached colonies float. Avoid pipetting to minimize iMEF carry-over.
7. Collect the supernatant containing the floating colonies into another well of 12-well plate.
8. Incubate the 12-well plate for 3–5 min in a humidified CO_2 incubator at 37 °C.

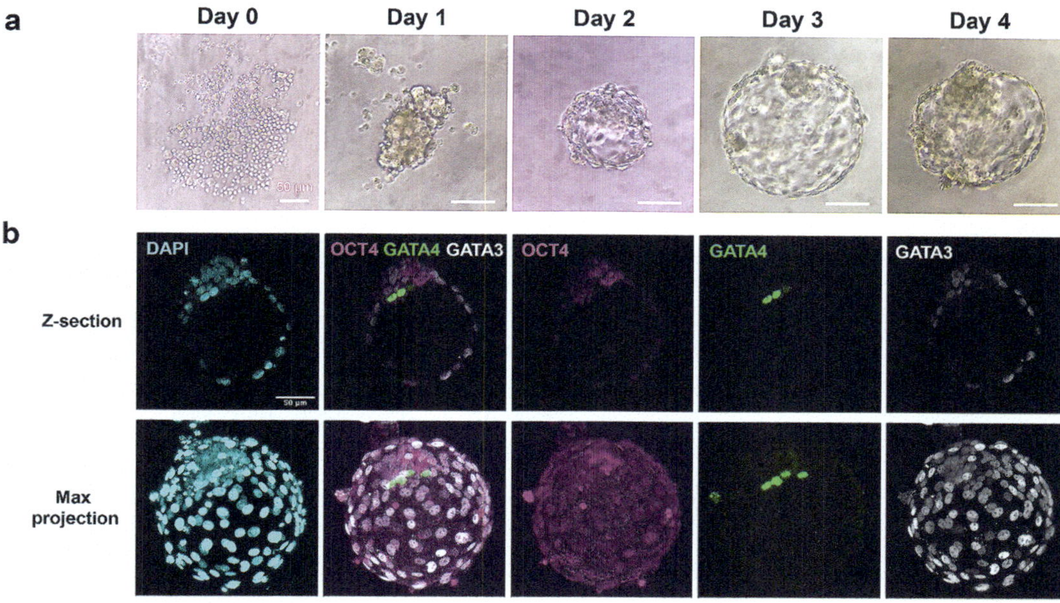

Fig. 2 Formation of human blastoids. (**a**) Cell aggregation morphology during blastoids formation. (**b**) Immunofluorescent staining for the naïve pluripotency marker (OCT4), hypoblast marker (GATA4), and trophectoderm marker (GATA3) during blastoids formation

9. If you want to passage the leftover colonies, add 1 mL of TrypLE to the 6-well plates and passage naïve hPSCs as described in Subheading 3.1.2, **steps 18–30** (optional).
10. Dissociate the cell clusters to a single-cell suspension by gently pipetting with a P1000 micropipette.
11. Add 2 mL of wash medium and gently mix using a 5 mL plastic pipette or a P1000 micropipette.
12. Transfer the cell suspension to a 15 mL centrifuge tube.
13. Count the cell using a hemocytometer or equivalent device to estimate total cell number (optional).
14. Centrifuge the 15 mL centrifuge tube for 3 min at 300× *g*.
15. Aspirate the supernatant.
16. Gently tap the tube to loosen the cell pellet.
17. Add 100–200 μL of PAY medium to the tube.
18. Resuspend the pellet by gently pipetting using a P200 micropipette.
19. Transfer the cell suspension to a 1.5 or 2.0 mL tube.
20. Count the cells again to confirm the total cell number.
21. Adjust the cell concentration to 2×10^3 cells/mL (for 100 cells per well), or as appropriate, in PAY. Preparing double the

expected volume is recommended to account for pipetting error.

22. Transfer the cell suspension to a reservoir.
23. Dispense 50 μL per well into a 96-well clear round bottom ultra-low attachment microplate using a multichannel pipette (*see* **Note 14**).
24. Centrifuge the plate at 300× *g* for 5 min with fast acceleration and slow deceleration.
25. Check cell clusters under a microscope to confirm that most cells gather at center of each well (*see* **Note 15**).
26. If the number of cells per well is below 100, add more cells by repeating **steps 23–25** until reaching 100–150 cells per well (*see* **Note 16**).
27. Incubate the plate in a humidified incubator at 37 °C with 5% O_2 and 5% CO_2. Set this day as Day 0.
28. On Day 2, prepare blastoid maturation medium and add 100 μL per well to a new U-bottom plate using a multichannel pipette (*see* **Note 4**). Incubate the plates in a humidified incubator at 37 °C with 5% O_2 and 5% CO_2 before transferring cell aggregates.
29. On Day 2 (*see* **Note 17**), transfer the cell aggregates to the plate containing blastoid maturation medium. Perform this under a stereo microscope in a laminar flow hood using a pulled glass capillary and mouth pipette. For alternative methods, *see* Subheading 3.2.2.
30. On Day 3, small blastoids should be observed (*see* Fig. 2a).
31. On Day 4, further developed blastoids should be apparent (*see* Fig. 2a, b).

3.2.2 Blastoids Formation Using a 24-Well Plate with Microwells

1. Prepare a 24-well plate with microwells (*see* **Note 18**) and add 2 mL of PAY medium per well.
2. Prepare a naïve hPSCs suspension as described in Subheading 3.2.1, **steps 1–18**.
3. Make a cell suspension with an appropriate cell density (100–200 cells per microwell) in PAY medium (*see* **Note 19**).
4. Transfer the cell suspension to the wells containing PAY medium.
5. Resuspend the cell suspension by gently pipetting using a P200 micropipette.
6. Immediately centrifuge the plate at 100× *g* for 3 min with fast acceleration and slow deceleration (*see* **Note 20**).
7. Check cell clusters under a microscope.

8. Incubate the plate in a humidified incubator at 37 °C with 5% O_2 and 5% CO_2. Set this day as Day 0.
9. On Day 2, gently remove the PAY medium using a P1000 and P200 micropipette. Be careful not to aspirate the cell aggregates (*see* **Note 21**).
10. Add 1–2 mL of blastoid maturation medium gently.
11. Incubate the plate in a humidified CO_2 incubator at 37 °C.
12. On Day 3, many small blastoids should be visible.
13. On Day 4, more developed blastoids should be observed.

4 Notes

1. 7% CO_2 and 5% O_2 condition can also be used for the culture of human naïve PSCs and the formation of human blastoids.
2. Low-attachment 96-well U-bottom plates with good shape and high quality are essential for making cell aggregates. For example, Corning Cat# 7007.
3. Aggregation well plates with multiple microwells per well can be used.
4. The same type of low-attachment 96-well U-bottom plates can be used to make cell aggregates. However, lower-cost alternative can be substituted.
5. Pasteur pipettes are pulled and bent over a small flame (Bunsen burner or spirit lamp) to achieve an internal diameter slightly wider than the expanded blastoids (around 50–200 μm). Alternatively, yellow tips and a multichannel pipette can be used.
6. It is vitally important to test the iMEF batch and seeding density before use. iMEF can be prepared by irradiation or mitomycin C treatment of primary MEF derived from E13.5 mouse fetuses. Propagate primary MEF in DMEM high glucose supplemented with 15% FBS, 1× Penicillin-Streptomycin-L-Glutamine, and 100 μM 2-mercaptoethanol for three passages, and then inactivate with irradiation or mitomycin C. Aliquot iMEF into cryovials with freezing medium, store at −80 °C for one day, and transfer to liquid nitrogen the next day. The numbers of cells per vial can be adjusted according to experimental needs. Each iMEF batch should be pre-tested for mycoplasma and well coverage. Optimal seeding density depends on how well the iMEF spread to cover the surface without becoming overly dense. iMEF seeded plates can be stored in the incubator for up to two weeks, with medium change once per week until use.

7. Adjust the amount of L-glutamine solution depending on whether DMEM/F-12 formulation contains L-glutamine. N2B27 medium prepared according to [14] or purchased from commercial suppliers (e.g., NDiff 277, Takara, Cat# Y40002) can be used. The quality of N2B27 medium is vital for successful naïve hPSCs culture.
8. Thaw Geltrex overnight on ice at 4 °C. Place thawed Geltrex on fresh ice and quickly aliquot into sterile 1.5 or 2.0 mL screw-cap tubes. Store stock Geltrex at −80 °C until use. Thaw aliquots on ice or at 4 °C and dilute 1:4 with cold DMEM/F-12 to prepare working solution. For feeder-free culture, dilute stock Geltrex 1:100 with cold PBS. Store this at 4 °C for up to one week.
9. Adjust the culture plate size and medium volume depending on the number and viability of cells in the thawed vial.
10. Y and Geltrex can be added to the cell suspension in PXGL medium before seeding or directly to the well after plating the cell in PXGL medium.
11. Most of the seeded cells should attach to the iMEF. If many dead cells are floating, check the concentration and quality of Y. If many viable cells remain unattached, check the Geltrex concentration and quality, and increase the volume added to the well if necessary.
12. If many viable colonies are floating in a well, transfer 1 mL of the culture medium along with the colonies into a 15 mL centrifuge tube. Centrifuge the tube for 3 min at 300× *g*, aspirate the supernatant, then resuspend the pellet in 1 mL of fresh PXGL medium, and return it to the well.
13. An alternative method for steps 2–7 is as follows. When iMEF are firmly attached to the well and human naïve PSCs have formed nicely round dome-shaped colonies, you can gently pipette the colonies using P1000 pipette and transfer floating colonies into a 15 mL centrifuge tube. Centrifuge at 300× *g* for 3 min, aspirate the supernatant and transfer the cell pellet with 1 mL of TrypLE into a 12-well plate.
14. For the initial experiment, it is recommended to use a relatively high number of cells (e.g., 200–300 cells per well) to evaluate the differentiation potential of the cells in use. Once the cells consistently form aggregates and cystic structures within five days in PAY medium, optimize the cell number per well accordingly.
15. If cells fail to aggregate at the center of the well, check the balance settings and carefully adjust the centrifuge parameters.
16. The optimal number of cells per well may vary depending on the cell lines. Optimization is required.

17. The optimal duration of TE induction may vary between cell lines. For initial experiments, treat cells with PAY medium for 3–5 days to assess their differentiation potential. Once the cells reliably form aggregates and cystic structures by Day 5, optimize the induction period accordingly.
18. It is essential to use low-attachment plates and to seed cells uniformly to generally well-shaped cell aggregates. For example, AggreWell™ 400 24-well plates (STEMCELL Technologies, Cat# 34411) or EZSPHERE®SP microplates 24-well (IWAKI, Cat# 4820-900SP) can be used. AggreWell™ plates require pretreatment with Anti-Adherence Rinsing Solution (500 μL/well, Cat# 07010). Be sure to remove air bubbles from microwells before seeding. Add rinsing solution or PAY medium and centrifuge for 5 min to remove bubbles. For AggreWell™ plates, aspirate the rinsing solution and wash once with 2 mL of PAY medium before seeding.
19. The number of microwells per well varies depending on the product. Adjust the cell number accordingly. For example, one well of an AggreWell™ 400 24-well plate contains approximately 1,200 microwells per well, so a suitable seeding density is $1.2–2.4 \times 10^5$ cells/well.
20. Centrifuge the plate immediately after seeding using proper balance. Uniform seeding into microwells is essential for generating well-formed aggregates.
21. When changing medium on Day 2, do not use a vacuum pump, as this may aspirate the cell aggregates.

Acknowledgments

This work was supported by Japan Society for the Promotion of Science (JSPS) FOREST Program (JPMJFR214Y to A.Y.) and KAKENHI (21K20651, 22K07886 to A.Y.)

References

1. Chazaud C, Yamanaka Y, Pawson T, Rossant J (2006) Early lineage segregation between epiblast and primitive endoderm in mouse blastocysts through the Grb2-MAPK pathway. Dev Cell 10:615–624
2. Plusa B, Piliszek A, Frankenberg S et al (2008) Distinct sequential cell behaviours direct primitive endoderm formation in the mouse blastocyst. Development 135:3081–3091
3. Rivron NC, Frias-Aldeguer J, Vrij EJ et al (2018) Blastocyst-like structures generated solely from stem cells. Nature 557:106–111
4. Yanagida A, Spindlow D, Nichols J et al (2021) Naive stem cell blastocyst model captures human embryo lineage segregation. Cell Stem Cell 28:1016–1022. e4
5. Karvas RM, Zemke JE, Ali SS et al (2023) 3D-cultured blastoids model human embryogenesis from pre-implantation to early gastrulation stages. Cell Stem Cell 30:1148–1165. e7
6. Guo G, von Meyenn F, Rostovskaya M et al (2017) Epigenetic resetting of human pluripotency. Development 144:2748–2763

7. Takashima Y, Guo G, Loos R et al (2014) Resetting transcription factor control circuitry toward ground-state pluripotency in human. Cell 158:1254–1269
8. Guo G, von Meyenn F, Santos F et al (2016) Naive pluripotent stem cells derived directly from isolated cells of the human inner cell mass. Stem Cell Rep 6:437–446
9. Guo G, Stirparo GG, Strawbridge SE et al (2021) Human naive epiblast cells possess unrestricted lineage potential. Cell Stem Cell 28:1040–1056.e6
10. Handyside AH (1978) Time of commitment of inside cells isolated from preimplantation mouse embryos. J Embryol Exp Morphol 45: 37–53
11. Nichols J, Gardner RL (1984) Heterogeneous differentiation of external cells in individual isolated early mouse inner cell masses in culture. J Embryol Exp Morphol 80:225–240
12. Spindle AI (1978) Trophoblast regeneration by inner cell masses isolated from cultured mouse embryos. J Exp Zool 203:483–489
13. Rostovskaya M (2022) Maintenance of human naïve pluripotent stem cells. Methods Mol Biol 2416:73–90
14. Mulas C, Kalkan T, von Meyenn F et al (2019) Defined conditions for propagation and manipulation of mouse embryonic stem cells. Development 146. https://doi.org/10.1242/dev.173146

Methods in Molecular Biology (2026) 3048: 61–70
DOI 10.1007/7651_2025_662

Published online: 05 August 2025

Use of miR-200-Mediated Reprogramming and Mechanical Cues to Generate Human Blastocyst Models

Sharon Arcuri, Georgia Pennarossa, Fulvio Gandolfi, and Tiziana A. L. Brevini

Abstract

Several early developmental events remain unclear due to ethical and technical limitations related to the use of natural embryos. To overcome this problem, over the last decade, different approaches aimed at the generation of in vitro blastocyst-like models have been developed.

Here, we describe a protocol that combines miR-200-mediated cell reprogramming and mechanical stimuli to create 3D spheroids phenotypically similar to natural embryos. Specifically, dermal fibroblasts are reprogrammed exploiting the miR-200 family property to induce a high plasticity state in somatic cells. Subsequently, miR-200-reprogrammed cells are either differentiated into trophoblast (TR)-like cells, using an ad hoc induction protocol, or encapsulated into polytetrafluoroethylene (PTFE) micro-bioreactors to generate inner cell mass (ICM)-like spheroids. Subsequently, TR-like cells and ICM-like spheroids are encapsulated and co-cultured in the same micro-bioreactor and then transferred to microwells to allow further differentiation and boost blastoid development.

The generated 3D artificial blastoids, phenotypically similar to natural embryos, could find a useful application in the field of assisted reproduction technologies to further characterize the events taking place during early embryogenesis as well as to study embryo disorders.

Key words Blastoids, Cellular reprogramming, ICM-like spheroids, Microbioreactors, miR-200s, TR-like cells

1 Introduction

Blastocyst formation constitutes a pivotal event in early mammalian embryogenesis, marking the transition from cleavage divisions to the beginning of gastrulation [1]. The blastocyst is characterized by the emergence of two distinct cell lineages: the trophoblast, forming the outer epithelial layer responsible for implantation and placenta formation, and the inner cell mass (ICM), which segregates into the epiblast and hypoblast contributing to the formation of the embryo proper and the yolk sac, respectively [2, 3]. However, different mechanisms taking place during the early stages of embryo development still remain unclear, due to ethical and practical limitations. To solve this problem, different approaches for the generation of blastocyst-like in vitro models, defined blastoids, have

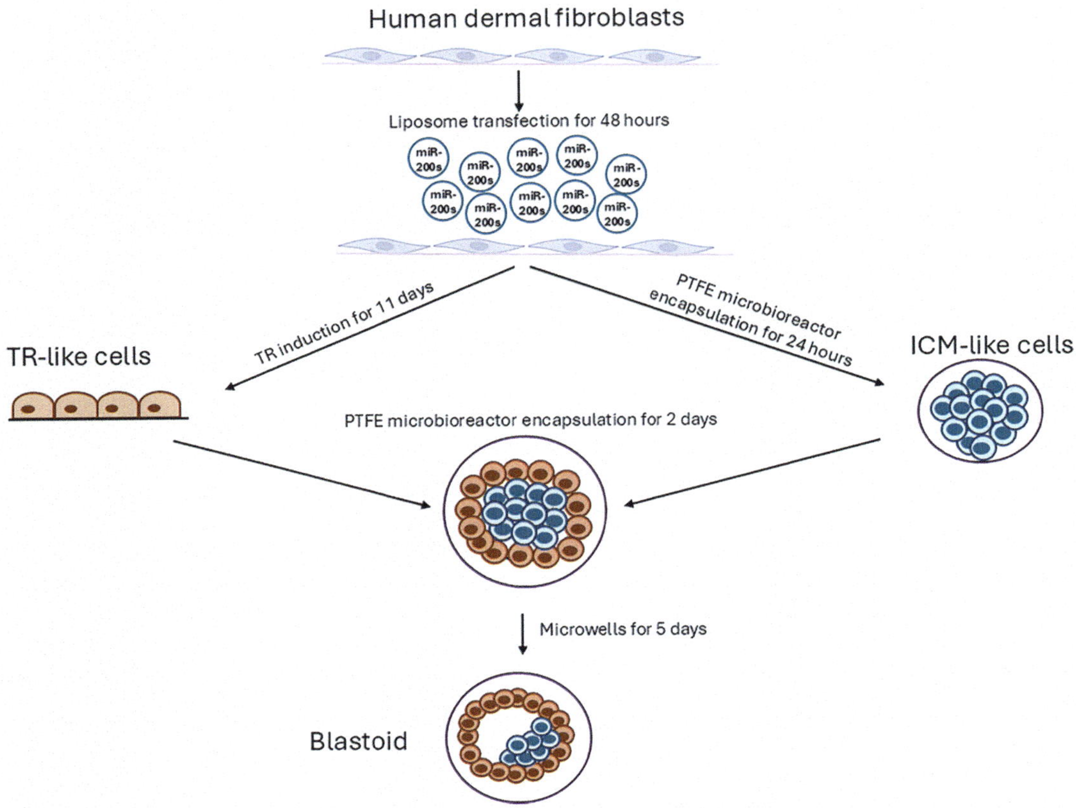

Fig. 1 Scheme describing blastoid generation from human dermal fibroblasts

been developed. In particular, the possibility to generate artificial embryoids starting from several stem cell types, including human [4–7], mouse [8], and bovine [9] embryonic stem as well as human-induced pluripotent stem cells (iPS) [10] and human and porcine chemical reprogrammed cells [11, 12] is reported. Interestingly, parallel studies also demonstrated the ability of miR-200 family to induce a transient window of increased cell plasticity, thanks to the involvement of Ten Eleven Translocation (TET) enzymes, which are known to be able to remove "epigenetic blocks" and reactivate previously silenced genes [13–16].

Here, a protocol that allows the generation of 3D blastocyst-like spheroids phenotypically similar to natural embryos (Fig. 1) is described. In particular, human dermal fibroblasts are reprogrammed using miR-200b/c and differentiated either into trophoblast (TR)-like cells [17, 18] or encapsulated in polytetrafluoroethylene (PTFE) microbioreactors to generate ICM-like spheroids [19]. Subsequently, TR-like cells are co-cultured with ICM-like organoids in the same micro-bioreactor to boost self-assembly and form artificial embryoids, which are then transferred to microwells, to encourage further differentiation.

2 Materials

2.1 Preparation of Mouse Embryonic Fibroblast (MEF)-Conditioned Medium

1. MEF cells.
2. Water bath.
3. Sterile 15 mL tubes.
4. Centrifuge.
5. T25 flasks.
6. 0.22 μm filters.
7. Optical inverted microscope.
8. CO_2 incubator.
9. Counting chamber.
10. Trypsin–EDTA solution.
11. Dulbecco's phosphate-buffered saline (PBS) containing 1% antibiotic/antimycotic solution.
12. MEF culture medium: 88% (v/v) Dulbecco's modified Eagle medium, 10% (v/v) fetal bovine serum (FBS), 2 mM (v/v) L-glutamine solution, 1% antibiotic/antimycotic solution.
13. MEF inactivating medium: 10 ug/mL Mytomicin C was added to 6 mL of MEF culture medium.
14. ESC culture medium without bFGF: 40% Ham's F-10 Nutrient mix, 40% DMEM low glucose, 10% Knock-out Serum Replacement, 5% FCS, 1% antibiotic/antimycotic solution, 2 mM L-glutamine, 1% Nucleoside Mix, 1% MEM Non-essential Amino Acids Solution, 0.1 mM 2-mercaptoethanol, 1 unit/mL LIF.

2.2 Transfection of Human Dermal Fibroblast with miR200b/c

1. Human skin dermal fibroblasts.
2. T25 flasks.
3. CO_2 incubator.
4. Water bath.
5. Sterile 15 mL tubes.
6. Cell counter.
7. Centrifuge.
8. 4-well multi-dishes.
9. Optical inverted microscope.
10. Trypsin–EDTA solution.
11. Dulbecco's phosphate-buffered saline (PBS) containing 1% antibiotic/antimycotic solution.

12. Fibroblast culture medium: 83% (v/v) Eagle's minimum essential medium, 15% (v/v) FBS, 2 mM (v/v) L-glutamine solution, 1% antibiotic/antimycotic solution.
13. 0.1% porcine gelatin: dissolve 0.1 g of porcine gelatin in 100 mL of sterile water. Autoclave and stock at +4 °C.
14. Stock solution of miR-200b (0.1 mM): add 50 μL of sterile water to 5 nmol of lyophilized predesigned mirVANA Mimics miR-200b.
15. Stock solution of miR-200c (0.1 mM): add 50 μL of sterile water to 5 nmol of lyophilized predesigned mirVANA Mimics miR-200c.
16. ESC culture medium: 40% Ham's F-10 Nutrient mix, 40% DMEM low glucose, 10% Knock-out Serum Replacement, 5% FCS, 1% antibiotic/antimycotic solution, 2 mM L-glutamine, 1% Nucleoside Mix, 1% MEM Non-essential Amino Acids Solution, 0.1 mM 2-mercaptoethanol, 1 unit/mL LIF, 5 ng/mL human basic fibroblast growth factor (bFGF).

2.3 Trophoblast Induction

1. Trophoblast induction medium: 10 mL of conditioned MEF inactivated medium, 10 μL of BMP4 (50 ng/mL), 10 μL A83–01 (1 μM), 10 μL PD173074 (0.1 μM) (*see* **Note 1**).
2. Tri-gas incubator.

2.4 Generation of Inner Cell Mass (ICM)-Like Spheroids

1. Water bath.
2. 35 mm bacteriology Petri dishes.
3. 100 mm Petri dishes.
4. Sterile 15 mL tubes.
5. Cell counter.
6. Centrifuge.
7. Optical inverted microscope.
8. PTFE powder.
9. 1000 μL pipette tip.
10. 200 μL pipette tip.
11. Trypsin–EDTA solution.
12. Sterile water.
13. Dulbecco's phosphate-buffered saline (PBS) containing 1% antibiotic/antimycotic solution.
14. ESC culture medium: 40% Ham's F-10 Nutrient mix, 40% DMEM low glucose, 10% Knock-out Serum Replacement, 5% FCS, 1% antibiotic/antimycotic solution, 2 mM L-glutamine, 1% Nucleoside Mix, 1% MEM Non-essential Amino

Acids Solution, 0.1 mM 2-mercaptoethanol, 1 unit/mL LIF, 5 ng/mL human basic fibroblast growth factor (bFGF).

15. CO_2 incubator.

2.5 Creation of Blastoids

1. Water bath.
2. 35 mm bacteriology Petri dishes.
3. 100 mm petri dishes.
4. Sterile 15 mL tubes.
5. Cell counter.
6. Sterile needle.
7. Centrifuge.
8. Optical inverted microscope.
9. Stereomicroscope.
10. PTFE powder.
11. 1000 μL pipette tip.
12. 200 μL pipette tip.
13. Microwells.
14. Trypsin–EDTA solution.
15. Dulbecco's phosphate-buffered saline (PBS) containing 1% antibiotic/antimycotic solution.
16. Sterile water.
17. G1-PLUS medium.
18. G2-PLUS medium.
19. PODS Activin A.
20. CO_2 incubator.

3 Methods

3.1 Ethical Statement

Ethical approval was not required because this study did not involve human participants. The NIGMS Human Genetic Cell Repository at the Coriell Institute for Medical Research (Camden, NJ, USA) was selected to purchase human dermal fibroblasts (GM02674, GM00495, and GM08402).

3.2 Preparation of MEF Conditioned Medium

1. When MEFs reach confluency, eliminate "MEF culture medium" from T25 flask.
2. Wash three times in sterile PBS containing 1% antibiotic/antimycotic solution.
3. Wash one time with 1 mL of trypsin–EDTA solution.

4. Add 0.6 mL of trypsin–EDTA solution and incubate at 37 °C for 3 min.
5. Add 5.4 mL of "MEF culture medium" supplemented with 10% FBS to neutralize trypsin–EDTA.
6. Gently and repeatedly pipette cells to detach them from the bottom of the culture dish.
7. Collect and transfer cellular suspension in a sterile 15 mL tube.
8. Calculate the volume of medium to resuspend cells at a concentration of 2×10^4 cells/cm^2, by counting cells using a chamber under an optical inverted microscope.
9. Centrifuge cell suspension at 300 g for 5 min.
10. Plate MEFs at a density of 2×10^4 in a new T25 flask and maintain for 24 h in CO_2 incubator.
11. Wash cells three times in sterile PBS containing 1% antibiotic/antimycotic solution.
12. Inactivate sub-confluent MEF monolayers using "MEF inactivating medium" for 3 h in CO_2 incubator.
13. Remove "MEF inactivating medium."
14. Add "ESC medium without b-FGF" and culture in CO_2 incubator.
15. Collect conditioned medium every 24 h.
16. Filter the collected MEF conditioned medium with 0.22 μm filters and store at −20 °C until use.

3.3 Transfection of Human Dermal Fibroblasts with miR-200b/c

1. Seed 0.5 mL of 0.1% porcine gelatine in each well of the 4-well multi-dishes and incubate for 20 min at room temperature.
2. At the end of the 20 min, remove the excess of gelatine and let the plate dry for 45 min under a sterile hood.
3. When human dermal fibroblasts reach confluency, eliminate "Fibroblast culture medium" from T25 flask.
4. Wash three times in sterile PBS containing 1% antibiotic/antimycotic solution and one time with 1 mL of trypsin–EDTA solution.
5. Add 0.6 mL of trypsin–EDTA solution and incubate at 37 °C for 3 min.
6. Add 5.4 mL of "Fibroblast culture medium" supplemented with 10% FBS to neutralize trypsin–EDTA.
7. Detach them from the bottom of the culture dish as described in Subheading 3.2, **step 6**.
8. Collect and transfer cellular suspension in a sterile 15 mL tube.

9. Calculate the volume of medium to resuspend cells at a concentration of 2.1×10^4 cells/cm^2, by counting cells using a chamber under an optical inverted microscope.
10. Centrifuge cell suspension at 300 g for 5 min.
11. Seed human dermal fibroblasts at a density of 2×10^4 in a 4-well multi-dishes and maintain for 24 h in CO_2 incubator.
12. Wash cells three times in sterile PBS containing 1% antibiotic/antimycotic solution.
13. Dilute 2 μL of 0.1 mM miR200b stock solution in 1.98 mL of OptiMEM medium to obtain 100 nM final concentration.
14. Dilute 2 μL of 0.1 mM miR200c stock solution in 1.98 mL of OptiMEM medium to obtain 100 nM final concentration.
15. Add 1 μL of 100 nM miR-200b (**step 13**) and 1 μL of 100 nM miR-200c (**step 14**) to 100 μL of OptiMEM medium (solution A).
16. Add 6 μL of Lipofectamine RNAiMAX to 100 μL of OptiMEM medium (solution B).
17. Mix solution A and solution B in a 1:1 ratio (solution C) and incubate for 5 min at room temperature (*see* **Note 2**).
18. Eliminate culture medium from the 4-well multi-dishes containing human dermal fibroblasts.
19. Wash three times in 1% PBS containing 1% antibiotic/antimycotic solution.
20. Add 200 μL of solution C (**step 17**) in three wells. Use one as a negative control.
21. Incubate at 37 °C in 5% CO_2 incubator for 48 h.
22. Remove solution C.
23. Rinse three times with sterile PBS containing 1% antibiotic/antimycotic solution.
24. Add 0.5 mL of "ESC culture medium" and incubate at 37 °C in CO_2 incubator for 3 h.
25. Wash cells three times in sterile PBS containing 1% antibiotic/antimycotic solution.
26. Proceed with trophoblast induction, Subheading 3.4, or Inner Cell Mass production, Subheading 3.5.

3.4 Trophoblast Induction

1. Remove washing PBS.
2. Add 0.5 mL of "Trophoblast induction medium" and incubate at 37 °C in tri-gas incubator for 11 days.
3. Refresh culture medium every other day.

3.5 Generation of ICM-Like Spheroids

1. Remove washing PBS.
2. Add 0.2 mL of trypsin–EDTA solution and incubate at 37 °C for 3 min.
3. Add 0.3 mL of "ESC culture medium" to neutralize trypsin–EDTA.
4. Collect cell suspension in a sterile 15 mL tube.
5. Count cells using a counting chamber under an optical inverted microscope. Calculate the volume of medium needed to resuspend cells in order to obtain 1×10^4 cells/30 μL.
6. Centrifuge cell suspension at 300 g for 5 min.
7. Resuspend the pellet in the calculated volume of "ESC culture medium" and pipette carefully.
8. Transfer PTFE powder onto a 35 mm bacteriology Petri dish.
9. Dispense 30 μL single droplet of "ESC culture medium" containing 1×10^4 cells onto the PTFE powder bed.
10. Gently rotate the 35 mm bacteriology Petri dish in a circular motion to completely cover the surface of the liquid drop with the PTFE powder particles and form a liquid marble micro-bioreactor.
11. Collect the generated PTFE micro-bioreactor using a 1000 μL pipette tip cut at the edge and lay it onto a clean 35 mm bacteriology Petri dish.
12. Place the 35 mm bacteriology Petri dish containing the micro-bioreactor within a 100 mm Petri dish containing sterile water to prevent evaporation.
13. Culture at 37 °C in 5% CO2 incubator for 24 h.

3.6 Creation of Blastoids

1. After 11 days of trophoblast induction (Subheading 3.4), eliminate trophoblast induction medium.
2. Wash three times in sterile PBS supplemented with 1% antibiotic/antimycotic solution.
3. Detach cells from the 4-well multi-dishes by adding 0.2 mL of trypsin–EDTA solution and incubating at 37 °C.
4. Add 0.3 mL of "Trophoblast induction medium" to neutralize trypsin–EDTA.
5. Dislodge cells by repeatedly and gently pipetting.
6. Collect cells in sterile 15 mL tubes, count them and calculate the volume of medium required to obtain 3×10^4 cells/30 μL.
7. Centrifuge cell suspension at 300 g for 5 min and resuspend in the calculated volume of G1-PLUS medium.
8. Break PTFE micro-bioreactors by puncturing them with a needle.

9. Collect the formed ICM-like spheroids using a 200 μL pipette tip under a stereomicroscope and transfer them in G1-PLUS culture medium.
10. Add new PTFE powder onto a 35 mm bacteriology Petri dish.
11. Generate a droplet of 40 μL containing 3×10^4 cells (30 μL) and one ICM-like spheroid (10 μL).
12. Dispense the 40 μL droplet onto the PTFE powder bed and generate new micro-bioreactors following the instruction described in Subheading 3.5, **step 10**.
13. Pick up the newly formed PTFE micro-bioreactors using a 1000 μL pipette tip cut at the edge and move onto a clean 35 mm bacteriology Petri dish.
14. Place 35 mm bacteriology Petri dish containing the micro-bioreactor within a 100 mm Petri dish containing sterile water to prevent evaporation and incubate at 37 °C in 5% CO_2 incubator for 48 h.
15. After 48-h incubation, break the micro-bioreactors, collect the generated epiBlastoid, and transfer it in microwells containing G2-PLUS medium.
16. Incubate for 24 h at 37 °C in 5% CO_2.
17. After 24 h, add 10^5 PODS Activin A in 150 μL of G2-PLUS medium.
18. Maintain at 37 °C in a 5% CO_2 incubator for 4 days.

4 Notes

1. Aliquot BMP4, A83–01, PD173074 just prior use.
2. Prepare miR-200b/c solution just prior use.

Acknowledgments

This study was supported by the Carraresi Foundation, PRIN 2022 Funded by the European Union—NextGenerationEU, PSR2022, and PSR2023.

Conflict of Interest Statement The authors declare no conflict of interest.

References

1. Rossant J, Tam PPL (2022) Early human embryonic development: blastocyst formation to gastrulation. Dev Cell 57:152–165. https://doi.org/10.1016/j.devcel.2021.12.022
2. Brevini TAL, Pennarossa G, Antonini S, Gandolfi F (2008) Parthenogenesis as an approach to pluripotency: advantages and limitations involved. Stem Cell Rev:127–135. https://doi.org/10.1007/s12015-008-9027-z
3. Brevini TAL, Pennarossa G, Vanelli A, Maffei S, Gandolfi F (2012) Parthenogenesis in non-rodent species: developmental competence and differentiation plasticity. Theriogenology 77:766–772. https://doi.org/10.1016/j.theriogenology.2011.11.010
4. Kagawa H, Javali A, Heidari Khoei H, Sommer TM, Sestini G, Novatchkova M, Scholte op Reimer Y, Rivron N (2022) Protocol for human blastoids modeling blastocyst development and implantation. J Vis Exp. https://doi.org/10.3791/63388-v
5. Kagawa H, Javali A, Khoei HH et al (2022) Human blastoids model blastocyst development and implantation. Nature:600–605. https://doi.org/10.1038/s41586-021-04267-8
6. Rivron NC, Frias-Aldeguer J, Vrij EJ, Boisset JC, Korving J, Vivié J, Truckenmüller RK, Van Oudenaarden A, Van Blitterswijk CA, Geijsen N (2018) Blastocyst-like structures generated solely from stem cells. Nature 557:106–111. https://doi.org/10.1038/s41586-018-0051-0
7. Weatherbee BAT, Cui T, Zernicka-Goetz M (2021) Modeling human embryo development with embryonic and extra-embryonic stem cells. Dev Biol 474:91–99. https://doi.org/10.1016/j.ydbio.2020.12.010
8. Li R, Zhong C, Yu Y et al (2019) Generation of blastocyst-like structures from mouse embryonic and adult cell cultures. Cell 179:687–702. https://doi.org/10.1016/j.cell.2019.09.029
9. Pinzón-Arteaga CA, Wang Y, Wei Y et al (2023) Bovine blastocyst-like structures derived from stem cell cultures. Cell Stem Cell 30:611–616. https://doi.org/10.1016/j.stem.2023.04.003
10. Niemann H, Seamark B (2021) Blastoids: a new model for human blastocyst development. Signal Transduct Target Ther 6:239. https://doi.org/10.1038/s41392-021-00663-8
11. Pennarossa G, Arcuri S, De Iorio T, Ledda S, Gandolfi F, Brevini TAL (2023) Combination of epigenetic erasing and mechanical cues to generate human epiBlastoids from adult dermal fibroblasts. J Assist Reprod Genet 40:1015–1027. https://doi.org/10.1007/s10815-023-02773-4
12. Arcuri S, Pennarossa G, Ledda S, Gandolfi F, Brevini TAL (2024) Use of epigenetic cues and mechanical stimuli to generate blastocyst-like structures from mammalian skin dermal fibroblasts. Methods Mol Biol:161–173. https://doi.org/10.1007/7651_2023_486
13. Pennarossa G, Arcuri S, Gandolfi F, Brevini TAL (2024) Generation of artificial Blastoids combining miR-200-mediated reprogramming and mechanical cues. Cells 13:628. https://doi.org/10.3390/cells13070628
14. Balzano F, Cruciani S, Basoli V, Santaniello S, Facchin F, Ventura C, Maioli M (2018) MiR200 and MiR302: two big families influencing stem cell behavior. Molecules 23:282. https://doi.org/10.3390/molecules23020282
15. Pennarossa G, De Iorio T, Arcuri S, Gandolfi F, Brevini TAL (2023) Synergistic effect of miR-200 and young extracellular matrix-based bio-scaffolds to reduce signs of aging in senescent fibroblasts. Stem Cell Rev Rep 19:417–429. https://doi.org/10.1007/s12015-022-10438-5
16. Arcuri S, Pennarossa G, De Iorio T, Gandolfi F, Brevini TAL (2023) 3D ECM-based scaffolds boost young cell Secretome-derived EV rejuvenating effects in senescent cells. Int J Mol Sci 24:8285. https://doi.org/10.3390/ijms24098285
17. Arcuri S, Pennarossa G, Gandolfi F, Brevini TAL (2021) Generation of trophoblast-like cells from Hypomethylated porcine adult dermal fibroblasts. Front Vet Sci 8:1–10. https://doi.org/10.3389/fvets.2021.706106
18. Arcuri S, Gandolfi F, Somigliana E, Brevini TAL (2021) A two-step protocol to erase human skin fibroblasts and convert them into trophoblast-like cells. Methods Mol Biol 2273:151–158. https://doi.org/10.1007/978-1-0716-1246-0_10
19. Pennarossa G, Manzoni EFM, Ledda S, deEguileor M, Gandolfi F, Brevini TAL (2019) Use of a PTFE micro-bioreactor to promote 3D cell rearrangement and maintain high plasticity in epigenetically erased fibroblasts. Stem Cell Rev Rep 15:82–92. https://doi.org/10.1007/s12015-018-9862-5

Methods in Molecular Biology (2026) 3048: 71–82
DOI 10.1007/7651_2025_652

Published online: 15 July 2025

Generation of the Nonintegrated Human Bilaminar Embryo Model (Bilaminoid) from Naïve Pluripotent Stem Cells

Takumi Okubo and Yasuhiro Takashima

Abstract

Human embryogenesis has been problematic to study due to technical and ethical issues. Recently, a human embryo model generated from pluripotent stem cells (PSCs) to mimic human embryogenesis, which has attracted attention as an invaluable tool for studying human embryonic development.

We have successfully developed a method to efficiently induce the differentiation of naïve human PSCs, which correspond to preimplantation epiblasts, into extraembryonic cells of the blastocyst. Furthermore, by combining these cells, we developed a novel nonintegrated human embryo model called "bilaminoid" that reproduces development from preimplantation to peri-gastrulation stages. Here, we describe a detailed protocol for the bilaminoid formation.

Key words Bilaminoid, Epiblast, Human embryo development, Hypoblast, Naïve human pluripotent stem cells, Trophoblast

1 Introduction

Human embryogenesis begins with a fertilized egg and develops into a blastocyst before implantation. The blastocyst contains the first three cell lineages: trophectoderm (future placenta), epiblast (future fetus body), and hypoblast (future yolk sac), all of which are essential for early development and body formation. After implantation, the epiblast forms the amniotic cavity by interacting with two types of extraembryonic tissue (trophectoderm and hypoblast). Then, the anterior-posterior axis forms while gastrulation initiates in the posterior epiblast.

Understanding the mechanisms of human embryogenesis is crucial not only for understanding human embryonic development but also for the development of regenerative medicine. However, much remains unclear because post-implantation human embryogenesis remains challenging to study due to ethical issues. Methods are being developed to culture human embryos in vitro until the post-implantation stage, but the number of human embryos available for research is limited [1–3]. Recently, stem cell-based human embryo models have been reported and have attracted attention as

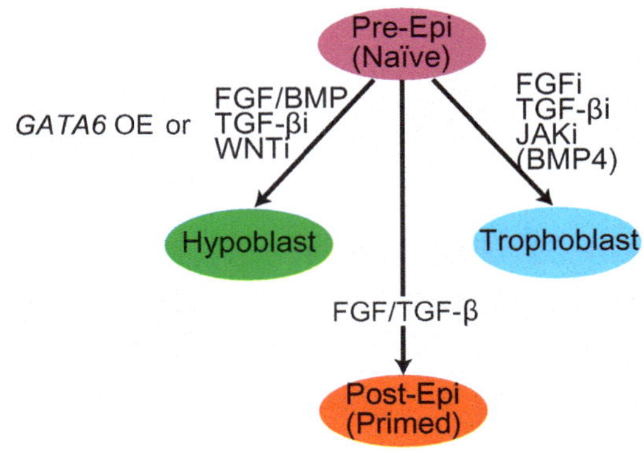

Fig. 1 Schematic for differentiating naïve human PSCs toward hypoblast-like cells and trophoblast-like cells. Naïve human PSCs can be efficiently induced into hypoblast-like cells by *GATA6* overexpression or compounds such as FGF4 and BMP4 or trophoblast-like cells using MEK and TGF-β inhibitors

a powerful tool to study the mechanisms of human embryogenesis [4].

Various stem cell-based human embryo models reported by several independent groups are classified into integrated and nonintegrated embryo models [5–17]. Integrated embryo models have all the component cells of the embryo and are more likely to form a structure that closely resembles a human embryo. By contrast, nonintegrated embryo models lack some cell types. As such, they do not fully mimic a human embryo and have the advantage of being less bound by ethical regulations than integrated embryo models.

We have developed methods to generate naïve human pluripotent stem cells (PSCs)—corresponding to the pre-implantation epiblast—and extraembryonic cells that compose the blastocyst (Fig. 1) [17–19]. In addition, a novel nonintegrated human embryo model (bilaminoid) has been developed by combining these cells [17]. The bilaminoid can reproducibly mimic development from preimplantation to peri-gastrulation (postimplantation) stages. Because each cell lineage is derived from a different cell, gene editing can be easily performed for each cell lineage, making it suitable for studying interactions and functions between cell lineages.

2 Materials

2.1 Cell Lines

1. MEF cells (*see* **Note 1**).
2. DR4-MEF cells (*see* **Note 1**).
3. Human PSCs (ESCs/iPSCs) (*see* **Note 2**).

2.2 Reagents and Equipment

1. Dulbecco's phosphate-buffered saline (PBS) (Nacalai Tesque, #14249-24).
2. DMEM (high glucose) (Nacalai Tesque, #08458-16).
3. FBS.
4. Ndiff227(Takarabio, #Y40002).
5. PD0325901(Tocris, #4192).
6. CHIR99021(Sigma, #SML1046).
7. Recombinant human LIF (Peprotech, #300-05).
8. Go6983(Tocris, # 2285/10).
9. FGF4 (Peprotech, 100-31).
10. Heparin (Wako, #081-00131).
11. Doxycycline (Dox; Sigma, #D9891).
12. A83–01 (Tocris, #517-75771).
13. BMP4 (R&D, #314-BP).
14. Y-27632 (Cayman Chemical, #CAY-10005583-50).
15. JAK inhibitorI (Merck, #420099).
16. DMEM/F12 (Nacalai Tesque, #08460-95).
17. BSA (Wako, # 012-23881).
18. Accutase (Sigma, #A6964).
19. Gelatin (Sigma, #G1890).
20. Geneticin (Thermo Fisher Scientific, #10131035).
21. iMatrix511 silk (Matrixome, #892021),
22. Elplasia (Corning, #4441).
23. Transwell (Corning, #3470).
24. Neon Transfection System (Thermo Fisher Scientific, #MPK5000).
25. Plasmid vector (Addgene #80474, pBase expression vector).

2.3 Culture Medium Composition

1. MEF medium: supplement DMEM (high glucose) with 10% FBS and 2-ME; can be stored at 4 °C for up to 4 weeks.
2. Naïve PSC maintain medium (t2iLGo): supplement Ndiff227 with 1 μM PD0325901, 1 μM CHIR99021, 10 ng/mL recombinant human LIF, 3 μM Go6983; can be stored at 4 ° C for up to 1 week.
3. Hypoblast induction medium (*GATA6* overexpression): supplement Ndiff227 with 25 ng/mL FGF4, 1 μg/mL Heparin, 0.1 μM Dox (D0-D2); can be stored at 4 °C for up to 1 week.

4. Trophoblast induction medium: supplement Ndiff227 with 2 μM PD0325901, 2 μM A83–01, 10 ng/mL BMP4 (D0-D1), 10 μM Y-27632 (D0-D1), 1 μM JAK inhibitorI (D1-D2); can be stored at 4 °C for up to 1 week.
5. Wash buffer: supplement DMEM/F12 with 0.1% BSA.

2.4 Characterization of Cellular Identity

1. Biotinylated anti-PDGFRA antibodies (R&D, #BAF322).
2. Streptavidin-APC (BioLegend, #405207).
3. Anti-CEACAM1 + 5 antibodies (Abcam, #ab91213).
4. Alexa Fluor 488 Donkey anti-Mouse (Thermo Fisher Scientific, #A32766).
5. DAPI.

3 Methods

3.1 Preparation of MEF Feeder Plates

1. Coat 6-well plates with 0.1% gelatin solution for at least 15 min.
2. Aspirate gelatin solution.
3. Seed MEF feeder cells at a density of 1×10^6 cells per 6-well in MEF medium.
4. Place the 6-well plate in a 37 °C humidified incubator at 5% CO_2 and incubate until ready to use. MEF feeder plate should be incubated for at least 24 h after cell seeding before use and used within a week.

3.2 Maintenance of Human Naïve PSCs

Naïve human PSCs were established from primed human PSCs using a chemical resetting method [20] and cultured on MEF feeder cells under t2iLGo medium [18] in a 37 °C humidified incubator at 5% CO_2 and 5% O_2 (*see* **Note 3**).

1. Aspirate t2iLGo medium from the 6-well plate and wash twice with PBS.
2. Add 500 μL of Accutase and return the plate to the incubator for 5 ~ 10 min.
3. After confirming cell detachment, add 1 mL wash buffer.
4. Pipette carefully and collect cells into tubes.
5. Centrifuge at 300 × *g* for 3 min.
6. After removing the supernatant, resuspend in t2iLGo medium and seed onto a new MEF feeder plate at a 1:3 ~ 1:5 ratio.
7. Change medium daily and passage every 3 ~ 5 days.

3.3 Preparation of Dox-Inducible GATA6-Overexpressing Human Naïve PSCs (GATA6-Naïve)

GATA6-Naïve is generated by incorporating a sequence that induces Dox-dependent *GATA6* overexpression into naïve human PSCs via the *piggyBac* transposon system. The plasmid vectors used are a *piggyBac* transposon plasmid (Addgene (#80474)) with the human *GATA6* and the pBase expression vector and are introduced into naïve human PSCs by electroporation.

1. Aspirate t2iLGo medium from the 6-well plate and wash twice with PBS.
2. Add 500 μL of Accutase and return the plate to the incubator for 5 ~ 10 min.
3. After confirming cell detachment, add 1 mL wash buffer.
4. Pipette carefully and collect cells into tubes.
5. Centrifuge at 300 × *g* for 3 min.
6. After removing the supernatant, resuspend in PBS.
7. Centrifuge at 300 × *g* for 3 min.
8. After removing the supernatant, introduce the plasmid vector using the Neon Transfection System (*see* **Note 4**).
9. Seed with t2iLGo medium containing Y-27632 (10 μM) on DR4-MEFs.
10. Add Geneticin (×250) to t2iLGo medium two days later.
11. Continue drug selection for about 2 weeks while passaging every 3 ~ 5 days.
12. Thereafter, maintain GATA6-Naïve cells with the same medium as wild-type naïve human PSCs described in Subheading 3.2.

3.4 Confirmation of Differentiation Efficiency into Hypoblast-Like Cells

Before starting bilaminoid creation, it is necessary to check the differentiation efficiency of GATA6-Naïve cells into hypoblast-like cells. The Dox concentration is crucial and should be optimized.

1. Prepare gelatine-coated dishes to exclude MEFs as described in Subheading 3.1.
2. Coat culture dishes with iMatrix511-silk (0.6 μg/cm^2) and incubate for about 0.5 h.
3. Aspirate t2iLGo medium from the 6-well plate and wash twice with PBS.
4. Add 500 uL of Accutase and return the plate to the incubator for 5 ~ 10 min.
5. After confirming cell detachment, add 1 mL wash buffer.
6. Pipette carefully and collect cells into tubes.
7. Centrifuge at 300 × *g* for 3 min.

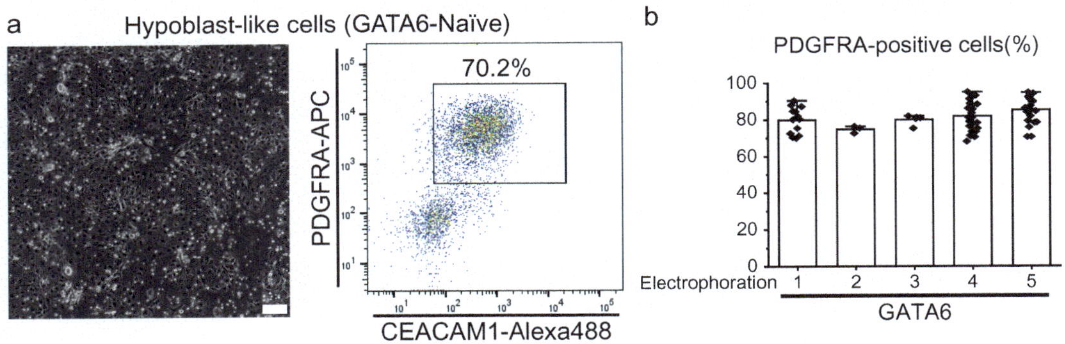

Fig. 2 Differentiation of human naïve PSCs into hypoblast-like cells by *GATA6* overexpression. (**a**) Brightfield image of hypoblast-like cells after three days of induction, expressing hypoblast cell surface markers: PDGFRA and CEACAM1. Scale bars, 100 μm. (**b**) Five GATA6-Naïve lines were independently established. For all lines, around 70–80% of induced cells are hypoblast-like cells (PDGFRA positive cells) when using 0.1 μM Dox. (Modified from ref. [17])

8. After removing the supernatant, resuspend cell pellets in t2iLGo medium containing Y-27632 (10 μM).
9. Seed cells with t2iLGo medium containing Y-27632 (10 μM) on gelatine-coated dishes.
10. Place the dishes in a 37 °C humidified incubator at 5% CO_2 and 5% O_2 and incubate for 1.5 ~ 2 h.
11. Verify that MEF feeder cells have adhered to the dish and collect the supernatant in a tube.
12. Centrifuge at 300 × *g* for 3 min.
13. After removing the supernatant, resuspend cell pellets in t2iLGo medium.
14. Count cells using a hemocytometer.
15. Seed 1 × 10^5 cells/cm^2 onto an iMatrix511-silk-coated dish in hypoblast induction medium containing Dox. Dox concentrations in the range of 0.01 ~ 1 μM should be considered.
16. Next day, change to hypoblast induction medium containing Dox.
17. On Day2, change to hypoblast induction medium.
18. On Day3, check the differentiation efficiency into hypoblast-like cells (PDGFRA+CEACAM1-double positive cells) by flow cytometry (Fig. 2) (*see* **Note 5**–**7**).

3.5 Generation of Human Embryo Models (Bilaminoids) from Human Naïve PSCs

Bilaminoids are generated by combining and aggregating two types of naïve human pluripotent stem cells (wild-type and GATA6-Naïve) (Fig. 3). Bilaminoids reproducibily mimic development from preimplantation to peri-gastrulation (postimplantation) stages (Fig. 4a).

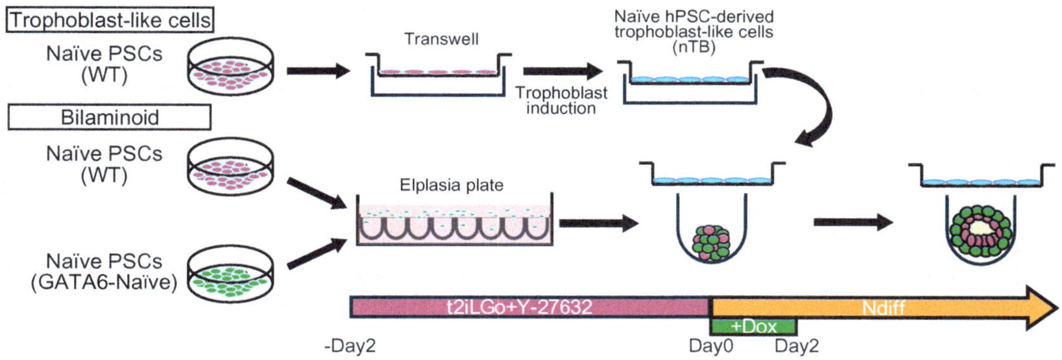

Fig. 3 Overview of protocol for forming bilaminoid models. Bilaminoids are formed by mixing two types of naïve human PSCs (wild-type and GATA6-Naïve). Cells are first aggregated under t2iLGo medium, with Dox added to overexpress *GATA6* for hypoblast-like cell induction. When co-culturing with trophoblast-like cells, co-cultures are initiated with Dox treatment

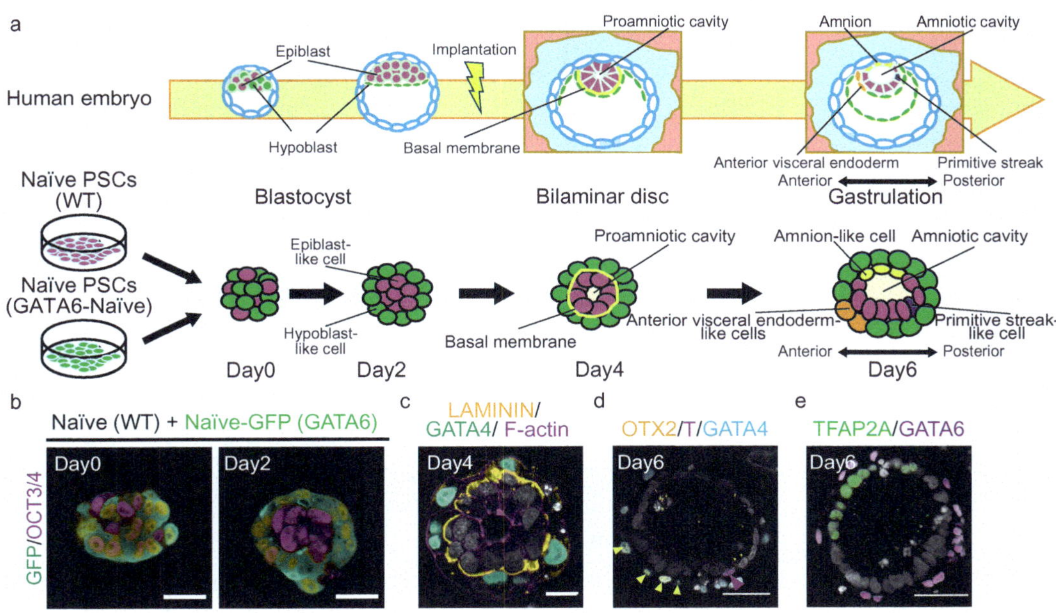

Fig. 4 Transition of bilaminoid formation. (**a**) Schematic diagrams of human embryogenesis and bilaminoid formation. Bilaminoids mimic human embryogenesis from pre-implantation to the peri-gastrulation stage. (**b**) Immunostaining and confocal imaging of a bilaminoid (Day0 and Day2) with GFP (green; GATA6-Naïve) and OCT3/4 (magenta; epiblast). (**c**) Immunostaining and confocal imaging of a bilaminoid (Day4) with LAMININ (yellow; basal membrane), GATA4 (green; hypoblast), and F-actin (magenta; apicobasal polarity). (**d**) Immunostaining and confocal imaging of a bilaminoid (Day6) with OTX2 (yellow; anterior visceral endoderm), GATA4 (blue; hypoblast), and T (magenta; primitive streak). Yellow arrows indicate OCT3/4- and GATA4-double-positive cells; the magenta arrow indicates a T-positive cell. (**e**) Immunostaining and confocal imaging of a bilaminoid (Day6) with TFAP2A (green; amnion) and GATA6 (magenta; hypoblast). Scale bars, 20 μm (b and g), 50 μm (e and y). (Modified from ref. [17])

1. Prepare gelatine-coated dishes to exclude MEFs as described in Subheading 3.1.
2. Prepare Elplasia plates. Add t2iLGo medium containing Y-27632 (10 μM) to the wells to be used. Then, centrifuge Elplasia plates at 300 × *g* for 5 min to remove any remaining air bubbles in microwells.
3. Aspirate t2iLGo medium from the 6-well plate and wash twice with PBS.
4. Add 500 μL of Accutase and return plates to the incubator for 5 ~ 10 min.
5. After confirming cell detachment, add 1 mL wash buffer.
6. Pipette carefully and collect cells into tubes.
7. Centrifuge at 300 × *g* for 3 min.
8. After removing the supernatant, resuspend cell pellets in t2iLGo medium containing Y-27632 (10 μM).
9. Seed cells in t2iLGo medium containing Y-27632 (10 μM) on gelatine-coated dishes.
10. Place dishes in a 37 °C humidified incubator at 5% CO_2 and 5% O_2 and incubate for 1.5 ~ 2 h.
11. Verify that MEF feeder cells have adhered to the dish and collect the supernatant in a tube.
12. Centrifuge at 300 × *g* for 3 min.
13. After removing the supernatant, resuspend cell pellets in t2iLGo medium containing Y-27632 (10 μM).
14. Count cells using a hemocytometer.
15. Seed the two cell types in t2iLGo medium containing Y-27632 (10 μM) into Elplasia plates by mixing 0.9×10^4 (wild-type) and 3.6×10^4 (*GATA6*-overexpressing) cells per well. Approximately 50 cells (10 wild-Naïve +40 GATA6-Naïve) can be seeded per microwell.
16. Shake the dish to distribute cells evenly and allow cells to settle for 5–10 min in incubator.
17. Centrifuge at 300 × *g* for 5 min.
18. Confirm by microscopy that about 50 cells per microwell have been seeded.
19. Place Elplasia plates in a 37 °C humidified incubator at 5% CO_2 and 5% O_2 and incubate for 48 h (2 Days).
20. Day0: Two types of naïve human PSCs form cell aggregations (*see* **Note 8**).

    Carefully remove the supernatant using a micropipette and slowly add Ndiff and Dox. Repeat twice to replace t2iLGo medium with induction medium (*see* **Note 9**).

21. Day1: Remove the supernatant and slowly add Ndiff and Dox.
22. Day2: Remove the supernatant and add Ndiff slowly.

    GATA6-Naïve cells migrate to the periphery of aggregations and surround epiblast-like cells (OCT3/4-positive cells) (Fig. 4b).
23. Day3: From Day3 onwards, change medium daily with Ndiff (*see* **Note 10**).
24. Day4: Hypoblast-like cells surround epiblast-like cells and form a basal membrane. Epiblast-like cells form a proamniotic cavity-like structure (Fig. 4c).
25. Day6: Cells expressing primitive streak makers emerge with anterior-posterior body axis formation (Fig. 4d). Some epiblast-like cells spontaneously differentiate into amnion-like cells (Fig. 4e).

3.6 Induction of Trophoblast-Like Cells and Co-culture with Bilaminoids

Co-culturing with trophoblast-like cells increases bilaminoid formation efficiency and enables extended culture. Trophoblast-like cells are cultured on Transwell inserts and used for co-culture with bilaminoids (Fig. 3). Trophoblast-like cells are induced from naïve human PSCs [19].

1. In parallel with Subheading 3.5, initiate the induction of trophoblast-like cells from naïve human PSCs.
2. Prepare gelatine-coated dishes to exclude MEFs as described in Subheading 3.1.
3. Coat Transwell inserts with iMatrix511-silk (0.15 μg/cm^2) and incubate at least for 1 h.
4. Aspirate t2iLGo medium from 6-well plates and wash twice with PBS.
5. Add 500 uL of Accutase and return the plate to the incubator for 5 ~ 10 min.
6. After confirming cell detachment, add 1 mL wash buffer.
7. Pipette carefully and collect cells into tubes.
8. Centrifuge at 300 × g for 3 min.
9. After removing the supernatant, resuspend cell pellets in t2iLGo medium containing Y-27632 (10 μM).
10. Seed cells with t2iLGo medium containing Y-27632 (10 μM) on gelatine-coated dishes.
11. Place dishes in a 37 °C humidified incubator at 5% CO_2 and 5% O_2 and incubate for 1.5 ~ 2 h.
12. Verify that MEF feeder cells have adhered to the dish and collect the supernatant in a tube.
13. Centrifuge at 300 × g for 3 min.

14. After removing the supernatant, resuspend cell pellets in fresh t2iLGo medium.
15. Count cells using a hemocytometer.
16. Seed cells in trophoblast induction medium(+BMP4 + Y-27632) on iMatrix511-silk coated Transwell inserts.
17. Next day, change medium to trophoblast induction medium (+JAKi).
18. On Day2 of induction, remove the medium, wash with PBS, and start co-culture with bilaminoids.

4 Notes

1. To stop the proliferation of MEFs, irradiate MEFs with 30 Gy using a Gammacell 40 Exactor irradiator (Best Theratronics).
2. We typically use H9 human ESCs for bilaminoid formation but confirmed that 585B1 and 1390D3 human iPSCs can also be used.
3. For bilaminoid formation, we typically use naïve human PSCs maintained in t2iLGo medium but confirmed that it is possible to generate bilaminoids using cells maintained in PXGL [20] medium.
4. Use Neon Program #20 (Voltage: 1150, Pulse width: 30, Pulse no: 2).
5. If the induction efficiency of differentiation into hypoblast-like cells is low, re-examine the concentration of Dox treatment. The Dox concentration is crucial for efficient induction; concentrations too high or too low will reduce the differentiation efficiency. The treatment time is also critical and should be adjusted for optimal induction.
6. If the induction efficiency does not increase by increasing the Dox concentration, repeat drug selection to ensure that no non-transfected cells remain. If this does not improve the efficiency, increase plasmid amounts, and perform electroporation again.
7. In most cases, 0.1 μM Dox should efficiently induce differentiation, with around 70–80% of cells differentiating into hypoblast-like cells (Fig. 2b).
8. If naïve PSCs do not form aggregations on Day0, they could have been damaged by the manipulation up to cell seeding. Perform cell manipulation, from MEF removal to seeding, more carefully and quickly.
9. Be careful not to move aggregations from microwells.

10. If aggregations collapse after Day0, the Dox concentration may be too high; repeat with a reduced Dox concentration. In some cases, extending the time of Y-27632 treatment to Day4 may improve the aggregation formation efficiency.

Acknowledgments

We thank Dr. Kelvin Hui for critical reading and English editing of the manuscript. This work was supported by JSPS (JP23K18340, JP25H01348) and AMED (JP21gm1310011, JP23bm1323001, JP24wm0625118).

References

1. Deglincerti A et al (2016) Self-organization of the in vitro attached human embryo. Nature 533:251–254. https://doi.org/10.1038/nature17948
2. Shahbazi MN et al (2016) Self-organization of the human embryo in the absence of maternal tissues. Nat Cell Biol 18:700–708. https://doi.org/10.1038/ncb3347
3. Xiang L et al (2020) A developmental landscape of 3D-cultured human pre-gastrulation embryos. Nature 577:537–542. https://doi.org/10.1038/s41586-019-1875-y
4. Shahbazi MN et al (2024) Early human development and stem cell-based human embryo models. Cell Stem Cell 31:1398–1418. https://doi.org/10.1016/j.stem.2024.09.002
5. Zheng Y et al (2019) Controlled modelling of human epiblast and amnion development using stem cells. Nature 573:421–425. https://doi.org/10.1038/s41586-019-1535-2
6. Yu L et al (2021) Blastocyst-like structures generated from human pluripotent stem cells. Nature 591:620–626. https://doi.org/10.1038/s41586-021-03356-y
7. Liu X et al (2021) Modelling human blastocysts by reprogramming fibroblasts into iBlastoids. Nature 591:627–632. https://doi.org/10.1038/s41586-021-03372-y
8. Yanagida A et al (2021) Naive stem cell blastocyst model captures human embryo lineage segregation. Cell Stem Cell 28:1016–1022. https://doi.org/10.1016/j.stem.2021.04.031
9. Kagawa H et al (2022) Human blastoids model blastocyst development and implantation. Nature 601:600–605. https://doi.org/10.1038/s41586-021-04267-8
10. Oldak B et al (2023) Complete human day 14 post-implantation embryo models from naive ES cells. Nature 622:562–573. https://doi.org/10.1038/s41586-023-06604-5
11. Weatherbee BAT et al (2023) Pluripotent stem cell-derived model of the post-implantation human embryo. Nature 622:584–593. https://doi.org/10.1038/s41586-023-06368-y
12. Pedroza M et al (2023) Self-patterning of human stem cells into post-implantation lineages. Nature 622:574–583. https://doi.org/10.1038/s41586-023-06354-4
13. Moris N et al (2020) An in vitro model of early anteroposterior organization during human development. Nature 582:410–415. https://doi.org/10.1038/s41586-020-2383-9
14. Liu L et al (2023) Modeling post-implantation stages of human development into early organogenesis with stem-cell-derived peri-gastruloids. Cell 186:3776–3792. https://doi.org/10.1016/j.cell.2023.07.018
15. Ai Z et al. Dissecting peri-implantation development using cultured human embryos and embryo-like assembloids. Cell Res 33:661–678. https://doi.org/10.1038/s41422-023-00846-8
16. Hislop J et al (2024) Modelling post-implantation human development to yolk sac blood emergence. Nature 622:367–376. https://doi.org/10.1038/s41586-023-06914-8
17. Okubo T et al (2024) Hypoblast from human pluripotent stem cells regulates epiblast

development. Nature 622:357–366. https://doi.org/10.1038/s41586-023-06871-2

18. Takashima Y et al (2014) Resetting transcription factor control circuitry toward ground-state pluripotency in human. Cell 158:1254–1269. https://doi.org/10.1016/j.cell.2014.08.029
19. Io S et al (2021) Capturing human trophoblast development with naive pluripotent stem cells in vitro. Cell Stem Cell 28:1023–1039. https://doi.org/10.1016/j.stem.2021.03.013
20. Guo G et al (2017) Epigenetic resetting of human pluripotency. Development 144:2748–2763. https://doi.org/10.1242/dev.146811

Methods in Molecular Biology (2026) 3048: 83–94
DOI 10.1007/7651_2026_697

Published online: 23 April 2026

Modeling Somitogenesis and Axial Development In Vitro with Axioloids

Keun-Tae Kim, Sofiane Hamidi, and Cantas Alev

Abstract

Axioloids are three-dimensional (3D) structures derived from pluripotent stem cells (PSCs) that model key aspects of human somitogenesis and early axis development in vitro. These mesoderm-based aggregates recapitulate essential morphogenetic features of the segmentation process, including axial elongation, sequential formation of epithelial somites with proper rostrocaudal patterning, and oscillatory activity of the segmentation clock. They further reproduce the spatiotemporal organization characteristic of early axial development, such as opposing FGF/WNT and retinoic acid signaling gradients and anteroposterior HOX gene expression patterns similar to those observed in vivo. Together, these attributes highlight axioloids as a robust and versatile platform for studying human axial development and congenital disorders of the spine and axial skeleton. In this chapter, we provide a concise, step-by-step protocol for generating axioloids, along with key developmental readouts and troubleshooting strategies to address common problems and issues encountered. This protocol provides a reproducible framework for producing and characterizing axioloids as an in vitro model of human axial development and disease.

Key words Axial development, Disease modeling, Extracellular matrix (ECM), Human pluripotent stem cells (hPSCs), In vitro embryo model, Presomitic mesoderm (PSM), Retinoid signaling, Segmentation clock, Somite epithelialization, Somitogenesis

1 Introduction

Somitogenesis, the periodic formation of somites along the anterior–posterior (AP) axis of the embryonic tail, is a fundamental process underlying the segmented body plan of vertebrates [1, 2]. While extensively studied and characterized in model organisms such as mouse, chick, and zebrafish, the molecular and morphogenetic mechanisms associated with human somitogenesis remain poorly understood, mainly due to ethical and technical limitations on research involving early-stage human embryos [3]. Recent advances in pluripotent stem cell (PSC)-based in vitro systems have enabled the alternative reconstitution of early developmental processes [4–8], including of the segmentation clock, a molecular oscillator that governs somitogenesis, using human PSC-derived presomitic mesoderm (PSM) [4, 6, 9]. However, these models often lack spatial organization or the ability to generate morphologically distinct anatomical structures including epithelial somites.

To overcome these limitations, we recently established a 3D PSC-derived mesoderm-based model of axial development termed "axioloid," which recapitulates key features of human somitogenesis in vitro. Axioloids exhibit synchronized segmentation clock oscillations, robust axial elongation, and sequential formation of epithelial somite-like structures with defined rostrocaudal polarity [10]. These structures share both molecular signatures and morphological characteristics of the developing embryonic trunk and tail, including spatially organized gradients of FGF, WNT, and retinoic acid (RA) signaling. Notably, we identified a novel role for retinoid signaling in segment stabilization and somite epithelialization. Retinoid signaling acts in coordination with extracellular matrix (ECM) components to promote somite epithelialization. Beyond developmental modeling, axioloids offer also a platform for disease modeling. Using patient-like PSCs carrying pathogenic mutations in, e.g., *HES7* and *MESP2*, we observed disruptions in somite morphology and rostrocaudal patterning, mimicking previous results observed in mouse models of congenital vertebral segmentation defects.

Here, we provide a comprehensive step-by-step protocol for the generation of human axioloids from PSCs, including key cell lines, reagents, culture steps, and troubleshooting strategies in case problems are encountered. This protocol offers an easy-to-follow guide for the robust and reproducible in vitro derivation of axioloids, a powerful novel system for studying human axial development and disease in a controlled and reproducible manner.

2 Materials

A list of all materials and brief descriptions of methods necessary to perform the suggested experiments for reproducibly deriving axioloids from human PSCs are mentioned below:

2.1 List of Reagents

2.1.1 Cell Culture Reagents

- 0.5 mmol/L-EDTA/PBS solution (Nacalai, Cat#: 13567-84).
- Albumin, bovine serum, F-V, pH 5.2 (Nacalai, Cat#: 01863-77).
- StemPro™ Accutase™ cell dissociation reagent (Thermo Fisher Scientific, Cat#: A11105-01).
- TrypLE™ Select (Thermo Fisher Scientific, Cat#: 12563-011).
- D-Luciferin potassium salt (Bioworld, Cat#: 40400076-1).
- Penicillin-streptomycin (Thermo Fisher Scientific, Cat#: 15140122).
- StemFit® AK02N medium (Reprocell, Cat#: AJ100).

- StemFit® AK03N medium (Reprocell, Cat#: SF031-001).
- NDiff 227 (Takara, Cat# Y40002).
- Matrigel® Growth Factor Reduced (GFR) Basement Membrane Matrix (Corning, Cat#: 356231).
- iMatrix-511 (Nippi, Cat#: 892021).

2.1.2 Small Molecules and Cytokines

- Y27632 (Nacalai, Cat#: 1267954).
- Recombinant human bFGF protein (Peprotech, Cat#: 100-18B).
- SB431542 (Selleck Chemicals, Cat#: S1067).
- CHIR99021 (Nacalai, Cat#: 1876444).
- Retinol (Sigma-Aldrich, Cat#: R7632).
- All trans-retinal (Sigma-Aldrich, Cat#: R2500).
- Retinoic acid (Sigma-Aldrich, Cat#: R2625).

2.1.3 Equipment

- Electronic pipette (Eppendorf, Cat#: 4861000040).
- Mechanical 12-channel pipette (Eppendorf, Cat#: 3122000060).
- Six-well sterile cell culture plate (WATSON, Cat#: 197-06CPS).
- 96-well flat-bottom sterile suspension culture plate (WATSON, Cat#: 197-96CS or 197-96CIS).
- 96-well glass-bottom EZVIEW plate (Iwaki, Cat#: 5899-096).
- 24-well cell imaging plates (Eppendorf, Cat#: 0030741005).
- PrimeSurface 96U plate (Sumitomo Bakelite, Cat#: MS-9096U).
- Cell counting chamber slides (OLYMPUS).
- Automated Cell Counter (OLYMPUS, Cat#: model R1).

2.2 Media

All induction and culture media should be prepared freshly on the day of use by adding individual small molecules and growth factors into the base medium immediately prior to application. Repeated freeze-thaw cycles of media and media components can significantly reduce the induction efficiency of axioloids.

2.2.1 Culture Media

StemFit® AK02N (Reprocell) medium is constituted of three different components termed A, B, and C that need to be mixed prior to use. The protocol is as follows:

- AK02N refers to the medium used for hiPSC maintenance; it is obtained by mixing three reagents provided by the company, A + B + C in addition to Penicillin-Streptomycin (50 U/mL).

- AK02N-C is used for differentiation of hiPSCs and is obtained by mixing the components A and B only while omitting the component C, in addition to Penicillin-Streptomycin (50 U/mL).

2.2.2 Induction Media

The standard axioloid induction medium consists of AK02N-C base medium supplemented with bFGF (20 ng/mL) and CHIR99021 (5 μM). Please note that the optimal concentration of CHIR99021 and bFGF may require adjustment depending on the specific PSC line (iPSC or ESC line) and initial cell condition and density. Alternative basal media, such as AK03N or N2B27, can also support axioloid induction, although their use may necessitate additional optimization steps including optimization of small molecule and growth factor concentrations (*see* **Notes 1** and **2**).

2.2.3 Aggregation Media

Aggregation media is based on AK02N-C, supplemented with bFGF (20 ng/mL), 10 μM of Y27632, 10 μM SB431542, and the identical CHIR99021 concentration used during the initial induction step; see also (Subheading 2.2.2).

2.2.4 Embedding Media

Embedding media is based on AK02N-C with 10% Matrigel and supplemented with retinol (10 μM) or all trans-retinal (1 μM) or retinoic acid (100 nM) (*see* **Notes 3–5**).

2.3 Cell Lines

The following human iPSC lines were used for axioloid generation in this protocol:

- 409B2 hiPSC (RIKEN BRC, HPS0076).
- 201B7 hiPSC (RIKEN BRC, HPS0063).
- 201B7-Luc a *HES7*-luciferase reporter based on 201B7 [6, 10].

Please note that optimal concentrations of small molecules, growth factors, and other related procedures may require adjustments depending on the properties of the specific cell line used (*see* **Note 6**).

3 Methods

3.1 Human iPSC Culture

3.1.1 Coating of Cell Culture Dishes and Plates for Feeder-Free iPSC Culture

Conditions described hereafter have been adjusted for a single well of a six-well culture plate (9.6 cm^2). The different volumes used need to be adjusted accordingly depending on the surface of the utilized culture vessel.

1. Add 1.5 mL of PBS containing iMatrix-511 (0.5 μg/cm^2) into each well of a six-well culture plate.
2. Incubate the plate for at least 1 h at 37 °C, or overnight at 4 °C.

3. Aspirate iMatrix-511 containing PBS and add 1.5 mL of StemFit® AK02N containing Y27632 (10 μM) (*see* **Note 7**).
4. Place the plate into a humidified 5% CO_2 incubator at 37 °C until further use.

3.1.2 Passaging Human iPSCs

1. Remove the culture media from the well.
2. Rinse cells with at least 1 mL of PBS.
3. Remove PBS and add 300 μL of 0.5X TrypLE™ Select solution (1:1 diluted with 0.5 mmol/L EDTA-PBS) and put the plate back into the incubator.
4. After 4 min, take out the plate and check the cells under a microscope to see if cells are detached. Otherwise, tap or move plates back-and-forth to evenly distribute 0.5X TrypLE™ Select solution in each well and incubate for an additional 2 min.
5. Aspirate TrypLE Select solution and rinse with at least 1 mL of PBS.
6. Add 0.5 mL of PBS into each well and delicately detach the cells using a P1000 pipette (*see* **Note 8**).
7. Transfer the cell suspension into a 1.5 mL microcentrifuge tube containing 0.5 mL of AK02N media.
8. Proceed to count the cell number and assess viability.
9. Transfer 1.3×10^4 cells/well into the previously prepared iMatrix-511-coated six-well culture plate containing 1.5 mL of StemFit® AK02N with 10 μM of Y-27632.
10. Gently distribute cells by slowly rocking back and forth and place the plate in a humidified 37 °C CO_2 incubator.
11. Change media to AK02N without Y-27632 the next day.

3.2 Induction of Axioloids from Human PSCs

On the 5th day after initial passage, cells are ready to be used for axioloid induction. This corresponds to the -24 h (24 h prior to cell aggregation) time point (Fig. 1) (*see* **Note 9**).

3.2.1 Primitive Streak Induction in 2D

1. Aspirate the maintenance AK02N medium from each well.
2. Replace with 1.5 mL of induction medium.
3. Place the plate in a humidified 37 °C, 5% CO_2 incubator.

3.2.2 3D Aggregation of hPSC-Derived Mesodermal Cells

1. Prepare the required volume of aggregation medium (50 μL per 96-well plate and 1 mL for cell resuspension; account for volume loss by preparing an extra 1 to 2 mL).
2. Aspirate the induction medium and rinse the cells with at least 1 mL of PBS.
3. Add 300 μL of Accutase solution to each well and gently distribute the Accutase solution to cover the entire well.

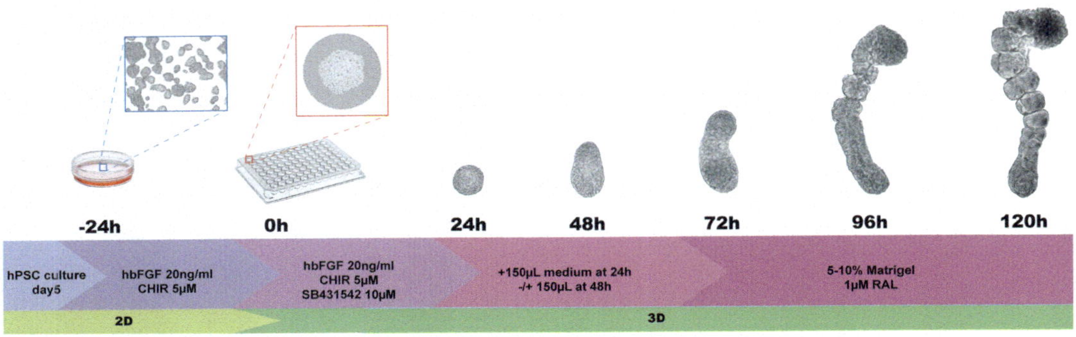

Fig. 1 Schematic representation of the axioloid induction protocol, illustrating the sequential steps required for successful generation of axioloids. The timeline highlights key medium changes and factor additions at critical time points necessary for proper mesoderm induction and differentiation

4. Incubate plate in a humidified 37 °C 5% CO_2 incubator for 4 min (*see* **Notes 10** and **11**).
5. Add 700 μL of PBS to the well.
6. Collect the cells by gently pipetting with a P1000 pipette; cells should detach very easily.
7. Transfer cells to a 1.5 mL microcentrifuge tube, containing 0.5 mL of AK02N-C medium and spin down at 800 rpm for 4 min.
8. Remove the supernatant and resuspend the cell pellet with 1 mL of aggregation media; pipette gently 10–15 times with a P1000.
9. Count the cell number and prepare a cell suspension solution of 1.0×10^4 cells/mL (in aggregation media) (*see* **Note 12**).
10. Distribute 50 μL of the cell suspension solution to each well of a 96-well U-bottom low-attachment plate (e.g., 500 cells per each well) (*see* **Note 13**).
11. Spin down the cells in the 96-well U-bottom low-attachment plate at 800 rpm for 2 min to accelerate the sedimentation of the cells at the bottom of each well.
12. Check under the microscope to confirm that the cells have sedimented at the bottom of each well. If not, spin down again for an additional 2 min (*see* **Note 14**).
13. The next day (Day 1 after aggregation), add 150 μL of AK02N-C media per well (*see* **Note 15**).
14. On the following day (Day 2 after aggregation), take out 150 μL of media from each well (using a multi-channel pipette) and add 150 μL fresh AK02N-C medium.

3.2.3 Embedding Axioloids

1. Prepare the required volume of embedding media.
2. Prepare the embedding plate by coating the wells of a 96-well flat-bottom suspension culture plate with 80 μL of a solution of 1% BSA in PBS. This step is critical as it prevents potential attachment of the axioloid structures to the bottom of each well.
3. Aspirate the BSA solution and place the plate on ice for a few minutes to chill it down before adding 80 μL of embedding medium per well using an electronic pipette.
4. Carefully transfer axioloids one-by-one from the U-bottom plate to a well of a 96-well flat-bottom suspension culture plate containing embedding medium (*see* **Note 16**).
5. Place the plate in a humidified 37 °C 5% CO_2 incubator.
6. After 24 h, you will see elongated axioloids containing 4–5 somites, 48 h later you should see 8–10 somites (Fig. 2a) with proper rostro caudal polarity and epithelization of somites (Figs. 2b, c) (*see* **Notes 3** and **4**).

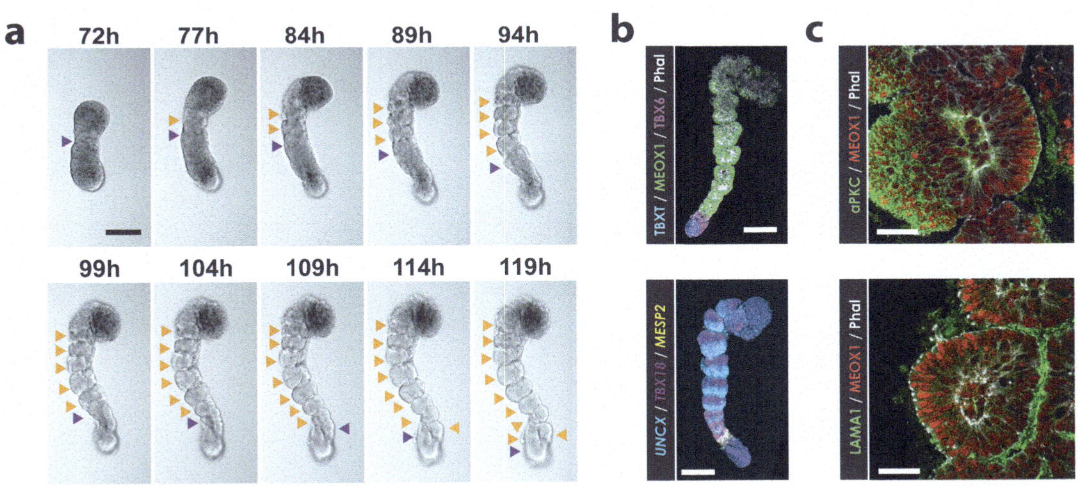

Fig. 2 Morphological and molecular characterization of axioloids. (**a**) Representative brightfield images capturing the overall morphology of a developing axioloid between 72 and 120 h of culture. The structures exhibit self-organization with defined axial features. Magenta and orange arrowheads indicate areas where somitogenesis is ongoing or completed, respectively. (**b**) Representative IF image (top) and HCR image (bottom) of 120 h axioloid. (Top) PSM marked with TBXT and TBX6 and somitic mesoderm marked with MEOX1; (bottom) *MESP2* for anterior PSM (aPSM), *UNCX*, and *TBX18* marking rostrocaudal polarity in each forming somite. Scale bars, 200 μm. (**c**) Characterization of somite formation and the acquisition of apicobasal polarity, a hallmark of somite maturation. Scale bars, 50 μm

3.3 Molecular Characterization of Axioloids

3.3.1 Live Imaging of Axioloids

1. One day prior to imaging, precoat a glass-bottom 96-well plate with 1.5% PVA (polyvinyl alcohol), aspirate, and dry overnight in a clean bench.
2. Transfer axioloids into a glass-bottom 96-well plate with one axioloid per well and start imaging.

3.3.2 Luciferase Activity Measurements in Axioloids

1. Add the required amount of embedding medium containing 100 μM of D-Luciferin to each well of a PVA-coated glass-bottom 96-well plate (EZVIEW) or a 24-well film bottom plate for microscopic imaging or bioluminescence quantification.
2. Transfer axioloids and analyze bioluminescence activity using proper analytical instrument. Here, we used either an IX83 (Olympus) microscope equipped with iXon EMCCD camera (Andor) and stage top incubator (Tokai Hit) to record time-lapse images of both bright-field and bioluminescence (Fig. 3a, b), or a Kronos HT (Atto) cell culture bioluminescence detection system (Fig. 3c).
3. Measure luciferase activity.

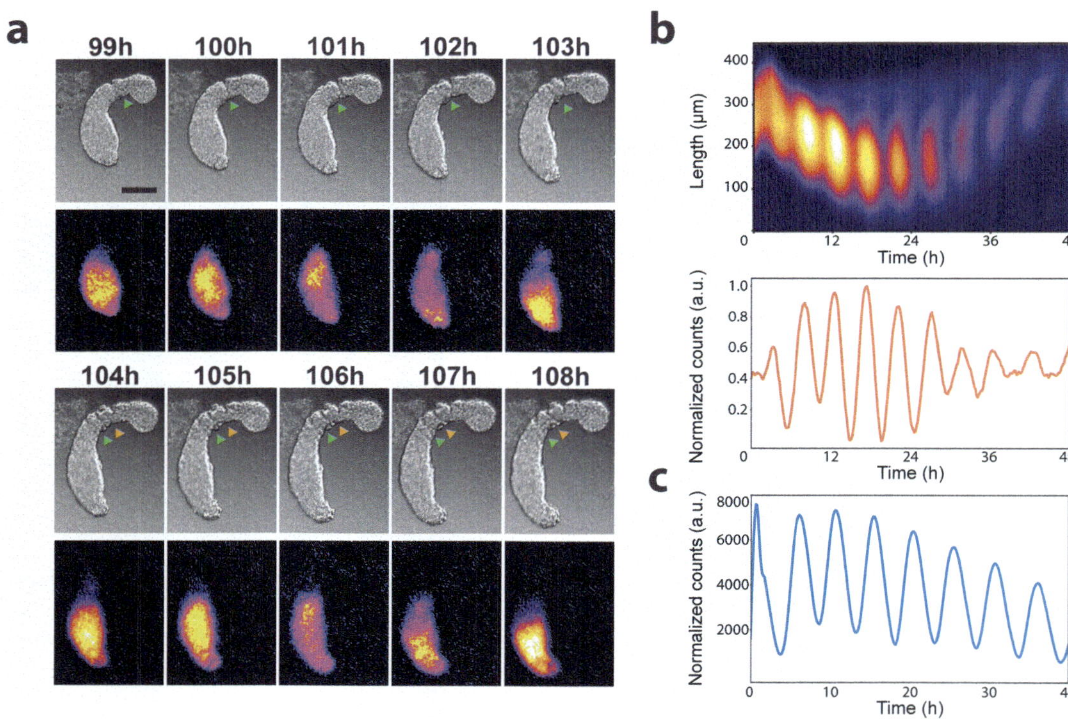

Fig. 3 Live imaging and quantification of HES7 oscillations in axioloids. (**a**) Representative images from IX83-based live imaging set up, showing luminescence of HES7-Luciferase (HES7-Luc) reporter activity in axioloids; (**b**) corresponding (top) kymograph representation of *HES7* oscillations illustrating wave propagation and periodicity of *HES7* gene expression and (bottom) measurement of HES7-Luc oscillation patterns over time. Green and orange arrowheads indicate areas where somitogenesis is ongoing or completed, respectively; (**c**) independent quantification of HES7-Luc oscillatory behavior in axioloids using the Kronos HT system. Scale bars, 200 μm

3.3.3 Fixation of Axioloids for Immunofluorescence or HCR/In Situ Hybridization

1. Axioloids should be collected using P10 or P2 pipettes set to 2–10 μL with a cut tip. Collect axioloids into a 1.5 mL tube containing 1 mL of ice cold 0.5% BSA-PBS solution.
2. Wash at least three times using the same solution to limit Matrigel carryover.
3. Proceed with fixation following specific protocols suited for downstream applications.

4 Notes

When precise and stable procedures for cell culture and passaging, cell counting, and proper media preparation including small molecule and growth factor storage and resuspension are implemented, axioloid induction is usually robust and reproducible with minimal operator-dependent variability. Below are the most common problems and troubleshooting issues mentioned which can be encountered when inducing axioloids. Possible solutions and adequate troubleshooting strategies are briefly mentioned.

1. *Failed or suboptimal induction of axioloids*: The most critical factors known to influence overall axioloid induction are (1) the identity of the used pluripotent cell line, (2) initial state of cells and cell density, (3) concentrations of used small molecules and growth factors, and (4) overall aggregate size. The balance between neural and mesoderm specification can be used as a readout to finetune the induction conditions (Fig. 4a) with basic rules being as follows:

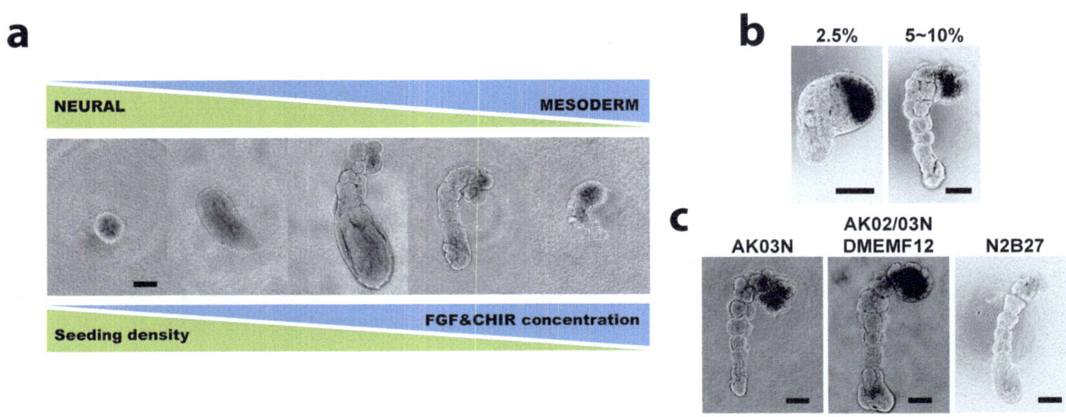

Fig. 4 Representative images of suboptimal morphologies of axioloids. (**a**) Bright-field images of axioloids illustrating the effect of seeding density and varying bFGF & CHIR99021 concentrations during axioloid induction; (**b**) bright-field images of axioloids demonstrating the impact of various basal media formulations used for paraxial mesoderm induction and/or embedding. Scale bars, 200 μm

As mentioned in Subheading 3.2.1, an optimal initial PSC density is essential for efficient and reproducible induction of axioloids. High initial cell density will result in a neural bias, whereas low density will result in an overall bias toward the mesodermal fate.

If bFGF or CHIR99021 concentrations are suboptimal, the outcome including size and morphological features of axioloids will vary largely. Low initial concentrations of FGF and WNT signaling will result in a neural bias, whereas high initial FGF and WNT concentrations will result in a mesodermal bias. Based on this, please titrate and optimize small molecule concentrations before performing actual aggregation and embedding experiments.

Depending on the used PSC line, if less than 500 cells are aggregated, it will usually result in a mesodermal bias, whereas more than 500 cells will result in a more neural bias.

Related to the previous point, and as already mentioned in the Subheading 3.2.2, cell viability after single-cell dissociation is critical in ensuring an optimal outcome.

2. *Compatibility of other basal media for axioloid induction*: Given the limited availability of the AK02N medium in the world, we have tested several alternative media on the market and found that axioloids can be induced using more broadly available media including NDiff227 (N2B27) or AK03N.

   A more economically viable alternative that we have tested consists of using AK02/03 N or NDiff227 (N2B27) for the initial 2 days of induction before switching to DMEM/F12 media supplemented with 1X NEAA, 1X ITS, 10 mM HEPES, and 50 U/mL Penicillin-Streptomycin (Fig. 4b).

3. *No axial elongation of axioloids after embedding*: During the embedding step, optimal Matrigel concentrations are critical for generating well-elongating and epithelialized somites in vitro. We recommend using 5–10% Matrigel for embedding axioloids, otherwise outcomes will vary, including inefficient elongation and less epithelialization. Other ECM mixtures and components available on the market have yielded suboptimal results in our hands.

4. *No epithelialized somites in axioloids after embedding*: Proper retinoid signaling is crucial for somite epithelialization. Ensure that retinol, retinal, or retinoic acid are present at adequate concentrations in the embedding medium. Ensure that Matrigel and retinoid containing embedding medium is freshly prepared and that concentrations are optimized; refer to Subheading 2.2.4.

5. Embedding media needs to be prepared freshly and should be kept on ice to avoid Matrigel reticulation.

6. *Cell line variations*: The working concentrations of CHIR99021 and bFGF may differ depending on the used pluripotent cell line. The above-mentioned concentrations are based on 409B2 and 201B7 human iPSC lines. For other lines, we highly recommend titrating and optimizing CHIR99021 and bFGF concentrations to identify the ideal conditions needed for reproducible axioloid induction. Common criterion for success is based on the morphology of axioloids at 120 h of culture with (1) well-elongated structures containing (2) well-epithelialized somites (see also **Note 1**).
7. Add media immediately after iMatrix solution removal to prevent iMatrix-511 from drying.
8. Cells should detach very easily, without the need to apply high pressure with a P1000 pipette. If high pressure is applied, it will likely result in suboptimal viability of detached cells [11].
9. Before proceeding to further steps and inducing axioloids, carefully check the overall quality of iPSCs, i.e., colonies should be round and epithelialized without differentiation, and overall cell density should be around 60–70%. If the cell density is suboptimal, adjust the initial seeding density during the next cell passage (*see* **Note 1**).
10. After 2 min of incubation, please check level of cell dissociation and redistribute Accutase by gently taping the plate. Cells should be rounded but not fully detached from the culture plate.
11. After 4 min, if the cells seem not to properly detach, incubate 2 more minutes. Do not incubate more than 8 min in total as viability will decrease.
12. At this stage, cell viability is critical for successful axioloid induction. We recommend viability higher than 80%. If the viability is below this threshold, please use another batch with a viability of more than 80% (*see* **Note 1**).
13. We recommend using an electronic pipette to avoid variations between wells.
14. Cells which are not sedimented after 4 min of centrifugation are most likely dead or dying cells.
15. Check under a microscope; cells must have formed small round aggregates with very little to no dead cells surrounding them.
16. Use a P10 or P2 pipette set to 2–3 μL to pick axioloids from each well. If needed, cut the end of the tips. Avoid placing axioloids on the border of the wells. If needed, reposition axioloids by gently tapping the 96-well plate.

Acknowledgments

We thank all members of the Alev lab, foremost Dr. Sirajam Munira, for their contributions in generating the data used in this book chapter.

Conflict of Interests S.H. and C.A. together with Kyoto University have filed a patent for the generation and application of axioloids, i.e., "Axioloid: a stem cell-based model of human axial development" (WO2023191099A1). K.K. declares no competing interests.

References

1. Maroto M, Bone RA, Dale JK (2012) Somitogenesis. Development 139(14):2453–2456. https://doi.org/10.1242/dev.069310
2. Hubaud A, Pourquie O (2014) Signalling dynamics in vertebrate segmentation. Nat Rev Mol Cell Biol 15(11):709–721. https://doi.org/10.1038/nrm3891
3. Diaz-Cuadros M, Pourquie O (2021) In vitro systems: a new window to the segmentation clock. Develop Growth Differ 63(2): 140–153. https://doi.org/10.1111/dgd.12710
4. Matsumiya M, Tomita T, Yoshioka-Kobayashi-K, Isomura A, Kageyama R (2018) ES cell-derived presomitic mesoderm-like tissues for analysis of synchronized oscillations in the segmentation clock. Development 145(4). https://doi.org/10.1242/dev.156836
5. Chu LF, Mamott D, Ni Z, Bacher R, Liu C, Swanson S, Kendziorski C, Stewart R, Thomson JA (2019) An in vitro human segmentation clock model derived from embryonic stem cells. Cell Rep 28(9):2247–2255 e2245. https://doi.org/10.1016/j.celrep.2019.07.090
6. Matsuda M, Yamanaka Y, Uemura M, Osawa M, Saito MK, Nagahashi A, Nishio M, Guo L, Ikegawa S, Sakurai S, Kihara S, Maurissen TL, Nakamura M, Matsumoto T, Yoshitomi H, Ikeya M, Kawakami N, Yamamoto T, Woltjen K, Ebisuya M, Toguchida J, Alev C (2020) Recapitulating the human segmentation clock with pluripotent stem cells. Nature 580(7801):124–129. https://doi.org/10.1038/s41586-020-2144-9
7. van den Brink SC, Alemany A, van Batenburg V, Moris N, Blotenburg M, Vivie J, Baillie-Johnson P, Nichols J, Sonnen KF, Martinez Arias A, van Oudenaarden A (2020) Single-cell and spatial transcriptomics reveal somitogenesis in gastruloids. Nature 582(7812):405–409. https://doi.org/10.1038/s41586-020-2024-3
8. Sanaki-Matsumiya M, Matsuda M, Gritti N, Nakaki F, Sharpe J, Trivedi V, Ebisuya M (2022) Periodic formation of epithelial somites from human pluripotent stem cells. Nat Commun 13(1):2325. https://doi.org/10.1038/s41467-022-29967-1
9. Diaz-Cuadros M, Wagner DE, Budjan C, Hubaud A, Tarazona OA, Donelly S, Michaut A, Al Tanoury Z, Yoshioka-Kobayashi K, Niino Y, Kageyama R, Miyawaki A, Touboul J, Pourquie O (2020) In vitro characterization of the human segmentation clock. Nature 580(7801):113–118. https://doi.org/10.1038/s41586-019-1885-9
10. Yamanaka Y, Hamidi S, Yoshioka-Kobayashi K, Munira S, Sunadome K, Zhang Y, Kurokawa Y, Ericsson R, Mieda A, Thompson JL, Kerwin J, Lisgo S, Yamamoto T, Moris N, Martinez-Arias A, Tsujimura T, Alev C (2023) Reconstituting human somitogenesis in vitro. Nature 614(7948):509–520. https://doi.org/10.1038/s41586-022-05649-2
11. Takahashi K, Okubo C, Nakamura M, Iwasaki M, Kawahara Y, Tabata T, Miyamoto Y, Woltjen K, Yamanaka S (2022) A stress-reduced passaging technique improves the viability of human pluripotent cells. Cell Rep Methods 2(2):100155. https://doi.org/10.1016/j.crmeth.2021.100155

Methods in Molecular Biology (2026) 3048: 95–107
DOI 10.1007/7651_2026_693

Published online: 31 March 2026

From Stem Cells to Post-gastrulation Amnioids (PGAs): A Protocol to Model Human Extra-Embryonic Development

Borzo Gharibi and Silvia D. M. Santos

Abstract

Successful embryonic development is a collaborative process: the embryo relies on the support of extra-embryonic and maternal tissues. However, understanding early human development is challenging due to the ethical and technical limitations. The generation of stem cell-based models that mimics embryonic and/or extra-embryonic tissues provides in vitro platforms to study human embryogenesis. We recently developed a 3D embryonic stem cell–derived model that mimics the post-gastrulation amnion and extra-embryonic tissues. These structures, which we call post-gastrulation amnioids (PGAs), capture early human amnion development up to the formation of a mature fluid-filled amniotic sac-like membrane composed of two cell layers of inner amniotic ectoderm and an outer extra-embryonic mesoderm. PGAs gradually expand in culture and exhibit functional traits of human amnion. PGAs are highly reproducible and can be scaled for high throughput studies. Here, we describe a step-by-step protocol for PGA generation and their morphological characterization by immunofluorescence staining as well as a co-culture assay to begin to understand interactions between extra-embryonic and embryonic cells.

Key words 3D models, Amnion, Early development, Extra-embryonic tissue, Human embryonic stem cells, Post-gastrulation

1 Introduction

Embryonic development unfolds through a coordinated interplay between the embryo and the extra-embryonic and maternal tissues that sustain it. The amnion and yolk sac, which together form the fetal membranes, play central roles in supporting and protecting the developing embryo (Fig. 1a). The amnion forms a fluid-filled membrane around the fetus that provides mechanical protection and nutritional support throughout the pregnancy [1] . The yolk sac supplies nutrition early in development and initiates hematopoiesis prior to full placenta development. [2] Increasing evidence suggests that extra-embryonic tissues might directly interact with the developing embryo to coordinate cellular specification and patterning through signaling pathways [3, 4]. However, our knowledge of how these tissues originate and function is still limited, mainly due to technical and ethical restriction associated with studying early human development [5–7].

The derivation of human embryonic stem cells (hESCs) and the generation of induced pluripotent cells (iPSc) have opened avenues for shedding light on human extra-embryonic tissue development. Differentiation of pluripotent cells in 2D allowed for derivation of definitive extra-embryonic cell types including extra-embryonic mesoderm, amniotic ectoderm, and yolk sac [8–10]. More recently, developments in generation of 3D stem cell–based models have recapitulated some aspects of extra-embryonic tissue development [11–19]. These 3D models mainly capture the extra-embryonic developmental stage around gastrulation. However, the development and functional maturation of extra-embryonic tissues extend beyond gastrulation, and until recently, there have been no robust models that recapitulate these later stages with high reproducibility and in isolation from embryonic lineages. This chapter provides a protocol for generation of post-gastrulation amnioids (PGAs), a primed pluripotent stem cell–derived extra-embryonic model that recapitulates the post-gastrulation amnion [20]. The protocol is well suited for both primed embryonic stem cells (hESC) and induced pluripotent stem cells (iPSCs).

To generate PGAs, cells cultured in 2D on Matrigel are subject to stimulation in a stepwise manner with a pulse of low level of BMP4 for 24 h to encourage differentiation toward amniotic ectoderm linage, followed by stimulation with CHIR, a WNT agonist, for further 24 h to promote differentiation of the mesodermal lineage (Fig. 1b). A 3D aggregate is generated by seeding the cells in suspension in Essential-6 media (E6) in a U-bottom low-adherence 96-well plate. Within the first couple of days, the cells form two distinct compartments, one positive for amniotic ectoderm markers such as TFAP2C and TFAP2A and another positive for GATA6, a marker of mesodermal fate (Fig. 1c, d). Over the next days, the amniotic ectoderm domain forms an amniotic-like cavity structure that rapidly expands in size resembling fluid-filled amniotic sac-like structures (Fig. 1e, f). PGA closely recapitulates cellular and structural composition of human amniotic sac composed of two cell layers of inner amniotic ectoderm and an outer extra-embryonic mesoderm (Fig. 1f) and a basement membrane that thickened as the PGA grows in size. This structural composition provides mechanical support for PGAs to expand to a size of over 1.5 cm in diameter. PGAs are highly reproducible and can be used for in vitro investigation of human amniogenesis, amnion function, and the embryonic:extra-embryonic interaction.

Here, we provide a detailed protocol for PGA generation, follow-up characterization of PGAs by immunofluorescence, and downstream application for studying signaling interactions between extra-embryonic and embryonic cells.

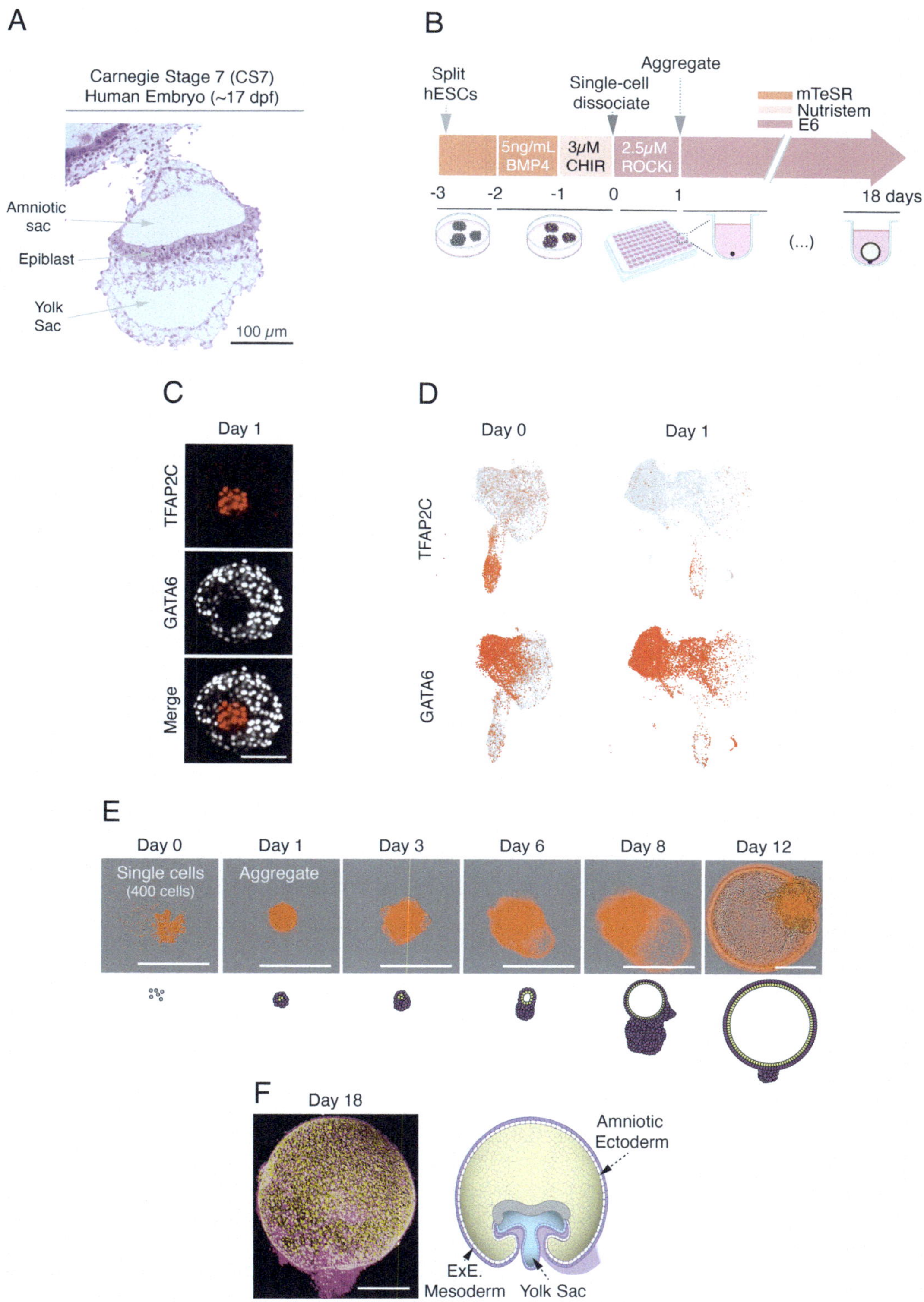

Fig. 1 Overview of PGA generation. (**a**) Section of the human embryo at Carnegie stage 7, showing extra-embryonic tissues, i.e., the amnion and the yolk sac. Sourced from Human Developmental Cell Atlas. Scale bar: 100 μm. (**b**) Schematic showing step-by-step protocol for generation of PGAs. (**c**) Representative

2 Materials

2.1 Equipment

1. Tissue culture laminar flow hood.
2. CO_2 humidified incubator.
3. Vacuum pump.
4. T25 tissue culture flasks.
5. 6- and 24-well tissue culture plates.
6. 96-well U-bottom Cell-Repellent Microplates.
7. 15 mL conical tubes.
8. Pipettes: P2, P10, P200, and P1000.
9. Sterile-filtered tips: P2, P10, P200, and P1000.
10. Multichannel pipette: P20–200.
11. Pipette Boy.
12. Reservoir.
13. Timer.
14. Parafilm.
15. Sterile scissors.
16. Automated cell counter.
17. Benchtop centrifuge.
18. Inverted phase-contrast microscope.
19. Incucyte fluorescence microscope.

2.2 Reagents

1. mTeSR1 (STEMCELL Technologies #85850).
2. TeSR E6 (STEMCELL Technologies #5946).
3. Nutristem hPSC XF (SARTORIUS #01-0005).
4. DMEM-F12 (ThermoFisher Scientific #11320033).
5. Gentle Cell Dissociation Reagent (STEMCELL Technologies #100-0485).
6. StemPro™ Accutase™ Cell Dissociation Reagent (ThermoFisher Scientific #A1110501).
7. Dulbecco's phosphate buffered saline (Merck Life Science #D8537).

Fig. 1 (continued) images of PGAs at day 1, showing two distinct compartments positive for the amniotic ectoderm marker TFAP2C (red) and positive for the extra-embryonic mesoderm marker GATA6 (gray). Scale bar: 100 μm. (**d**) UMAPs showing the expression of TFAP2C and GATA6 at day 0 and day 1 of PGA formation. (Adapted from Gharibi et al. [20]) (**e**) Representative images showing day 0 to day 12 of the PGA formation from H1 H2B-mCherry hESCs. Scale bar: 500 μm. (**f**) (Left) 3D light-sheet image of PGA (GATA6 positive, magenta, and TFAP2C positive, yellow) and (right) sagittal sections of human embryo at around day 28. (Illustration adapted from Sadler and Feldkamp [21]. Scale bar: 500 μm)

8. HyClone Water, molecular biology grade (ThermoFisher Scientific #10247783).
9. Recombinant human BMP4 (LifeTechnologies #PHC9534).
10. CHIR99021 (Sigma Aldrich #SML1046-5MG).
11. Y27632 (TOCRIS #1254).
12. Matrigel Growth Factor Reduced (GFR) LDEV-Free (Corning #354230).
13. Bovine Serum Albumin (Sigma Aldrich #A3294).
14. Fetal bovine serum (GIBCO #10270-106).
15. 4′,6-diamidino-2-phenylindole (DAPI) (ThermoFisher Scientific #D1306).
16. Lipofectamine™ RNAiMAX (ThermoFisher Scientific #13778030).
17. Formalin solution (PFA) (Sigma Aldrich #HT501128-4 L).
18. Triton X-100 (Sigma Aldrich #T8787).
19. TWEEN® 20 (Sigma Aldrich #9005-64-5).

2.3 Preparation of Solutions

1. Differentiation Media A: Prepare a stock of BMP4 at 50 μg/mL 4 mM HCl, aliquot in single-use volumes, and store at −20 °C. For a 5 ng/mL working concentration, add 1 μL per 10 mL of mTeSR media (1:10,000).
2. Differentiation Media B: Prepare a stock of CHIR99021 at 30 mM in DMSO, aliquot in single-use volumes, and store at −20 °C. Aliquot NutriStem media and store at −20 °C. For a 3 μM working concentration, add 1 μL per 10 mL of NutriStem medium (1:10,000).
3. ROCK inhibitor: Prepare a stock of Y-27632 at 10 mM in DMSO, aliquot in single-use volumes and store at −20 °C. For a 5 μM working concentration, add 5 μL per 10 mL of E6 media (1:2000).
4. Permeabilization solution: 0.3% Triton in PBS. Store at room temperature.
5. Blocking solution: 10% FBS/1% BSA in PBS. Store at 4 °C.
6. Primary antibody solutions: 1% Tween-20/3% BSA in PBS. Store at 4 °C.
7. Washing solution: 0.1% Tween-20 in PBS. Store at room temperature.

3 Methods

3.1 Preparation of Matrigel Plates

1. Aliquot growth factor-reduced Matrigel and store at −80 °C.
2. Thaw aliquots on ice for around 30 min.
3. Dilute Matrigel 1:100 in ice-cold DMEM/F12.
4. Add diluted Matrigel to culture plates at 50% of the final growth volume (1 mL per 1 well of a 6-well plate).
5. Swirl gently to coat the entire surface.
6. Incubate plates at room temperature for 1 h (*see* **Note 1**).

3.2 Routine Culture of hESCs

1. Thaw hESCs and culture for at least 1 week prior to use in PGA formation.
2. Passage cells every 3–4 days using Gentle Cell Dissociation Reagent (GCDR) in 6-well plates.
3. Warm mTeSR1 medium at room temperature for 1 h.
4. Wash hESCs once with PBS.
5. Add 400 μL of GCDR per well and incubate at room temperature for 5–6 min.
6. Aspirate GCDR and add mTeSR1 medium.
7. Gently dislodge colonies using a 5 mL serological pipette.
8. If needed, pipette gently to break cell aggregates to around 100–200 μm in size.
9. Seed cells at a 1:8 dilution into Matrigel-coated plates.
10. Incubate at 37 °C with 5% CO_2 in a humidified incubator.
11. After overnight incubation, inspect the cells under a microscope.
12. Replace medium with 2 mL fresh mTeSR1. Change mTeSR1 medium on a daily basis (*see* **Note 2**).

3.3 Preparation of hES Cells

1. Culture human ESCs in mTeSR1 on Matrigel-coated 6-well plates to reach 60–70% confluency (Fig. 2a).
2. Gently dissociate cells using GCDR. Break up the aggregates into small clumps of 15–20 cells with a 5 mL serological pipette.
3. Plate the dissociated cells at dilutions of 1:10 into a Matrigel-coated 6-well plate (*see* **Note 3**).
4. Distribute the colonies evenly by rocking the plate back and forth and side to side.
5. Incubate for 24 h at 37 °C in a 5% CO_2 humidified incubator.
6. Assess colony morphology and confluency under a microscope to make sure that colonies are of appropriate size and confluency (Fig. 2a).

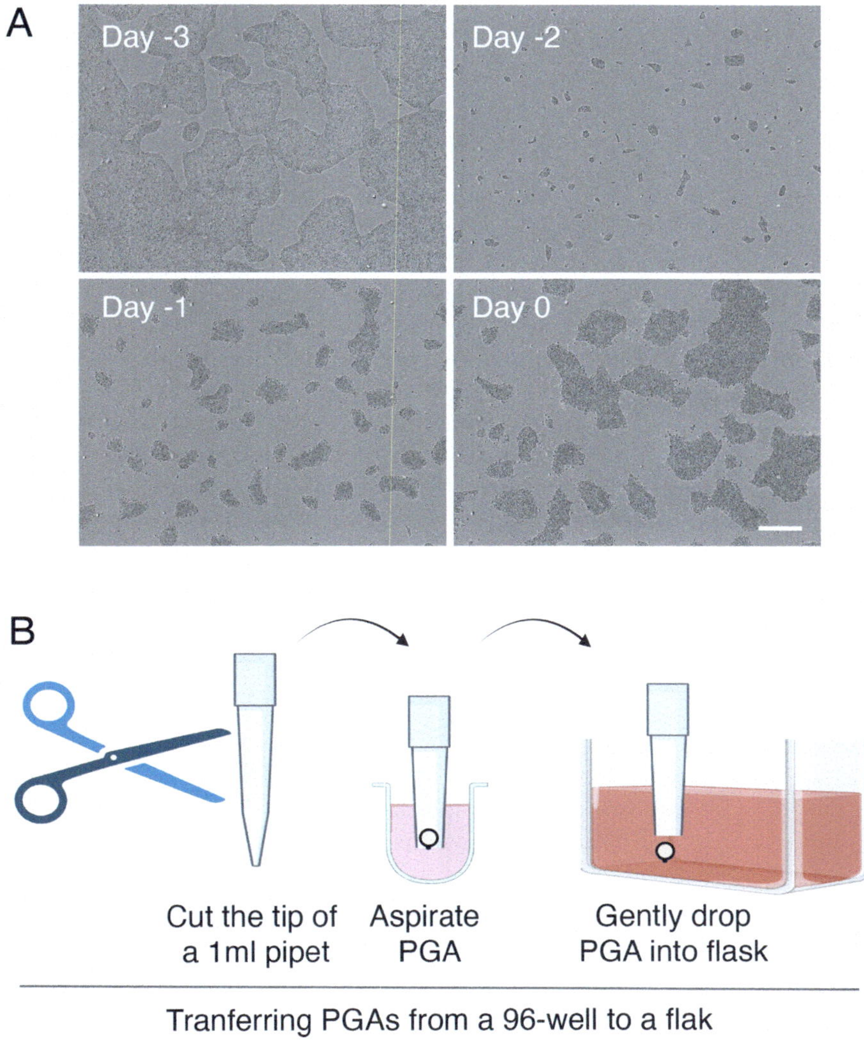

Fig. 2 hESC colony morphology and long-term culture of PGAs. (**a**) Representative images showing ideal colony size and confluency during step-wise stimulation of H1 hESCs during PGA generation. Day-3 (before plating of small colonies), day-2 (before stimulation with BMP4), day-1 (before stimulation with CHIR), and day 0 (before single cell seeding into 96-well plate) are shown. Scale bar: 500 μm. (**b**) Schematic showing transfer of PGAs from a 96-well plate to a T25 flask for long-term culture. Sterilize scissors using 70% ethanol and cut a P1000 tip around 1 cm above the tip. Carefully aspirate the media along with PGA by holding the pipette in vertical position. Insert the tip inside the media within the T25 flask and slowly release the PGA into the media

3.4 Induction of Step-Wise Differentiation

1. Aspirate the media and stimulate the cells by adding BMP4 (5 ng/mL final) in mTeSR1 (Differentiation media A).
2. Incubate for 24 h at 37 °C and 5% CO_2.

3. Assess the colonies after 24 h of BMP4 treatment. At this point, colonies show signs of peripheral differentiation with flattened morphology (Fig. 2a) (*see* **Note 4**).
4. Wash the cells once with PBS and stimulate with 3 μM CHIR99021 in NutriStem medium (Differentiation medium B).
5. Incubate for 24 h at 37 °C and 5% CO_2.
6. After the 48 h of step-wise stimulation, colonies exhibit dense central regions and differentiating edges (Fig. 2a).

3.5 Generation of Single-Cell Suspension

1. Aspirate medium from the well, wash with 2 mL PBS and add 400 μL of GCDR per well. Incubate at 37 °C, 5% CO_2 for 12 min, swirling the plate every 4 min to maintain uniform distribution throughout the incubation. Alternatively, StemPro™ Accutase™ Cell Dissociation Reagent (A1110501) can be used for single-cell dissociation. It is important that all colonies are completely dissociated into single cells.
2. Add 1 mL pre-warmed E6 media to each well. Pipette (P1000) the cells around 10 to 12 times to generate a single-cell suspension.
3. Transfer the cell suspension to a 15 mL tube containing an additional 3 mL of E6 media.
4. Centrifuge for 5 min at 1000 RPM.
5. Aspirate the supernatant, and resuspend the cell pellet in 500 μL of pre-warmed E6 media.
6. Count the cells using an automated counter such as Countess automated cell counter (Invitrogen).

3.6 Aggregate Formation in 96-Well Plate

1. Make a cell suspension of 10 cells/μL in E6 medium supplemented with 5 μM Y-27632 (ROCK inhibitor). Prepare 4 mL per full 96-well plate.
2. Using a P20–200 multichannel pipette, dispense 40 μL of the cell suspension into each well of a 96-well U-bottom cell-repellent plate.
3. Seal the plate with Parafilm and centrifuge at 700 RPM for 2 min at room temperature.
4. Incubate at 37 °C and 5% CO_2 for 24 h.

3.7 PGA Culture and Maintenance

1. On Day 1 (24 h after plating), check the formation of single spherical aggregates (*see* **Note 5**).
2. Add 150 μL of pre-warmed E6 medium to each well and return to the incubator.
3. On Day 2 (48 h post-plating), replace medium with fresh, pre-warmed 150 μL E6 medium.

4. Continue culture in the same E6 media for up to 16 days.
5. Monitor aggregates daily for the emergence of amniotic cavity-like structures.
6. Between Days 18 and 20, transfer aggregates either to a 24-well plate or directly into a T25 flask.
7. Equilibrate E6 medium (1 mL/well for 24-well plate, 30 mL/flask) at 37 °C for 30 min before transfer.
8. To transfer, cut 1 cm above the tip of a sterile P1000 tip. Insert the tip vertically into the well and gently aspirate the aggregate with surrounding medium (Fig. 2b).
9. Carefully transfer the PGA into a flask in vertical position (Fig. 2b).
10. Incubate at 37 °C and 5% CO_2.
11. For media changes, aspirate media while retaining 10 mL to avoid loss of PGAs. Add 30 mL of fresh, equilibrated E6 media.
12. Culture 1 PGA per well (24-well) or 6 PGAs per T25 flask. Perform weekly media changes for the first 3 weeks, followed by twice weekly changes thereafter.

3.8 Immunostaining of PGAs

1. Transfer PGAs to 2 mL Eppendorf tubes using a P1000 pipette tip with the end cut to avoid mechanical damage.
2. Allow PGAs to settle at the base of the tube for 3–5 min. Carefully aspirate the supernatant, leaving 50–100 μL of liquid to avoid accidental aspiration of PGAs (*see* **Note 6**).
3. Wash twice with PBS to remove residual medium.
4. Fix by adding 1.8 mL of 4% PFA to each tube.
5. Incubate at room temperature for 40 min.
6. Remove PFA and wash three times with PBS, incubating for 5 min each wash on a seesaw shaker (*see* **Note 7**).
7. Permeabilize the cells by adding 1.5 mL of 0.3% Triton X-100 in PBS (*see* **Note 8**).
8. Incubate for 40 min at room temperature with rocking.
9. Remove the permeabilization buffer and add 1.5 mL of blocking buffer (10% FBS, 1% BSA in PBS). Incubate for 2 h at room temperature.
10. Wash twice with 1% Tween-20 in PBS.
11. Prepare primary antibody solutions at the appropriate concentration in 1% Tween-20/3% BSA in PBS. Make 1.5× the final concentration of antibody to compensate for dilution by residual buffer.

12. Remove blocking buffer and add 200 μL of primary antibody solution per tube.
13. Incubate overnight at 4 °C on a seesaw shaker.
14. Wash three times with 0.1% Tween-20 in PBS, 10 min per wash.
15. Prepare secondary antibody in 1% Tween-20/3% BSA and add 200 μL per tube.
16. Incubate for 2 h at room temperature on a seesaw shaker.
17. Perform nuclear staining by adding a nuclear dye.
18. Incubate for 1 h at room temperature on a seesaw shaker.
19. Wash once with 0.1% Tween-20 in PBS for 10 min on a seesaw shaker.
20. Wash three times with molecular biology-grade water to remove antibody residues.
21. Store samples overnight at 4 °C in molecular biology-grade water before imaging (*see* **Note 9**).

3.9 Co-culture of Extra-Embryonic and Pluripotent Cells

To study signaling interactions between extra-embryonic and embryonic cells, a co-culture of PGA and pluripotent cells can be established (Fig. 3a, b). To track the fate of each cell population within the co-culture, the cells can be fluorescently tagged and visualized using fluorescent proteins. This can be achieved either by knocking in a fluorescent protein using CRISPR–Cas9 or by integrating a fluorescent protein using a lentiviral strategy.

1. Prepare extra-embryonic progenitor PGA cells by step-wise stimulation according to Subheadings 3.3 and 3.4.
2. Culture fluorescently tagged pluripotent stem cells in mTeSR media to around 70% confluency.
3. Perform single-cell dissociation of both cell types as described in Subheading 3.5.
4. Combine extra-embryonic progenitor cells with tagged pluripotent stem cells at ratio of 50:50% (*see* **Note 10**) (Fig. 3a, b).
5. Prepare a mixed cell suspension at 10 cells/μL in E6 medium supplemented with 5 μM ROCK inhibitor (Y-27632).
6. Dispense 40 μL of the cell suspension into each well of a Cell Repellent-Treated U-bottom 96-well plate.
7. Continue with spheroid formation steps as described in Subheading 3.6.
8. After 24 h of incubation post single-cell plating, replace the media with 150 μL of pre-warmed E6 media.

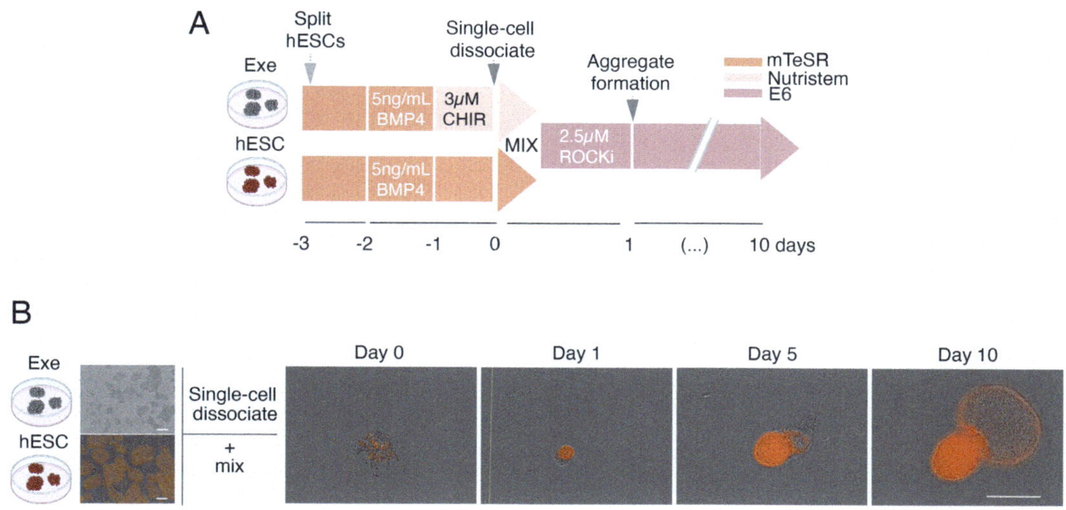

Fig. 3 Co-culture of extra-embryonic and pluripotent cells. (**a**) Schematic of the step-by-step protocol for generation of PGA/pluripotent cell co-culture. (**b**) Representative images showing PGA co-culture from day 0 to day 10. H2B-mCherry hESC cells and WT cells from PGA protocol at day 0 were mixed at ratio of 50:50%. Scale bar: 500 μm

9. Incubate the cells for further 24 h and replace the media with 150 μL of E6 media. Thereafter, maintain the cells in the same media.
10. Samples can be collected at various timepoint for analysis.

4 Notes

1. Matrigel-coated plates can be stored at 4 °C for up to 1 week.
2. Maintain hESCs up to a maximum of 15 passages for PGA applications.
3. Multiple seeding dilutions help achieve the right colony size and confluency for stimulation. Parallel seeding of 1:8, 1:10, and 1:12 is recommended.
4. Response to BMP4 may vary among ES and iPSC lines; optimization of BMP4 concentration may be needed at this stage. 5 ng/mL BMP4 is optimal for H1 and GLR hES lines, while 10 ng/mL increased the efficiency for H9 hES and KOLF2 iPSC lines.
5. 24 h aggregates should form spheroids of around 150–200 μm. Low efficiency or loss of PGA formation may occur with incorrect aggregate size.

6. Make sure samples are settled at the base of tube before removing any media. It is critical to allow around 50–100 μL of liquid at the base of the tube with each media change.
7. Samples can be stored in PBS at 4 °C after fixation.
8. Amniotic sac-like structures may collapse or deflate after permeation.
9. It is critical to wash and store the PGAs in water to reinflate the structures to their original size and shape.
10. Varying cell ratio can be used to optimize co-culture of PGA: embryonic cells.

References

1. Dobreva MP, Pereira PN, Deprest J, Zwijsen A (2010) On the origin of amniotic stem cells: of mice and men. Int J Dev Biol 54:761–777
2. Ross C, Boroviak TE (2020) Origin and function of the yolk sac in primate embryogenesis. Nat Commun 11:3760
3. Yang R, Goedel A, Kang Y, Si C, Chu C, Zheng Y, Chen Z, Gruber PJ, Xiao Y, Zhou C (2021) Amnion signals are essential for mesoderm formation in primates. Nat Commun 12:5126
4. Okubo T, Rivron N, Kabata M, Masaki H, Kishimoto K, Semi K, Nakajima-Koyama M, Kunitomi H, Kaswandy B, Sato H (2023) Hypoblast from human pluripotent stem cells regulates epiblast development. Nature 626: 357–366
5. Rugg-Gunn PJ, Moris N, Tam PP (2023) Technical challenges of studying early human development. Development 150:dev201797
6. Cavaliere G (2017) A 14-day limit for bioethics: the debate over human embryo research. BMC Med Ethics 18:1–12
7. Hyun I, Wilkerson A, Johnston J (2016) Embryology policy: revisit the 14-day rule. Nature 533:169–171
8. Pham TXA, Panda A, Kagawa H, Ertekin C, Georgolopoulos G, van Knippenberg SS, Allsop RN, Bruneau A, Chui JS-H, Vanheer L (2022) Modeling human extraembryonic mesoderm cells using naive pluripotent stem cells. Cell Stem Cell 29:1346–1365.e1310
9. Mackinlay KM, Weatherbee BA, Souza Rosa V, Handford CE, Hudson G, Coorens T, Pereira LV, Behjati S, Vallier L, Shahbazi MN (2021) An in vitro stem cell model of human epiblast and yolk sac interaction. elife 10:e63930
10. Rostovskaya M, Andrews S, Reik W, Rugg-Gunn PJ (2022) Amniogenesis occurs in two independent waves in primates. Cell Stem Cell 29:744–759.e746
11. Shao Y, Taniguchi K, Townshend RF, Miki T, Gumucio DL, Fu J (2017) A pluripotent stem cell-based model for post-implantation human amniotic sac development. Nat Commun 8: 208
12. Zheng Y, Xue X, Shao Y, Wang S, Esfahani SN, Li Z, Muncie JM, Lakins JN, Weaver VM, Gumucio DL (2019) Controlled modelling of human epiblast and amnion development using stem cells. Nature 573:421–425
13. Shao Y, Taniguchi K, Gurdziel K, Townshend RF, Xue X, Yong KMA, Sang J, Spence JR, Gumucio DL, Fu J (2017) Self-organized amniogenesis by human pluripotent stem cells in a biomimetic implantation-like niche. Nat Mater 16:419–425
14. Hislop J, Song Q, Keshavarz FK, Alavi A, Schoenberger R, LeGraw R, Velazquez J, Mokhtari T, Taheri MN, Rytel M (2023) Modeling post-implantation human development to yolk sac blood emergence. Nature 626: 367–376
15. Nakanoh S, Sham K, Ghimire S, Mohorianu I, Rayon T, Vallier L (2024) Human surface ectoderm and amniotic ectoderm are sequentially specified according to cellular density. Science. Advances 10:eadh7748
16. Weatherbee BA, Gantner CW, Iwamoto-Stohl LK, Daza RM, Hamazaki N, Shendure J, Zernicka-Goetz M (2023) Pluripotent stem cell-derived model of the post-implantation human embryo. Nature 622:584–593
17. Oldak B, Wildschutz E, Bondarenko V, Comar M-Y, Zhao C, Aguilera-Castrejon A, Tarazi S, Viukov S, Pham TXA, Ashouokhi S (2023) Complete human day 14 post-implantation embryo models from naïve ES cells. Nature 622:562–573

18. Pedroza M, Gassaloglu SI, Dias N, Zhong L, Hou T-CJ, Kretzmer H, Smith ZD, Sozen B (2023) Self-patterning of human stem cells into post-implantation lineages. Nature 622: 574–583
19. Liu L, Oura S, Markham Z, Hamilton JN, Skory RM, Li L, Sakurai M, Wang L, Pinzon-Arteaga CA, Plachta N (2023) Modeling post-implantation stages of human development into early organogenesis with stem-cell-derived peri-gastruloids. Cell 186:3776–3792.e3716
20. Gharibi B, Inge OCK, Rodriguez-Hernandez I, Driscoll PC, Dubois C, Jiang M, Howell M, Skehel JM, Macrae JI, Santos SDM (2025) Post-gastrulation amnioids as a stem cell-derived model of human extra-embryonic development. Cell 188:3757–3774.e3720. https://doi.org/10.1016/j.cell.2025.04.025
21. Sadler TW, Feldkamp ML (2008) The embryology of body wall closure: relevance to gastroschisis and other ventral body wall defects. Am J Med Genet C: Semin Med Genet 148C:180–185

Methods in Molecular Biology (2026) 3048: 109–120
DOI 10.1007/7651_2025_643

Published online: 03 July 2025

Interaction Between Pluripotent Stem Cells and Amniotic Epithelium for the In Vitro Study of Human Embryogenesis

Daniela Ávila-González, Axel Castro-Abrego, Jonathan Salazar-Alonso, Omar Martínez Alarcon, Alejandro Martinez-Juarez, Anayansi Molina Hernández, Nestor Emmanuel Diaz-Martinez, Wendy Portillo, and Nestor Fabián Diaz

Abstract

During the peri-implantation stage in primate embryos, the epiblast coexists with amniotic epithelium, suggesting a pivotal role for this tissue in embryogenesis. This interaction has attracted considerable interest in developmental biology and regenerative medicine due to its potential influence on embryonic patterning and tissue specification. In this protocol, we examined whether co-culturing human amniotic epithelial cells (hAEC) with human embryonic stem cells (hESC) enhances the latter's capacity for differentiation toward human primordial germ cell-like cells (hPGCLC).

Key words Amnion, Epiblast, Germ line induction, Human primordial germ cells

1 Introduction

In vertebrates amniotes, embryonic development occurs within the amniotic membrane, which offers protection and stability during development. This membrane is primarily composed of the mesoderm and amniotic ectoderm, separated by a basal lamina [1]. Among mammals, amnion formation varies in both the mechanism and stages of development. In rodents, amniogenesis occurs through the folding of a single amniochorionic fold, which grows laterally and expands from the posterior aspect of the embryo during the gastrula stage of the medial primitive streak (E6.5). In primates, by contrast, amniogenesis initiates during implantation, when a subset of epiblast cells adopts a squamous phenotype, migrating to form a cavity between itself and the epiblast. This process established a polarized embryonic sac, with a columnar epiblast at the base and squamous amnion dorsally [2, 3]. During the peri-implantation stage, the epiblast coexists with the emerging amniotic epithelium generating significant interest in primate amniogenesis. Studies on human embryos cultured in vitro and three-dimensional embryonic models derived from stem cells suggest that the amnion may function as an organizational or inductive center for other lineages [4]. For instance, recent findings indicate that the amnion significantly influences embryonic mesoderm formation [5, 6]. Furthermore, research point to a critical relationship between the amnion and pluripotency, supporting germline specification and suggesting that primordial germinal cells (PGC) formation in primates may directly arise from the dorsal amnion [7].

An alternative hypothesis "germinal pluripotency" proposes that the germline originates from an intermediate population within the epiblast, which subsequently gives rise to the amnion, PGCs, embryonic and extraembryonic mesoderm, and trophectoderm [8].

Our previous research demonstrated that the human amniotic epithelium, isolated from term placentas, supports the derivation of human embryonic stem cells (hESC) when cultured on a monolayer of these cells, resulting in a transcriptomic profile distinct from conventional primed conditions [9]. In this chapter, we describe the optimized conditions for maintaining hESC on a feeder layer of human amniotic epithelial cells (hAEC) and explore their effect in the co-culture, facilitating the derivation of human primordial germ cell-like cells (hPGCLC).

2 Materials

2.1 Cells

1. hESC lines: H9 (WA09, WiCell; NIH registration no. 0062; RRID: CVCL_9773) and Amicqui-1 (INPERe001-A, RRID:CVCL_UL74). These hESC lines were previously authenticated by STR analysis and confirmed to be free of mycoplasma contamination.
2. hAEC: Human amniotic epithelial cells were isolated from term fetal membranes, as approved by the Ethics and Research Committees of the National Institute of Perinatology of Mexico (INPer) under project number 2019-1-40.

2.2 Solutions and Culture Media

1. hAEC medium: 450 mL high-glucose DMEM (Gibco, 12430–054), 50 mL inactivated fetal bovine serum (qualified for stem cells), 5 mL non-essential amino acids 100× (Gibco, 11140–050), 5 mL antibiotic-antimycotic 100× (Gibco, 15240–062), 5 mL sodium pyruvate 100 nM (Gibco, 11360–070), 500 μL 2-mercaptoethanol 1,000× (Gibco, 21985–023), and 10 ng/mL recombinant Epidermal Growth Factor (EGF) (Peprotech, AF-100–15).
2. hESC medium: 395 mL DMEM/F-12 (Gibco, 11330–032), 20% knockout serum replacement (KSR) (Gibco, 1082-8028), 5 mL non-essential amino acids 100×, and 910 μL

2-mercaptoethanol, supplemented with 10 ng/mL recombinant fibroblast growth factor 2 (FGF2) (Peprotech, 100-18B).

3. GK15 medium: Glasgow medium (Sigma-Aldrich, G6148), 15% KSR (75 mL), 5 mL non-essential amino acids 100×, 5 mL GlutaMAX 100× (Gibco, 35050061), 5 mL sodium pyruvate, and 500 μL of 2-mercaptoethanol 1,000×.
4. Pretreatment medium: GK15 medium supplemented with 3 μM CHIR99021 (Tocris, 4423), 50 ng/mL Activin A (R&D Systems, 338-AC), and 10 μM ROCK inhibitor (StemCell Technologies, 72302).
5. Germ induction medium: GK15 medium supplemented with 200 ng/mL BMP4 (R&D Systems, 314 BP), 20 ng/mL human LIF, 100 ng/mL SCF (R&D Systems, 255-SC), 50 ng/mL EGF, and 10 μM ROCK inhibitor.
6. Other solutions and enzymes: Sterile saline solution, 1× PBS/0.5 mM EDTA solution (prepared by adding 500 μL of 0.5 M EDTA stock [Ambion, AM9260G] to 500 mL of 1× PBS), 0.25% trypsin/EDTA (Gibco, 2520-0056), 0.05% trypsin/EDTA (Gibco, 2530-0054), and StemPro® Accutase solution (Gibco, A1110501).

3 Methods

3.1 Isolation of Amniotic Epithelial Cells from Term Membrane

1. Patient selection: Select patients with a gestational age of 37–40 weeks who are eligible for scheduled elective cesarean delivery to obtain term placentas. Patients must be free from chronic diseases or active infections. The study must be approved by relevant local and international Ethics Committees and comply with applicable regulations. Informed consent must be obtained from patients before placenta collection.
2. Placenta collection: In the operating room, clamp the umbilical cord to prevent blood flow to the remaining tissue. Place the entire placenta, with the umbilical cord attached, in a sterile container and add sufficient saline to keep the tissue hydrated. Close the container and transport the placenta to the laboratory at room temperature for processing.
3. Storage: If dissection cannot be performed within 15 min of collection, store the placenta refrigerated, ensuring that no more than 2 h elapse from collection to the start of dissection.
4. Preparation for dissection: Under sterile conditions in a biosafety cabinet at room temperature, remove the placenta from the container and place it on a stainless-steel tray with the umbilical cord facing up.

5. Membrane separation: Remove blood clots from the fetal membranes surfaces using sterile gauze. Mechanically separate the amniotic membrane from the chorion.
6. Amniotic membrane dissection: Using dissecting forceps, carefully hold the amniotic membrane. Employ a scalpel to separate the membrane from the chorion, taking care to avoid cutting placental vessels. Stretch the membrane with forceps as necessary to facilitate separation.
7. Washing: Place the amniotic membrane in a sterile beaker and wash it several times with saline until the membrane appears translucent, removing all blood clots (*see* **Note 1**).
8. Final cleaning: If clots remain, spread the membrane on a sterile table and further clean with gauze. If blood clots persist, excise those sections with a scalpel.

3.2 Enzymatic Digestion of the Amniotic Membrane

1. Cut the clot-free, translucent amniotic membranes into three or four fragments. Cutting the membrane into smaller pieces ensures the enzyme solution can fully cover its surface.
2. Place the fragments into two sterile 50 mL centrifuge tubes, adding 20 mL of 0.25% trypsin/EDTA to each tube.
3. Manually shake each tube gently for 30 s.
4. Using tweezers, transfer the membrane fragments to new sterile 50 mL centrifuge tubes and add 30 mL of 0.05% trypsin/ EDTA.
5. Incubate the tubes for 40 min at 37 °C, with manual shaking every 10 min for 30 s.
6. After the first incubation, transfer the membrane fragments to new centrifuge tubes and 30 mL of 0.05% trypsin/EDTA.
7. Transfer the cells/trypsin solution (approximately 30 mL) into two new tubes (15 mL each), adding twice the volume of prewarmed hAEC medium (37 °C) per tube to inactivate trypsin. Store tubes on ice.
8. Perform a second incubation for 40 min at 37 °C, with manual shaking every 10 min for 30 s.
9. At the end of the second incubation, use dissecting forceps to hold one end of each membrane fragments. With a second pair, apply gentle pressure along the tissue to detach more epithelial cells.
10. Discard the digested membranes fragments into a biohazard container.
11. Collect the second cell/trypsin solution into separates tubes and inactivate it by adding twice the volume of hAEC medium.

12. Centrifuge the mixture at 200× *g* for 10 min at 4 °C. Discard the supernatant and add 10 mL of prewarmed hAEC medium to each tube. Gently pipette to dissolve pellets.
13. Filter the cell suspensions through 100 μm cell strainers (Corning®, 431752) to remove extracellular matrix residues and obtain individual cells.
14. Count the cells using a hemocytometer or automated cell counter.

3.3 Preparation of Amniotic Epithelial Cells as Feeder Layer

1. Seed hAECs onto 100 mm plates at a density of 3×10^4 cells/cm^2 in pre-warmed AEC medium.
2. Supplement with 10 ng/mL of human EGF, then incubate the plates at 37 °C in normoxic conditions (5% CO_2) within a humidified incubator.
3. Add EGF daily and replace the medium every three days (*see* **Note 2**).
4. Remove the hAEC medium and wash cell twice with a 1× PBS/0.5 mM EDTA solution.
5. Incubate with PBS/ EDTA solution for 15 min at 37 °C. Remove the solution and add 1.5–2 mL of 0.05% trypsin/ EDTA, then incubate for an additional 5–8 min at 37 °C.
6. Inactivate the trypsin by adding twice the volume of hAEC medium.
7. Collect the cell suspension and centrifuge at 200× *g* for 5 min at 4 °C.
8. Count the cells, which are now in passage 1 (P1), for immediate use as a feeder layer or freeze in liquid nitrogen for future use.
9. For feeder layer use, seed cells at a density of $5 \times 10^4/cm^2$ in 6- or 4-well plates with hAEC medium supplemented with EGF (Fig. 1A).
10. For cryopreservation, add 100 μL of StemMACS™ Cryo-Brew per 5×10^5 cells and store cryovials containing 5×10^5 cells for later thawing (sufficient for a 4-well plate) or 3×10^6 cells for a 6-well plate (*see* **Note 3**).

3.4 Maintenance and Passage of hESC on an hAEC Feeder Layer

1. Thaw hESCs directly onto the hAEC feeder layer in hESC medium supplemented with 10 ng/mL FGF2. By the third day, small hESC colonies should be visible.
2. Maintain cells on hAEC with daily medium changes and FGF2 supplementation. Passages are typically performed every 5–7 days when hESC colonies reach 70–80% confluency, comparable to the use of inactivated mouse embryonic fibroblasts (iMEF) layers (refer to Fig. 2).

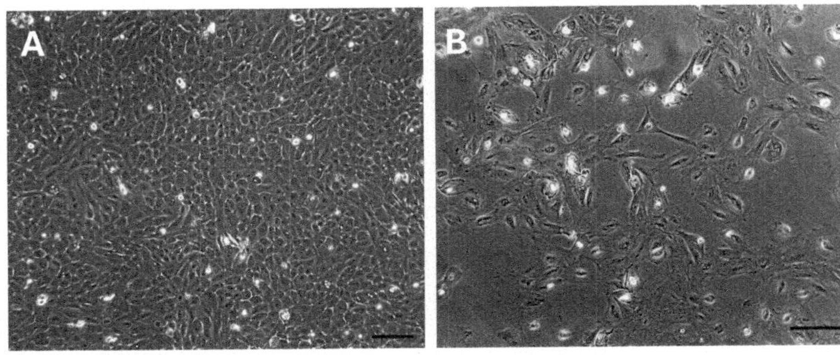

Fig. 1 Morphology of hAEC in vitro culture. (**a**) Representative micrograph of confluent hAEC isolated from the amniotic membrane at passage zero after 8 days in vitro. (**b**) Example of a primary hAEC culture showing a negative outcome, with senescent cells and fibroblast morphology. Scale bars: 50 μm

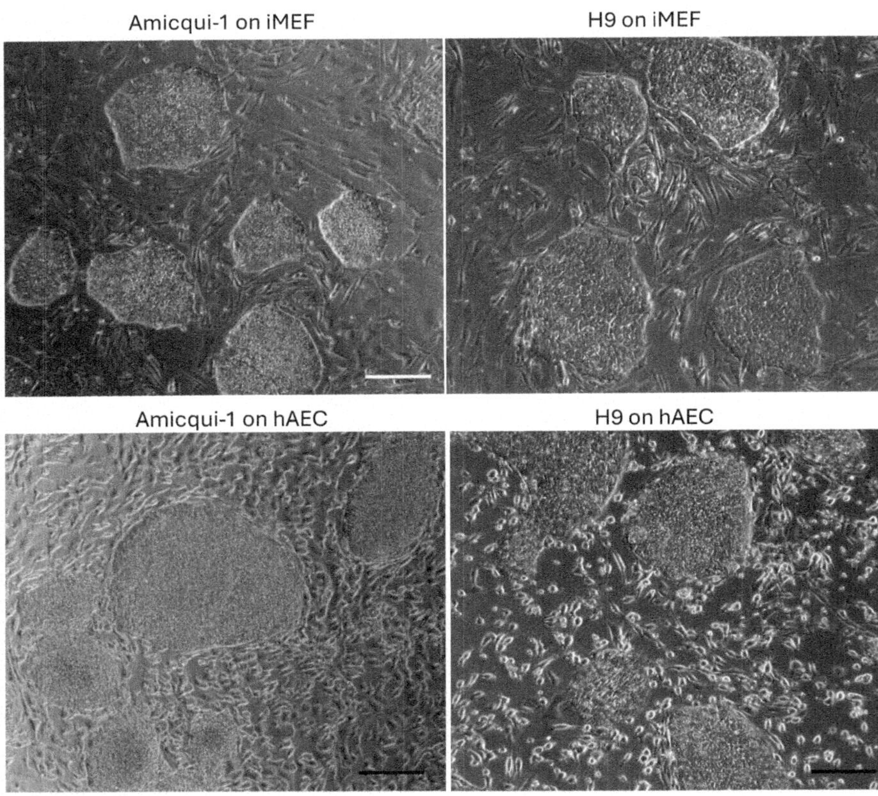

Fig. 2 Morphology of hESC colonies on feeder layers. Representative phase-contrast micrographs of Amiqui-1 and H9 hESC colonies maintained on either iMEF or hAEC feeder layers. Scale bars: 100 μm

3. For passage, remove the hESC medium and wash cells with warm hESC medium. Add 1 mL or 500 μL of prewarmed Accutase enzyme per well in a 6- or 4-well plate, respectively.
4. Incubate the cells at 37 °C for 5–7 min.

5. Rinse with Accutase to further hESC detachment (*see* **Note 4**).
6. Collect the hESC/Accutase mixture in a 15 mL Falcon tube, diluting with twice the volume of hESC medium to neutralize the Accutase.
7. Centrifuge at 200× *g* for 5 min at 4 °C.
8. For a 1:6 passage, seed the hESC onto a 6- or 24-well plate with a monolayer of hAEC (or, if unavailable iMEF).

3.5 hESC Differentiation Toward Germline Lineages

1. Detach hESC from hAEC or iMEF feeders layers using Accutase (*see* **Notes 5** and **6**).
2. Plate 200,000 cells per well in a 12-well plate pre-treated with Matrigel® hESC-Qualified (Corning®), using the appropriate dilution factor based on the batch. Use pretreatment medium for initial culture.
3. After 24 h, detach cells with Accutase and replate at 4,000 cells per well in 96-well clear round-bottom ultra-low attachment plates (Corning®, 7007) in germline induction medium (see **Note 7**).
4. Replace germline induction medium every three day.
5. On day 7, transfer cell aggregates to 4-well plates pre-coated with Matrigel for continued culture.
6. After 48 h, fix and analyze samples to identify germline markers via immunofluorescence. Quantify marker expression as described previously [10].

3.6 Observations

1. Using immunofluorescence, we confirmed that hESCs (Amicqui-1 and H9) expressed key pluripotency markers OCT4, SOX2, and NANOG, along with surface antigen SSEA-4 (Fig. 3), when co-cultured with amniotic epithelium. This expression pattern aligns with observations under conventional culture conditions with iMEF, consistent with findings reported by our group [9].
2. To evaluate whether hESC-hAEC co-culture facilitates competence for germline specification, we employed a differentiation protocol to induce hPGCLC in Amicqui-1 and H9 hESCs cultured on hAEC, comparing their differentiation efficiency to hESC on the iMEF feeder layer. On day 11 post-induction, double immunofluorescence was performed for TFAP2C+/SOX17+ (Fig. 4), SOX17+/NANOG+ (Fig. 5), and TFAP2C+/NANOG+ (Fig. 6). A significant increase in double-positive cells was observed in Amicqui-1 on hAEC compared to iMEF conditions (TFAP2C+/SOX17+, *adj* $p = 0.0256$; SOX17+/NANOG+, *adj* $p < 0.0001$; TFAP2C+/NANOG+, *adj* $p = 0.0014$). For H9 on hAEC, a significantly greater number of SOX17+/NANOG+ double-positives was observed (*adj*

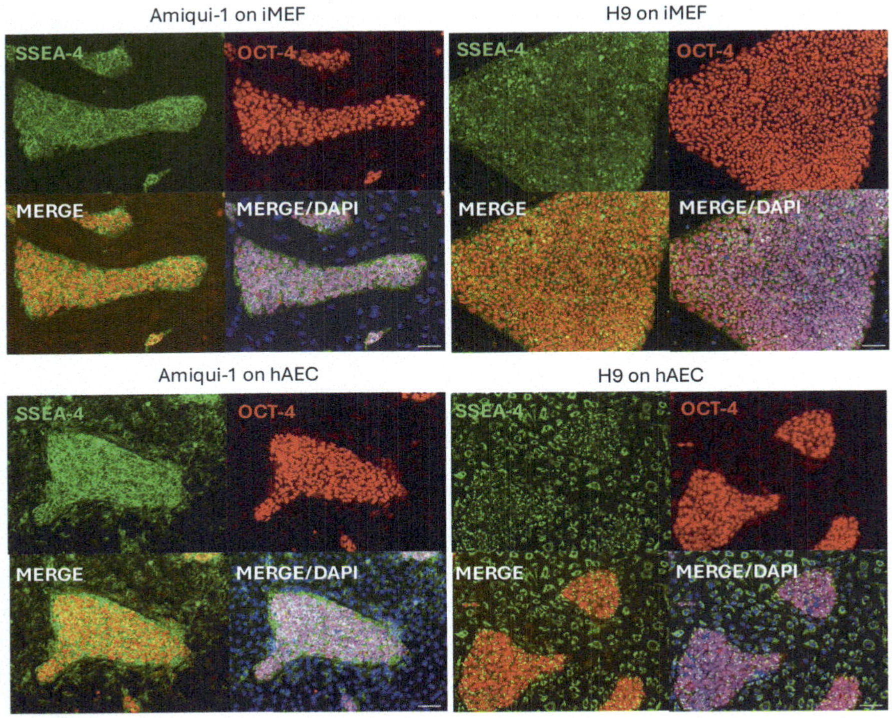

Fig. 3 Immunofluorescent characterization of hESC colonies on feeder layers. Representative epifluorescence micrographs of Amicqui-1 and H9 hESC cultured on inactivated mouse embryonic fibroblast (iMEF) or human amniotic epithelial cell (hAEC), showing double immunostaining for OCT-4/SSEA-4. Scale bars: 50 μm

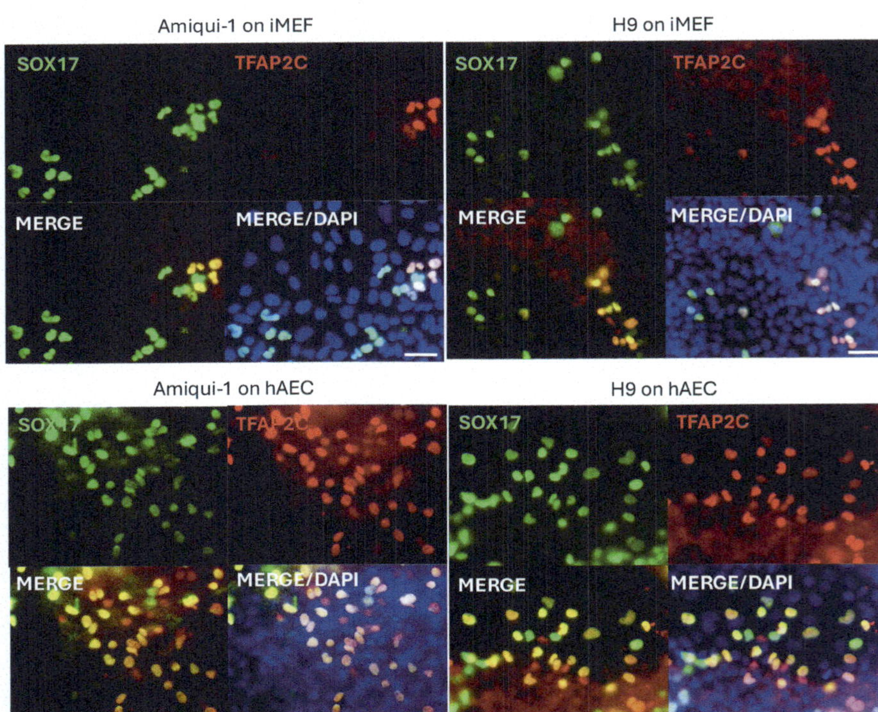

Fig. 4 Detection of germline markers in hPGCLC derived from hESC. Immunofluorescence images showing SOX17+ (green) and TFAP2C+ (red) markers on day 11 of induction into human primordial germ cell-like cells (hPGCLC) derived from Amicqui-1 and H9 hESC, on both iMEF and hAEC feeders. Scale bars: 50 μm

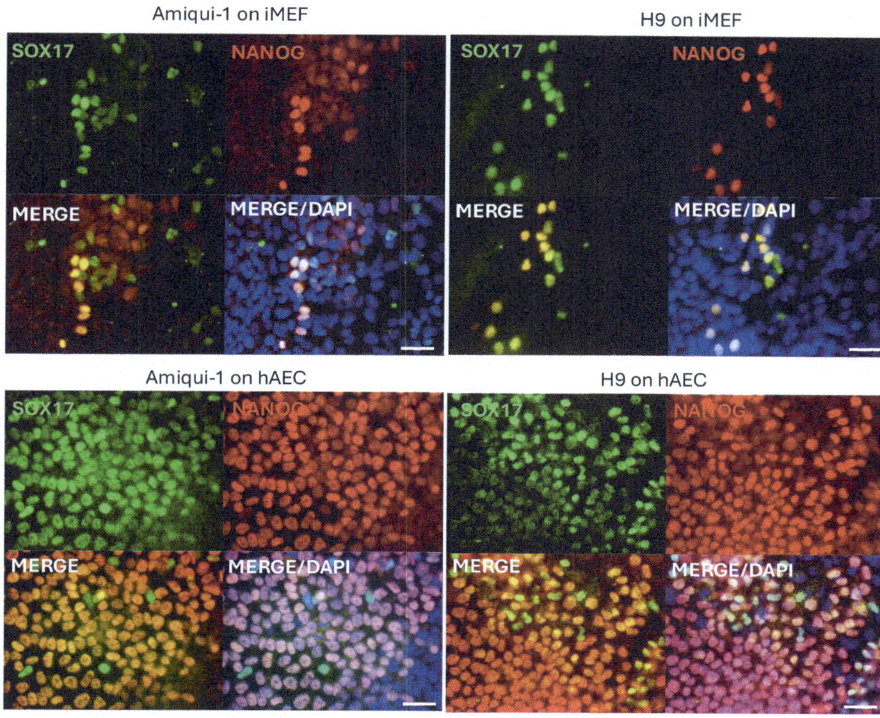

Fig. 5 Detection of germline markers in hPGCLC derived from hESC. Immunofluorescence images showing SOX17+ (green) and NANOG+ (red) markers on day 11 of induction into human primordial germ cell-like cells (hPGCLC) derived from Amicqui-1 and H9 hESC, on both iMEF and hAEC feeders. Scale bars: 50 μm

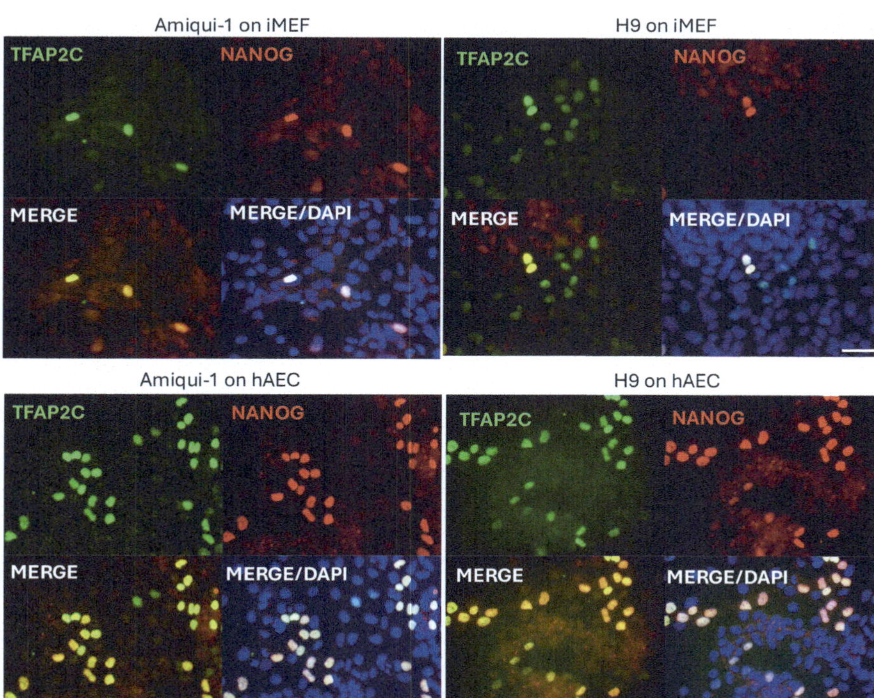

Fig. 6 Detection of germline markers in hPGCLC derived from hESC. Immunofluorescence images showing TFAP2C+ (green) and NANOG+ (red) markers on day 11 of induction into human primordial germ cell-like cells (hPGCLC) derived from Amicqui-1 and H9 hESC, on both iMEF and hAEC feeders. Scale bars: 50 μm

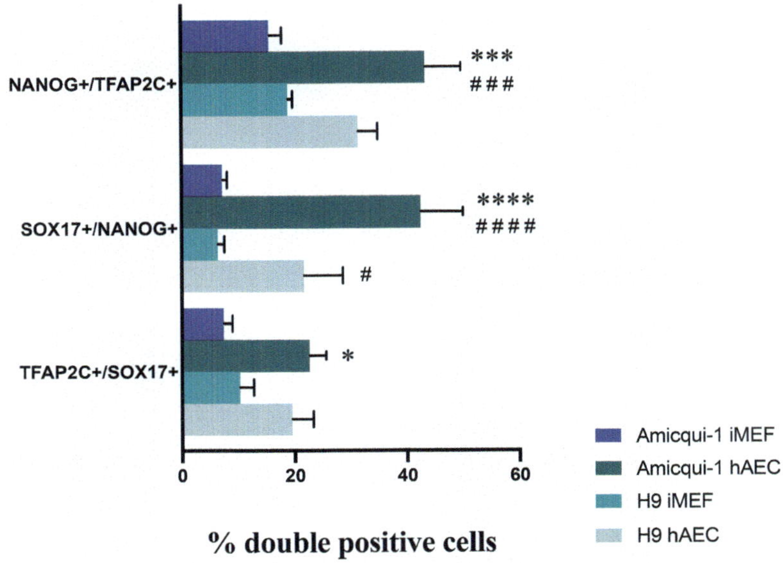

Fig. 7 Immunofluorescence analysis of hPGCLC derived from hESC
Quantification of immunofluorescence for TFAP2C+/NANOG+, SOX17+/NANOG+, and TFAP2C+/NANOG+ in hPGCLC derived from hESC on hAEC or iMEF feeders on day 11 after germinal induction. Data represent mean ± SE, with n = 3–4 biological replicates per group
Significant difference from Amicqui-1 on iMEF:
* $p < 0.05$, *** $p < 0.001$, **** $p < 0.0001$
Significant difference from H9 on iMEF:
#$p < 0.05$, # # # $p < 0.001$, # # # # $p < 0.0001$

$p = 0.0316$) (Fig. 7). This difference may be attributed to the shorter pre-induction culture period for H9 on hAEC (only two passages), in contrast to the minimum 10 passages for Amicqui-1 on hAEC, Amicqui-1 on iMEF, and H9 on iMEF.

3. This study underscore the significance of interactions between the amniotic epithelium and both embryonic and extraembryonic lineages, offering valuable insight into the mechanisms of early embryogenesis in primates, particularly in relation to cell fate determination and tissue organization.

4 Notes

1. Complete removal of erythrocytes is crucial, as residual blood cells interfere with trypsin function and reduce cell viability during epithelial cell isolation.
2. The cells should reach confluency within 8 days, forming an epithelial monolayer with cobblestone morphology. If

confluency is not achieved within 10 days, hAECs are not recommended for use as a feeder layer, as they may exhibit a fibroblastic morphology (see Fig. 1).

3. After thawing or direct passage, hAECs should reach confluency within 24–48 h, which is optimal for their use as a feeder layer.
4. hAECs typically remain adherent after 5–7 min of Accutase treatment.
5. Undifferentiated hESC cultures should be utilized for germline induction experiments, ensuring they exhibit the classical characteristics of hESC when grown on feeder layers: uniform, compact cells organized in flat colonies with well-defined edges (Fig. 2).
6. In addition to morphological assessment, verification of pluripotency markers at the protein level is essential prior to initiating experiments. At least two core pluripotency markers, such as OCT4 and NANOG, must be confirmed with nuclear localization (Fig. 3).
7. During the differentiation stage, it is important to seed an optimal number of cells in the wells of the 96-well plate. Higher cell densities can compromise the acquisition of germline markers, especially in hESC previously cultured on iMEF feeder layer.

Acknowledgments

This work was supported by INPER 2019-1-40 and 2024-1-24 as well as CONACYT A1-S-8450.

References

1. Garcia-Lopez G et al (2015) Human amniotic epithelium (HAE) as a possible source of stem cells (SC). Gac Med Mex 151(1):66–74
2. de Sousa C, Lopes SM, Roelen BAJ, Lawson KA, Zwijsen A (2022) The development of the amnion in mice and other amniotes. Philos Trans R Soc Lond Ser B Biol Sci 377(1865): 20210258. https://doi.org/10.1098/rstb.2021.0258
3. Dobreva MP, Pereira PN, Deprest J, Zwijsen A (2010) On the origin of amniotic stem cells: of mice and men. Int J Dev Biol 54(5):761–777. https://doi.org/10.1387/ijdb.092935md
4. Avila-Gonzalez D et al (2023) Pluripotent stem cells as a model for human embryogenesis. Cells 12(8). https://doi.org/10.3390/cells12081192
5. Yang R et al (2021) Amnion signals are essential for mesoderm formation in primates. Nat Commun 12(1):5126. https://doi.org/10.1038/s41467-021-25186-2
6. Zheng Y et al (2019) Controlled modelling of human epiblast and amnion development using stem cells. Nature 573(7774):421–425. https://doi.org/10.1038/s41586-019-1535-2
7. Sasaki K et al (2016) The germ cell fate of Cynomolgus monkeys is specified in the nascent amnion. Dev Cell 39(2):169–185. https://doi.org/10.1016/j.devcel.2016.09.007
8. Chen D et al (2019) Human primordial germ cells are specified from lineage-primed progenitors. Cell Rep 29(13):4568–4582. e4565.

https://doi.org/10.1016/j.celrep.2019.11.083

9. Avila-Gonzalez D et al (2022) The human amniotic epithelium confers a bias to differentiate toward the neuroectoderm lineage in human embryonic stem cells. elife 11. https://doi.org/10.7554/eLife.68035

10. Garcia-Lopez G et al (2019) Pluripotency markers in tissue and cultivated cells in vitro of different regions of human amniotic epithelium. Exp Cell Res 375(1):31–41. https://doi.org/10.1016/j.yexcr.2018.12.007

Methods in Molecular Biology (2026) 3048: 121–141
DOI 10.1007/7651_2025_641

Published online: 15 July 2025

Extended In Vitro Culture of Human Embryos

Sylwia M. Czukiewska, Felicitas Azpiroz, Susana M. Chuva de Sousa Lopes, and Mina Popovic

Abstract

Recent innovations in extended in vitro culture (IVC) systems have revolutionized our understanding of human peri-implantation development. Building on foundational animal studies, refined protocols now support human embryo culture beyond the blastocyst stage, providing unprecedented access to previously elusive developmental events. These systems have yielded critical insights into early morphogenetic processes, lineage specification, and tissue organization, significantly advancing developmental biology. Here, we provide our current protocol for the extended culture of human embryos, followed by immunofluorescence for lineage markers of interest. Unveiling human peri-implantation development also promises to improve reproductive medicine, potentially addressing challenges related to implantation failure, chromosomal instability in embryos, as well as congenital disorders. Insights gained from this research may pave way for novel therapeutic approaches and advancements in medically assisted reproduction.

Key words Embryo outgrowth, Human embryo, Implantation, In vitro culture, Peri-implantation development

1 Introduction

Over the past decade, advances in media formulations have significantly improved the culture conditions for human embryos from fertilization to the blastocyst stage [1, 2]. These innovations have markedly enhanced the efficiency of medically assisted reproduction (MAR), contributing to improved embryo selection, higher implantation rates, and overall increased pregnancy success. Nevertheless, culture systems used in clinical settings are inherently restricted to supporting embryos solely through the preimplantation stages of development. This limitation leaves the peri-implantation period, an essential phase marked by critical processes such as lineage specification and differentiation, largely unexplored.

To better understand early post-implantation (or peri-implantation) development, efforts have been made to develop in vitro systems that extend embryo culture beyond the blastocyst

Sylwia M. Czukiewska and Felicitas Azpiroz contributed equally as first-authors. Susana M. Chuva de Sousa Lopes and Mina Popovic contributed equally as senior-authors.

stage. While animal models laid an essential foundation for this work, they faced challenges in efficiency and lacked quantitative comparisons with in vivo development [3–8]. Refinements in these protocols have enabled the culture of mouse embryos for extended periods [9, 10] and have also been successfully optimized for studying peri-implantation development in human [11–15].

Extended in vitro culture (IVC) of human embryos allows blastocyst attachment, outgrowth formation, and culture for up to 14 days post-fertilization (dpf), capturing aspects of implantation, while omitting the requirement for maternal endometrial tissues [11, 15]. Notably, extended IVC allows embryo outgrowth organization representative of the peri-implantation stages of in vivo-developed human embryos, both structurally and based on lineage marker expression. For instance, human embryo outgrowths show a clear distinction between the epiblast, marked by POU5F1, and GATA6-positive hypoblast or primitive endoderm [11, 15]. Moreover, the embryo outgrowths form putative amniotic and yolk sac cavities; however, these are often not maintained, likely due to the flattened two-dimensional nature of IVC. Interestingly, blastocyst outgrowths morphologically resemble the post-inner cell mass intermediate (PICMI), a transient structure previously identified and characterized during human embryonic stem cell (hESC) derivation [12, 16]. The PICMI is marked by a flattened morphology with densely packed, epithelial-like cells and is thought to closely resemble the human peri-implantation epiblast. Excising and plating the PICMI from embryo outgrowths, 4–5 days after blastocyst plating, significantly enhance hESC derivation efficiency [12, 16].

While extended in vitro embryo culture platforms present powerful tools for studying the peri-implantation period, they do come with inherent limitations and technical challenges [17]. One significant drawback is that embryo outgrowths adopt a flattened, two-dimensional structure, which can lead to interactions between cells that would not typically be in close proximity in a natural three-dimensional environment. Additionally, IVC conditions, including medium composition, surface coating, pressure, oxygen concentration, may impact gene expression and alter cellular behavior. Advancements in IVC conditions and the use of refined extracellular matrices continue to offer improvements in the precision of these platforms in mimicking physiological conditions more accurately [18–20]. To date, extended IVC platforms have provided critical insights into the early morphogenetic events, including cellular dynamics, lineage specification trajectories, chromosomal instability, and tissue organization [13, 14, 18, 19, 21], offering new opportunities for understanding implantation failure, congenital disorders, and the molecular mechanisms underlying early human development. Ultimately, this evolving field promises to transform both basic research and clinical applications, providing powerful tools for unravelling the complexities of human embryogenesis. Here, we provide details to set up an extended IVC platform to investigate the human peri-implantation period.

2 Materials

2.1 Facilities

1. Clinical IVF laboratory (or Biosafety Level 2).
2. Liquid nitrogen storage room.
3. Microscopy imaging facility.

2.2 Equipment

1. IVF Workstation with integrated stereo microscope, heated stage, and laminar flow with HEPA filter (L124 IVF Workstation, Cooper Surgical).
2. Standard tissue culture laminar flow hood with vacuum aspirator.
3. CO_2 humidified incubator for mammalian cell culture with N_2 source or IVF benchtop incubator (MINC® Mini Incubator, Cook Medical).
4. Liquid nitrogen storage tank.
5. Inverted microscope (Nikon Eclipse Ti2 brightfield microscope equipped with Octax Eyeware, MedITEX IVF Software).
6. Inverted confocal microscope for immunofluorescence with appropriate fluorophore filters (Dragonfly 505, Andor).
7. Stereomicroscope.
8. MiniSpin Centrifuge (Eppendorf).
9. Cooling rack for handling of vitrification devices (Kitazato).
10. Embryo manipulation adjustable pipette handle (Cook® Flexipet® adjustable handle, Cook Medical, G18674).
11. Electronic pipette controller.
12. 1000 μL, 200 μL, 20 μL, 2 μL micropipettes.
13. Timer.
14. Lint-free lens cleaning tissues.

2.3 Disposables

1. Nunc™ cell-culture treated four-well dishes (ThermoFisher Scientific, 176740).
2. 60 mm culture dish (Vitrolife, 16002).
3. IVF-certified center-well dish (Vitrolife, 16005).
4. Flexible polycarbonate 300 μm pipettes for embryo manipulation (Cook® Flexipet®, Cook Medical, G18878).
5. 1000 μL, 200 μL, 20 μL, 2 μL sterile filtered pipette tips.

6. 5 mL, polypropylene, sterile MTC™ Bio DuoClick™ two position screw-cap culture tube (Sigma Aldrich, MTCT8734).
7. 15 mL Falcon polypropylene tubes.
8. μ-Slide 8 Well^{high} IbiTreat plates (Ibidi GmbH, IB-80826).
9. 0.22 μm disposable syringe filter units.
10. 250 mL Nalgene™ Rapid-Flow™ sterile disposable filter units (ThermoFisher Scientific, 568-0010).
11. 10 mL disposable syringes.
12. 5 mL, 10 mL serological pipettes, plugged.
13. 1.5 mL tubes (Eppendorf).
14. Disposable nitrile gloves.
15. Parafilm.
16. Permanent marker.

2.4 Reagents

2.4.1 Blastocyst Warming

1. IVF-grade disinfectant (Oosafe® surface disinfectant). Store at room temperature (RT).
2. Thawing Media Kit (Kitazato, VT102), including thawing solution (TS), diluent solution (DS), and washing solution (WS). Store at 4 °C.
3. Continuous single IVF culture medium, CSCM (Irvine Scientific, 90164). Store at 4 °C.
4. OVOIL Mineral oil (Vitrolife). Store at 4 °C (*see* **Note 1**).

2.4.2 Extended Embryo Culture

1. Sterile deionized water. Store at RT.
2. Ethanol, pure (Sigma Aldrich, 493546). Store at RT in a dry and well-ventilated place. Keep away from heat and sources of ignition. Refer to the manufacturer's safety data sheet for specific storage and disposal instructions.
3. Hydrochloric acid solution, volumetric, 1 M HCl (1 N) Honeywell™ Fluka™ (Fisher Scientific, 15676840). Store in a cool, dry, and well-ventilated area, away from heat sources. Refer to the manufacturer's safety data sheet for specific storage and disposal instructions.
4. Dimethyl sulfoxide (DMSO). Store at RT in the dark. Refer to the manufacturer's safety data sheet for specific storage and disposal instructions.
5. Acidic Tyrode's Solution (Sigma Aldrich, T1788). Store aliquots at −20 °C.
6. Advanced Dulbecco's Modified Eagle Medium/Ham's F-12, DMEM/F12 (ThermoFisher Scientific, 12634010). Store at 4 °C.

7. Fetal Bovine Serum (FBS), qualified, New Zealand (ThermoFisher Scientific, 10091148). Store aliquots at −20 °C (*see* **Note 2**).
8. L-glutamine (ThermoFisher Scientific, 25030). Store aliquots at −20 °C.
9. Penicillin–streptomycin, 10,000 U/mL (ThermoFisher Scientific, 15140122). Store aliquots at −20 °C.
10. Insulin-Transferrin-Selenium-Ethanolamine (ITS-X) 100x (ThermoFisher Scientific, 51500056). Store at 4 °C.
11. Sodium DL-Lactate Solution synthetic, syrup, BioXtra, suitable for mouse embryo cell culture, 60% (w/w) (Sigma Aldrich, L7900). Store at 4 °C.
12. 0.22% (v/v) sodium lactate. Store at 4 °C.
13. N-acetyl-L-cysteine (Sigma Aldrich, A7250): Reconstitute 1 g of N-acetyl-L-cysteine in 10 mL of sterile water (heat gently if required). Make a 1 mg/mL stock solution. Store at 4 °C.
14. Progesterone-water soluble (Sigma Aldrich, P7556): Dissolve 100 mg of progesterone in 100 mL of sterile water. Vacuum-filter. Store aliquots at −20 °C. For a 200 ng/mL final concentration, add 0.2 μL per mL of IVC base medium.
15. β-estradiol (Sigma Aldrich, E8875): Dissolve 250 mg estradiol in 250 mL of pure ethyl alcohol. Dilute 1 mL of the 1 mg/mL stock into 45 mL of pure ethanol to achieve a 21.79 μg/mL working stock. Store aliquots at −20 °C. For a 8 nM final concentration, add 1 μL of working stock per mL of IVC base medium.
16. Activin A (R&D Systems, 338-AC-050): First prepare a 4 mM HCl solution from 1 M HCl by adding 40 μL of 1 M HCl to 9.96 mL of sterile water. Reconstitute 10 μg of Activin A in 0.5 mL of the prepared 4 mM HCl solution. Store aliquots at −20 °C. For a 100 ng/mL concentration, add 5 μL per mL of IVC base medium.
17. ROCK inhibitor, Y-27632 (Enzo Life Sciences, ALX-270-333): Reconstitute 5 mg of Y27634 in 1.56 mL of DMSO. Store aliquots at −20 °C. For a 10 μM concentration, add 1 μL per mL of IVC base medium.
18. Sodium pyruvate (P4562, Sigma Aldrich): Reconstitute 5 g of sodium pyruvate in 90.9 mL of sterile water. Store aliquots at −20 °C. For a 1 mM final concentration, add 2 μL per mL of IVC base medium.
19. Insulin-like growth factor 1, IGF-1 (STEMCELL Technologies, 78078): Reconstitute 10 μg of IGF-1 in 100 μL of sterile water. Store aliquots at −20 °C. For a 100 ng/mL final concentration, add 1 μL per mL of IVC base medium.

20. IVC base medium: Advanced DMEM/F12 with 2 mM L-glutamine, 1% penicillin-streptomycin, 1X ITS-X (containing 10 mg/mL insulin, 5.5 mg/L transferrin, 0.0067 mg/L sodium selenite, and 2 mg/L ethanolamine), 25 μM N-acetyl-L-cysteine, and 0.22% (v/v) sodium lactate. Vacuum filter. Prepare 10 mL aliquots in 15 mL Falcon tubes. Store aliquots at −20 °C.
21. IVC Medium: Prepare prior to each experiment. IVC base medium with 20% (v/v) FBS, 8 nM β-estradiol, 200 ng/mL progesterone, 1 mM sodium pyruvate, 10 μM Y27632, and 100 ng/mL IGF-1.

2.4.3 Immunofluorescence Staining

1. Phosphate-buffered saline (PBS, 1X), sterile-filtered (Thermo Fisher Scientific, 10010056). Store at RT.
2. Sterile deionized water. Store at RT.
3. Donkey serum (Sigma Aldrich, D9663). Store aliquots at −20 °C.
4. Bovine Serum Albumin (BSA) solution, 30% in DPBS, sterile-filtered, BioXtra, suitable for cell culture (Sigma Aldrich, A9576).
5. Paraformaldehyde (PFA), 4% in PBS (ThermoFisher Scientific, J61899.AK). Refer to the manufacturer's safety data sheet for specific storage and disposal instructions.
6. Triton X-100 (Sigma Aldrich, T9284). Store at RT.
7. Tween 20 (Sigma Aldrich, P9416). Store at RT.
8. Permeabilization Solution: 1X PBS with 0.1% (v/v) Triton X-100. Store at RT for up to 1 month.
9. Blocking Solution: 1X PBS with 0.1% (v/v) Tween 20, 10% (vol/vol) donkey serum, 1% (vol/vol) BSA. Store at 4 °C for up to 2 weeks.
10. Washing Solution: 1X PBS with 0.1% (v/v) Tween 20. Store at RT up to 1 month.
11. Primary antibodies: Select lineage-specific protein markers of interest (Table 1). Prepare antibody dilutions in blocking solution as reported (Table 1) or as per manufacturer's instructions (*see* **Note 3**).
12. 1 mg/mL 4′,6-diamidino-2-phenylindole (DAPI) solution (ThermoFisher Scientific, 62248) (Table 1).
13. Secondary antibodies: Select adequate fluorescence-conjugated secondary antibodies for immunofluorescence (Table 1) diluted in blocking solution. Include in the same solution a nuclear marker (such as DAPI) and/or marker of the actin filaments (phalloidin) (Table 1) (*see* **Note 4**).

Table 1
Overview of antibodies

Cell type/structure	Primary antibody	Host	Company	Catalogue Number	Clone	Dilution	Reference
Epiblast	SOX2	Goat	Santa Cruz	sc-17319	D-17, polyclonal	1:200	Overeem et al. 2023
Epiblast/PGCs	POU5F1 (or OCT3/4)	Mouse	Santa Cruz	sc-5279	C-10	1:250	Popovic et al. 2019
Epiblast/PGCs	POU5F1 (or OCT3/4)	Goat	Santa Cruz	sc-8628	N-19, polyclonal	1:250	Popovic et al. 2019
Epiblast/PGCs	NANOG	Rabbit	R&D Systems	AF1997	Polyclonal	1:250	Popovic et al. 2019
Epiblast/PGCs	NANOG	Rabbit	ThermoFisher Scientific	PA1-097	Polyclonal	1:200	Meistermann et al. 2021
Epiblast/PGCs	NANOG	Mouse	BD Pharmingen	560482	N31-355	1:500	Deglincerti et al. 2016
Epiblast/PGCs	NANOG	Goat	R&D Systems	AF1997	Polyclonal	1:200	Shahbazi et al. 2020
Hypoblast	GATA4	Rat	ThermoFisher Scientific	14-9980-82	Polyclonal	1:200	Meistermann et al. 2021
Hypoblast/PGCs	SOX17	Goat	R&D Systems	AF1924	Polyclonal	1:250	Popovic et al. 2019
Hypoblast/extraembryonic mesoderm	GATA6	Goat	R&D Systems	AF1700	Polyclonal	1:200	Shahbazi et al. 2020
Hypoblast/extraembryonic mesoderm	GATA6	Rabbit	Santa Cruz	sc-9055	H-92, polyclonal	1:250	Popovic et al. 2019
Trophoblast (sub-types)	HCGB	Mouse	Abcam	ab9582	5H4-E2	1:100	Deglincerti et al. 2016

(continued)

Table 1
(continued)

Cell type/structure	Primary antibody	Host	Company	Catalogue Number	Clone	Dilution	Reference
Trophoblast (sub-types)	HLAG	Mouse	Abcam	ab52455	4H84	1:200	Xiang et al. 2020
Trophoblast (sub-types)	HLAG	Mouse	Abcam	ab7759	MEM-G/1	1:200	Shahbazi et al. 2020
Trophoblast (sub-types)	GATA2	Mouse	Merk	WH0002624M1	2D11	1:200	Meistermann et al. 2021
Trophoblast (sub-types)	NR2F2	Rabbit	Abcam	ab211776	EPR18442	1:100	Meistermann et al. 2021
Trophoblast (sub-types)/PGCs	TFAP2C (or AP-2 g)	Rabbit	Santa Cruz	sc-12762	6E4/4	1:100	Yang et al. 2021
Trophoblast (sub-types)/PGCs	TFAP2C (or AP-2 g)	Rabbit	Cell Signaling	2320	Polyclonal	1:200	Overeem et al. 2023
Trophoblast (sub-types)/amniotic ectoderm	KRT7	Rabbit	Abcam	ab181598	EPR17078	1:400	Deglincerti et al. 2016
Trophoblast (sub-types)/amniotic ectoderm	GATA3	Mouse	ThermoFisher Scientific	MA1-028	1A12-1D9	1:100	Deglincerti et al. 2016
Trophoblast (sub-types)/amniotic ectoderm	GATA3	Rat	ThermoFisher Scientific	14-9966-82	TWAJ	1:100	Pham et al. 2022
Trophoblast (sub-types)/amniotic ectoderm	GATA3	Goat	R&D Systems	AF2605	Polyclonal	1:200	Shahbazi et al. 2020
Primitive streak	TBXT (or T)	Rabbit	Santa Cruz	sc-20109	H-210, polyclonal	1:50	Xiang et al. 2020
Primitive streak	TBXT (or T)	Goat	R&D Systems	AF2085	Polyclonal	1:200	Xiang et al. 2020
Primitive streak	MIXL1	Rabbit	Merk	HPA005662	Polyclonal	1:100	Yang et al. 2021

PGCs	PDPN	Mouse	Abcam	ab256561	LpMab-7	1:200	Overeem et al. 2023
PGCs	PRDM1 (or BLIMP1)	Rat	ThermoFisher Scientific	14-5963-82	6D3	1:200	Overeem et al. 2023
Amniotic ectoderm	ISL1	Mouse	DSHB	39.4D5	39.4D5	1:300	Yang et al. 2021
Amniotic ectoderm	ISL1	Rabbit	Millipore	MABN1107	EP4182	1:300	Yang et al. 2021
Amniotic ectoderm/PGCs	TFAP2A (or AP-2a)	Mouse	Santa Cruz	sc-12726	3B5	1:200	Overeem et al. 2023
Amniotic ectoderm/extraembryonic mesoderm	HAND1	Goat	R&D Systems	AF3168	Polyclonal	1:200	Overeem et al. 2023
Extraembryonic mesoderm	CDX2	Mouse	BioGenex	MU392-UC	CDX2-88	1:250	Popovic et al. 2019
Extraembryonic mesoderm	POSTN	Rabbit	Abcam	ab14041	Polyclonal	1:100	Pham et al. 2022
Apical domain	PODXL	Mouse	R&D Systems	MAB1658	222328	1:500	Shahbazi et al. 2020
	Secondary antibody	**Host**	**Company**	**Catalogue Number**	**Clone**	**Dilution**	
	Anti-rabbit IgG AF-488	Donkey	ThermoFisher Scientific	A-21206	Polyclonal	1:500	
	Anti-rabbit IgG AF-594	Donkey	ThermoFisher Scientific	A-21207	Polyclonal	1:500	
	Anti-rabbit IgG AF-647	Donkey	ThermoFisher Scientific	A-31573	Polyclonal	1:500	
	Anti-mouse IgG AF-488	Donkey	ThermoFisher Scientific	A-21202	Polyclonal	1:500	
	Anti-mouse IgG AF-594	Donkey	ThermoFisher Scientific	A-21203	Polyclonal	1:500	

(continued)

Table 1
(continued)

Cell type/structure	Primary antibody	Host	Company	Catalogue Number	Clone	Dilution	Reference
	Anti-mouse IgG AF-647	Donkey	ThermoFisher Scientific	A-31571	Polyclonal	1:500	
	Anti-goat IgG AF-488	Donkey	ThermoFisher Scientific	A-11055	Polyclonal	1:500	
	Anti-goat IgG AF-594	Donkey	ThermoFisher Scientific	A-11058	Polyclonal	1:500	
	Anti-goat IgG AF-647	Donkey	ThermoFisher Scientific	A-21447	Polyclonal	1:500	
	Anti-sheep IgG AF-647	Donkey	ThermoFisher Scientific	A-21448	Polyclonal	1:500	
	Anti-rat IgG AF-594	Donkey	ThermoFisher Scientific	A-21209	Polyclonal	1:500	
Cell structure	**Product**		**Company**	**Cat**	**Clone**	**Dilution**	**Reference**
Actin filaments	Phaloidin (F-actin) AF-488		ThermoFisher Scientific	A12379	na	1:500	Shahbazi et al. 2020
Nucleus	DAPI		ThermoFisher Scientific	D3571	na	1:2000	Shahbazi et al. 2020

3 Methods

3.1 Research Using Supernumerary Human Embryos

In accordance with local and national regulations, approval from an Institutional Review Board or relevant Ethics Committee is required for the use of human preimplantation embryos in scientific research. Supernumerary embryos may be donated for research, subject to patient consent and regulatory compliance. These embryos, generated during conventional IVF or intracytoplasmic sperm injection (ICSI) cycles, are generally of good (morphological) quality. The term "supernumerary" indicates that these embryos are extra or additional, beyond what is required for immediate clinical purposes. These embryos may not have been selected for immediate transfer, or may have been deemed clinically unsuitable, for instance, following preimplantation genetic testing (PGT). Supernumerary embryos are typically vitrified at the blastocyst stage, on 5 dpf or 6 dpf. However, they can also be obtained at different developmental stages, cryopreserved using different protocols, and may or may not have undergone PGT.

Written informed consent must be obtained from all patients donating their embryos for scientific research purposes. All procedures should comply with the ethical standards of the institutional research committee and the 1964 Helsinki Declaration, as revised in 2024 [22, 23]. To ensure confidentiality, data associated with supernumerary embryos donated for research should be anonymized and systematically recorded for subsequent analysis.

3.2 Blastocyst Warming and Grading

All steps are to be carried out at RT in an IVF workstation or laminar flow cabinet, unless otherwise specified. Ensure to use aseptic tissue culture techniques throughout. Ensure that all solutions are prepared using tissue culture/embryo grade reagents and water (endotoxin free). All incubations should be performed in a humidified incubator at 37 °C, 5% CO_2 in hypoxic conditions (5% O_2), unless otherwise specified.

1. Disinfect the workspace surface using Oosafe® surface disinfectant.
2. Working in the IVF workstation, prepare a post-thawing culture dish (60 mm) by pipetting 25 μL drops of IVF culture medium (CSCM), overlayed with 3 mL of OVOIL (Fig. 1a) (*see* **Note 5**).
3. Label individual droplets and prepare as many dishes as required based on the number of embryos to be warmed.
4. Place the prepared culture dishes in the incubator to equilibrate overnight.

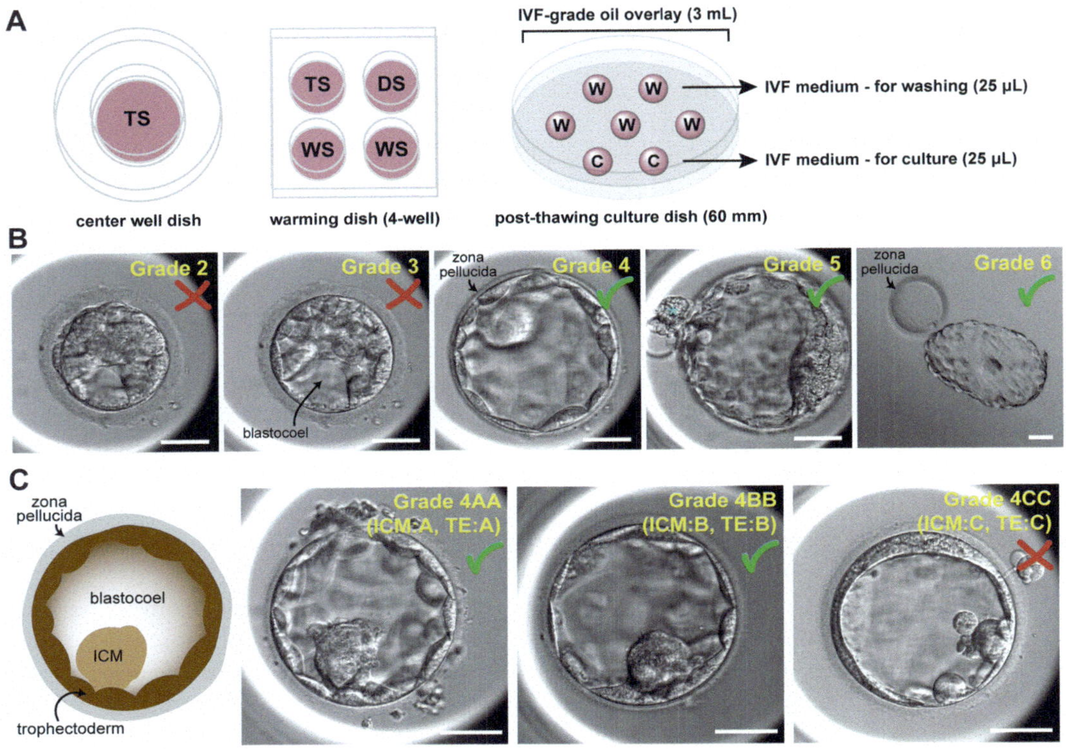

Fig. 1 Embryo warming and grading. (**a**) Dish layout for embryo warming and culture. Recommended layout of media droplets for embryo warming and culture following embryo warming. The central well dish contains 1 mL of thawing solution (TS). The warming dish (4-well dish) contains one well will TS, one well with diluent solution (DS), and two wells with washing solution (WS). The post-thawing culture dish contains drops of IVF medium overlayed with IVF-grade mineral oil. Drops can be used either for washing (W) or culture (C), accommodating up to five individual embryos per dish. (**b**) Blastocyst expansion grading. Grade 2 (early blastocyst): the blastocoel occupies less than half of the total volume of the embryo. Grade 3 (full blastocyst): the blastocoel occupies half or more of the total volume of the embryo. Grade 4 (expanded blastocyst): the blastocoel fills the entire embryo, accompanied by a thinning of the zona pellucida. Grade 5 (hatching blastocyst): the blastocoel is larger than the original embryo volume, accompanied by further thinning of the zona pellucida; the trophectoderm starts to herniate (cyan asterisk) through a natural breach in the zona pellucida. Grade 6 (hatched blastocyst): the blastocyst has completely emerged from the zona pellucida. Good-quality blastocysts (expansion Grade 4 or higher) are used for further culture. Scale bars are 50 μm. (**c**) Blastocyst inner cell mass (ICM) and trophectoderm (TE) grading. The first letter after the expansion grade refers to the ICM grade (A, B, C) followed by the TE grade (A, B, C). Left panel: cartoon of a blastocyst showing the ICM and TE. Middle left panel (Grade 4AA): ICM Grade A, the ICM is composed of many tightly packed cells; TE Grade A, the TE consists of many cells forming a cohesive epithelium. Middle right panel (Grade 4BB): ICM Grade B, the ICM has several cells grouped more loosely, suggesting a less compacted mass compared to Grade A; TE Grade B, the TE includes fewer cells forming a loose epithelium, showing less structural integrity than Grade A. Right panel (Grade 4CC): ICM Grade C, the ICM has few cells and is potentially compromised; TE Grade C, the TE has few large cells, indicating poor proliferation. Good-quality blastocysts (expansion Grade 4BB or higher) are used for further culture. Scale bars are 50 μm

The following steps outline the warming steps in accordance with the Kitazato Warming Kit (VT602) protocol, using a Cryotop vitrification device (*see* **Note 6**).

5. Pipette 1 mL of TS into the center well dish (Fig. 1a) and prewarm for 1 h at 37 °C. Ensure that the TS is prewarmed during this step.
6. Bring the DS and WS to RT.
7. Prepare a "warming dish" by dispensing 300 μL of TS into the first well, 300 μL of DS into the second well, and 300 μL of WS into the third and fourth wells of a Nunc™ cell-culture treated 4-well dish (Fig. 1a). The "warming dish" should be kept at RT.
8. Remove the Cryotop containing the embryo from the liquid nitrogen storage tank and place it carefully in a cooling rack filled with liquid nitrogen.
9. Following the manufacturer's instructions, open the Cryotop device by sliding or unscrewing the protective cover.
10. Once opened, immediately submerge the Cryotop strip into the prewarmed TS solution in the center well dish.
11. Aspirate the embryo using a 300 μm Flexipet® pipette tip and gently place it in TS in the first well of the "warming dish" for 1 min.
12. Using a 300 μm Flexipet® pipette tip, sequentially transfer the embryo through the following solutions in the "warming dish": DS, the second well for 3 min, WS, the third well for 5 min, and finally wash in WS, fourth well, for 1 min (*see* **Note 7**).
13. Carefully aspirate the blastocyst and gently place in the washing drop of the pre-equilibrated post-thawing culture dish (Fig. 1a).
14. Wash embryo three times in wash droplets.
15. Transfer blastocyst to individual culture droplet.
16. Assess blastocyst viability using a brightfield microscope (Nikon Eclipse Ti2).
17. Culture viable embryos for at least 2–4 h at 37 °C, 6% CO_2, and 5% O_2, to allow for re-expansion.

3.3 Extended In Vitro Culture of Human Embryos

Ensure all embryos are at 6 dpf on the day of plating for extended in vitro culture. For embryos that have been warmed on 5 dpf, allow them to develop for an additional day until they reach 6 dpf. For embryos that are already at 6 dpf, wait until they re-expand before proceeding.

1. On the day of plating (IVC day 0), prepare IVC medium.
2. Add 300 μL of IVC medium to each well of an μ-Slide 8 Well-high IbiTreat plate, according to the number of embryos to be cultured. For any unused wells, add 300 μL of sterile water to maintain humidity (*see* **Note 8**).
3. Place μ-Slide 8 Well^{high} IbiTreat plates in a humidified incubator at 37 °C, 6% CO_2, and 5% O_2 to equilibrate for a minimum of 1 h.
4. Using an inverted brightfield microscope (Nikon Eclipse Ti2), assess blastocyst quality using the Gardner and Schoolcraft grading system [24] and record the embryo grade for further reference.
5. Capture images of individual blastocysts for further reference using the Octax Eyeware and MedITEX IVF Software.
6. Plate only expanded, good-quality blastocysts, expansion Grade 4 or higher (Fig. 1b), and inner cell mass and trophectoderm, quality Grade B or higher (Fig. 1c).
7. Label each well of the μ-Slide 8-Well^{high} IbiTreat plates with the corresponding embryo sample ID according to the experimental setup.
8. For embryos still enclosed in the zona pellucida, expansion Grade 4 or 5 (Fig. 1b, c), perform zona pellucida removal (Subheading 3.3.1, **step 27**) before proceeding with extended culture (Subheading 3.3.2, **step 35**).
9. For fully hatched, zona-free blastocysts, expansion Grade 6 (Fig. 1b), proceed directly to initiate extended culture (Subheading 3.3.2, **step 35**).

3.3.1 Zona Pellucida Removal

1. Set Acidic Tyrode's solution for at least 30 min at 37 °C (*see* **Note 9**).
2. In a Nunc™ Cell-Culture Treated four-well dish, fill the first and third wells with pre-warmed Acidic Tyrode's solution. Add pre-equilibrated IVC medium to the second and fourth wells (*see* **Note 10**).
3. Working in the IVF workstation on a heated stage, transfer one blastocyst at a time using a 300 μm Flexipet® pipette tip into the well with pre-warmed Acidic Tyrode's solution (*see* **Note 11**).
4. Gently pipette the embryo up and down a few times, ensuring that it does not touch the bottom of the plate.
5. Observe under a stereomicroscope as the zona pellucida begins to dissolve. This process should take about 5–10 s. Continuously monitor the embryo to reduce exposure time to the Acidic Tyrode's solution.

6. If required, move the embryo to the next well with Acidic Tyrode's solution to aid in the complete dissolution of the zona pellucida.
7. Once the zona pellucida is completely dissolved, immediately transfer the embryo into the well with IVC medium and gently wash the blastocyst by pipetting up and down to neutralize any transferred Acidic Tyrode's solution.
8. Use a clean 300 μm Flexipet® pipette tip to transfer the zona pellucida-free embryo into the other IVC medium well. Gently wash the blastocyst by pipetting up and down and proceed to initiate extended culture (Subheading 3.3.2, **step 35**) (*see* **Note 12**).

3.3.2 Blastocyst Plating for Extended Culture

1. Use a clean 300 μm Flexipet® pipette tip to transfer the zona pellucida-free embryos into IVC medium, into well 4 of the Nunc™ Cell-Culture Treated four-well dish for an extra wash, gently pipetting up and down (*see* **Note 12**).
2. Plate the zona pellucida-free embryos in the equilibrated μ-Slide 8-Well^{high} IbiTreat plate (1 embryo/well; IVC day 0) and immediately place the plate in a humidified incubator at 37 °C, 6% CO_2 in hypoxic conditions (5% O_2).
3. On IVC day 2 (~ 48 h after plating), prepare IVC medium and place it in a humidified incubator at 37 °C, 6% CO_2, and 5% O2 to equilibrate for a minimum of 1 h.
4. Examine the embryos using a brightfield microscope (Nikon Eclipse Ti2). Assess and classify each embryo according to its attachment status (Fig. 2).
5. Capture images of each embryo using the Octax Eyeware, MedITEX IVF Software for future reference.
6. Working in the IVF workstation on a heated stage, carefully remove 100 μL of IVC medium from each well using a sterile pipette tip, while observing the embryo, and add 200 μL of pre-equilibrated IVC medium to each well (*see* **Note 13**).
7. Immediately place the μ-Slide 8-Well^{high} IbiTreat plate in a humidified incubator at 37 °C, 6% CO_2 in hypoxic conditions, and 5% O_2.
8. Repeat **steps 37–41** on IVC day 4 (~ 96 h after plating).
9. Repeat **steps 37–41** on IVC day 6 (~ 144 h after plating) (*see* **Note 14**).
10. Following final observation, embryo outgrowths can be fixed for further analysis using immunofluorescence (*see* **Note 15**).

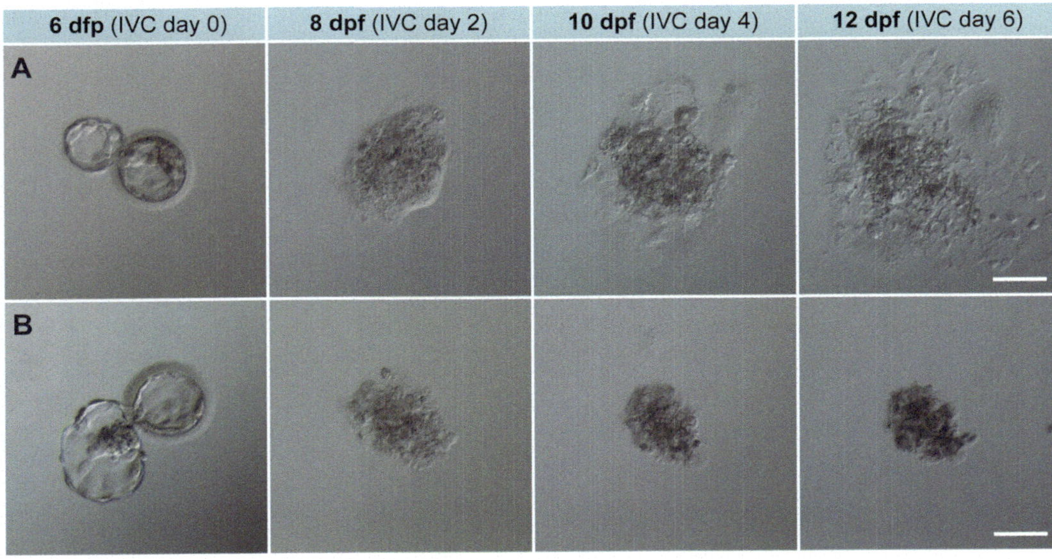

Fig. 2 Extended in vitro culture (IVC) of human blastocysts. (**a**) Good developmental progression of a representative (hatching) embryo from day 6 to 12 post fertilization (dpf), corresponding to IVC day 0 to IVC day 6. Embryo attached on IVC day 2 and remained attached during the extended culture forming an embryo outgrowth. Scale bar is 50 μm. (**b**) Poor developmental progression of one representative (hatching) embryo from day 6 to 12 dpf, corresponding to IVC day 0 to IVC day 6. Embryo attached on IVC day 2, but showed cell degeneration shortly thereafter. Embryo detached between IVC days 4 and 6. Scale bar is 50 μm

3.4 Immunofluorescence of Embryo Outgrowths

For immunofluorescence, select only attached and viable embryo outgrowths (Fig. 2a).

1. Working in the IVF workstation, carefully remove the IVC medium from the μ-Slide 8-Well^{high} IbiTreat plate using a sterile pipette tip while observing the embryo outgrowth under the stereomicroscope.
2. Add 300 μL PBS 1X to each well to wash the embryo outgrowths.

The following steps should be performed on a standard laboratory bench, using a stereomicroscope to monitor the embryo outgrowth. Care should be taken to ensure the embryo outgrowth is not detached during the procedure.

3. Remove PBS 1X from each well and add 300 μL 4% PFA to each well of the μ-Slide 8-Well^{high} IbiTreat plate for 30 min at RT.
4. Remove 4% PFA and add 300 μL washing solution to each well of the μ-Slide 8-Well^{high} IbiTreat plate for 10 min at RT.
5. Repeat this washing process twice.

6. This step can serve as a stopping point before proceeding with immunofluorescence. Add 300 μL PBS 1X to each well and store at 4 °C until future use (*see* **Note 16**).
7. Remove washing solution (or PBS 1X) and add 300 μL of permeabilization solution to each well of the μ-Slide 8-Well^{high} IbiTreat plate for 10 min at RT.
8. Remove permeabilization solution and add 300 μL of blocking solution to each well of the μ-Slide 8-Well^{high} IbiTreat plate for at least 2 h at RT (*see* **Note 17**).
9. Calculate total volume of primary antibody solution based on the number of wells and the recommended dilution (Table 1) (*see* **Note 18**).
10. Remove the blocking solution and add 200 μL of the prepared primary antibody solution to each well. Incubate overnight at 4 °C (*see* **Note 19**).
11. Remove the primary antibody solution from each well. Wash the embryo outgrowths carefully five times with 300 μL washing solution for 5, 10, 15, 20, and 20 min, respectively, at RT.
12. Calculate total volume of secondary antibody solution, including DAPI and or phalloidin, based on number of wells and the recommended dilution (Table 1). Work in the dark and cover secondary antibody solution with foil (*see* **Note 20**).
13. Remove the washing solution, add 200 μL of the prepared secondary antibody solution to each well and incubate overnight at 4 °C (*see* **Notes 21** and **22**).
14. Remove the secondary antibody solution from each well. Wash the embryo outgrowths by adding 300 μL of washing solution for 10 min at RT.
15. Repeat this washing process twice.
16. Remove washing solution and add 300 μL PBS 1X (*see* **Note 23**).
17. Proceed with confocal imaging (*see* **Note 24**).

4 Notes

1. It is critical to use mineral oil specifically designed and validated for embryo culture, as specified by the manufacturer. Mineral oil that is not certified for embryo culture can introduce contaminants, disrupt gas exchange, or alter the medium composition, potentially compromising embryo viability and development.
2. FBS exhibits considerable batch-to-batch variation, which can significantly impact experimental outcomes due to differences

in growth factors, hormones, and other bioactive components. To mitigate this and enhance reproducibility, it is recommended to test each new FBS batch on a small number of embryos, comparing performance against a baseline established from previous batches or from published data. Purchasing and testing large, single batches for consistency over the course of a study can further improve the reliability and reproducibility of experimental results.

3. Different primary antibodies may require specific fixation methods, dilutions, or incubation times. When using multiple primary antibodies together, it is critical that they are from different host species or alternatively have distinct isotypes.
4. When choosing a secondary antibody ensure that it matches the host of the primary antibody. When using multiple primary and hence secondary antibodies together, it is critical that the different secondary antibodies emit in different color ranges (wavelength). DAPI is a convenient nuclear counterstain as it binds to DNA. It has an absorption maximum at 358 nm and fluoresces blue at an emission maximum of 461 nm.
5. Incubate additional aliquots of CSCM medium and OVOIL in 5 mL sterile screw-cap culture tubes, in case additional dishes are required on the day of warming. Keep the caps loosened to maintain gas exchange.
6. Alternative warming protocols may be employed depending on the vitrification method. It is essential to follow the specific warming instructions as per manufacturer's instructions, according to the vitrification system used for cryopreserving the embryos. This will ensure optimal embryo viability upon recovery.
7. Continuous monitoring of the embryo with a stereo microscope is recommended.
8. μ-Slide 8 Well^{high} IbiTreat plates are sensitive to rapid temperature changes, which may cause damage to the plate or fracture the slide. To minimize this risk, maintain a constant temperature of 37 °C by performing all steps on a heated stage within the IVF workstation.
9. Ensure to maintain the Acidic Tyrode's solution at 37 °C as its efficiency reduces at lower temperatures.
10. This dish setup provides convenient access to both solutions during the procedure. The IVC medium neutralizes the Acidic Tyrode's solution through a buffering effect and the presence of FBS proteins.
11. Minimize the amount of culture medium carried over when transferring embryos into the Acidic Tyrode's solution. The medium can neutralize the Acidic Tyrode's solution, making it less efficient.

12. Using a sterile pipette tip and washing the blastocyst helps preventing the transfer of any mineral oil from the embryo culture dish into the μ-Slide 8-Well^{high} IbiTreat plates (Subheading 3.3.2).
13. Exercise precision when refreshing the medium to avoid disrupting the culture or detaching the embryos from the plate. Carefully remove and add the medium along the side wall of each well to minimize disturbances.
14. Following internationally recognized guidelines [25, 26], embryo culture should be stopped when the embryo reaches 14 dfp, corresponding to IVC day 8 (~192 h after plating).
15. In addition to immunofluorescence, embryo outgrowths can be used for a variety of applications, such as chromosomal analysis using next-generation sequencing (NGS) techniques and transcriptomics using single-cell RNA sequencing.
16. If storing for extended periods (>2 weeks), wrap the plate in parafilm and ensure all wells are filled with PBS 1X to prevent evaporation. Check regularly for evaporation and maintaining PBS 1X levels.
17. Embryo outgrowths can also be left in blocking solution overnight at 4 °C.
18. As an example, for immunofluorescence in 4 wells using three primary antibodies with a 1:200 dilution, add 4 μL of each of the three antibodies (total of 12 μL) to 788 μL of blocking solution, to prepare a total of 800 μL of primary antibody solution.
19. The use of appropriate controls during immunofluorescence is essential for the accurate interpretation of the results. To confirm the specificity of the primary antibodies used, consider including a "negative control" by omitting those primary antibodies and a "positive control" by testing the antibodies in different tissue/cell types or using different antibodies to recognize the same protein.
20. As an example, for immunofluorescence in four wells using three secondary antibodies with a 1:500 dilution, add 1.6 μL of each of the three antibodies (total of 4.8 μL) and 0.4 μL of DAPI to 794.8 μL of blocking solution to prepare a total volume of 800 μL of secondary antibody solution.
21. An incubation step of 2 h may be sufficient.
22. From this step on, the samples should be protected from light to prevent photobleaching of the fluorophore-conjugated secondary antibodies. Hence, cover the plate with aluminum foil or place it in a box to avoid exposure to light.

23. While embryo outgrowths can be stored at 4 °C, it is recommended to perform the imaging as soon as possible, ideally within a few days to ensure optimal quality of fluorescence signal. For extended storage, consult guidelines for preserving fluorescence signals.

24. Imaging should be performed using a confocal microscope, ensuring that the settings are optimized for the specific fluorophores used. Parameters such as laser power, gain, and exposure time should be adjusted according to the signal intensity to prevent photobleaching and achieve high-quality images. Each imaging setup may vary, so it is important to tailor the imaging conditions to the specific equipment and sample preparation used.

Acknowledgments

This work was supported by the Novo Nordisk Foundation (reNEW NNF21CC0073729 to SMC and SMCSL).

References

1. Glujovsky D, Quinteiro Retamar AM, Alvarez Sedo CR, Ciapponi A, Cornelisse S, Blake D (2022) Cleavage-stage versus blastocyst-stage embryo transfer in assisted rep roductive technology. Cochrane Database Syst Rev 5(5): CD002118. https://doi.org/10.1002/14651858.CD002118.pub6
2. Govindasamy N, Duethorn B, Oezgueldez HO, Kim YS, Bedzhov I (2019) Test-tube embryos – mouse and human development in vitro to blastocyst stage and beyond. Int J Dev Biol 63(3–5):203–215. https://doi.org/10.1387/ijdb.180379ib
3. Enders AC, Schlafke S, Hendrickx AG (1986) Differentiation of the embryonic disc, amnion, and yolk sac in the rhesus monkey. Am J Anat 177(2):161–185. https://doi.org/10.1002/aja.1001770205
4. Hsu Y-C (1972) Differentiation in vitro of mouse embryos beyond the implantation stag e. Nature 239(5369):200–202. https://doi.org/10.1038/239200a0
5. Hsu Y-C (1973) Differentiation <i>in vitro</i> of mouse embryos to the stage of early somite. Dev Biol 33(2): 403–411. https://doi.org/10.1016/0012-1606(73)90145-0
6. Hsu Y-C, Baskar J, Stevens LC, Rash JE (1974) Development in vitro of mouse embryos from the two-cell egg stage to t he early somite stage. Development 31(1):235–245. https://doi.org/10.1242/dev.31.1.235
7. Hsu YC (1979) In vitro development of individually cultured whole mouse embryos from blastocyst to early somite stage. Dev Biol 68(2):453–461. https://doi.org/10.1016/0012-1606(79)90217-3
8. Tachi C (1992) Partial characterization of macromolecular components in fetal bovine serum required for development of mouse blastocysts cultured in vitro. Develop Growth Differ 34(1):69–77. https://doi.org/10.1111/j.1440-169X.1992.00069.x
9. Aguilera-Castrejon A, Oldak B, Shani T, Ghanem N, Itzkovich C, Slomovich S, Tarazi S, Bayerl J, Chugaeva V, Ayyash M, Ashouokhi S, Sheban D, Livnat N, Lasman L, Viukov S, Zerbib M, Addadi Y, Rais Y, Cheng S, Stelzer Y, Keren-Shaul H, Shlomo R, Massarwa R, Novershtern N, Maza I, Hanna JH (2021) Ex utero mouse embryogenesis from pre-gastrulation to late organogenesis. Nature 593(7857):119–124. https://doi.org/10.1038/s41586-021-03416-3
10. Bedzhov I, Leung CY, Bialecka M, Zernicka-Goetz M (2014) In vitro culture of mouse blastocysts beyond the implantation stages. Nat Protoc 9(12):2732–2739. https://doi.org/10.1038/nprot.2014.186

11. Deglincerti A, Croft GF, Pietila LN, Zernicka-Goetz M, Siggia ED, Brivanlou AH (2016) Self-organization of the in vitro attached human embryo. Nature 533(7602):1–13. https://doi.org/10.1038/nature17948
12. O'Leary T, Heindryckx B, Lierman S, Van Bruggen D, Goeman JJ, Vandewoestyne M, Deforce D, De Sousa Lopes SMC, De Sutter P (2012) Tracking the progression of the human inner cell mass during embryonic stem cell derivation. Nat Biotechnol 30(3): 278–282. https://doi.org/10.1038/nbt.2135
13. Popovic M, Bialecka M, Fernandes MG, Taelman J, Jeught MVD, Sutter PD, Heindryckx B, Lopes SMCDS (2019) Human blastocyst outgrowths recapitulate primordial germ cell specific ation events. Mol Hum Reprod 25(9):519. https://doi.org/10.1093/molehr/gaz035
14. Popovic M, Dhaenens L, Taelman J, Dheedene A, Bialecka M, De Sutter P, De Sousa C, Lopes SM, Menten B, Heindryckx B (2019) Extended in vitro culture of human embryos demonstrates the complex nature of diagnosing chromosomal mosaicism from a single trophectoderm biopsy. Hum Reprod 34(4):758–769. https://doi.org/10.1093/humrep/dez012
15. Shahbazi MN, Jedrusik A, Vuoristo S, Recher G, Hupalowska A, Bolton V, Fogarty NME, Campbell A, Devito LG, Ilic D, Khalaf Y, Niakan KK, Fishel S, Zernicka-Goetz M (2016) Self-organization of the human embryo in the absence of maternal tissues. Nat Cell Biol 18(6):700–708. https://doi.org/10.1038/ncb3347
16. O'Leary T, Heindryckx B, Lierman S, Van Der Jeught M, Duggal G, De Sutter P, De Sousa C, Lopes SM (2013) Derivation of human embryonic stem cells using a post-inner cell mass intermediate. Nat Protoc 8(2):254–264. https://doi.org/10.1038/nprot.2012.157
17. Rossant J (2016) Human embryology: implantation barrier overcome. Nature 533(7602): 182–183. https://doi.org/10.1038/nature17894
18. Molè MA, Coorens THH, Shahbazi MN, Weberling A, Weatherbee BAT, Gantner CW, Sancho-Serra C, Richardson L, Drinkwater A, Syed N, Engley S, Snell P, Christie L, Elder K, Campbell A, Fishel S, Behjati S, Vento-Tormo-R, Zernicka-Goetz M (2021) A single cell characterisation of human embryogenesis identifies pluri potency transitions and putative anterior hypoblast centre. Nat Commun 12(1):3679. https://doi.org/10.1038/s41467-021-23758-w
19. Weatherbee BAT, Weberling A, Gantner CW, Iwamoto-Stohl LK, Barnikel Z, Barrie A, Campbell A, Cunningham P, Drezet C, Efstathiou P, Fishel S, Vindel SG, Lockwood M, Oakley R, Pretty C, Chowdhury N, Richardson L, Mania A, Weavers L, Christie L, Elder K, Snell P, Zernicka-Goetz M (2024) Distinct pathways drive anterior hypoblast specification in the implan ting human embryo. Nat Cell Biol 26(3):353–365. https://doi.org/10.1038/s41556-024-01367-1
20. Xiang L, Yin Y, Zheng Y, Ma Y, Li Y, Zhao Z, Guo J, Ai Z, Niu Y, Duan K, He J, Ren S, Wu D, Bai Y, Shang Z, Dai X, Ji W, Li T (2020) A developmental landscape of 3D-cultured human pre-gastrulation embryos. Nature 577(7791):537–542. https://doi.org/10.1038/s41586-019-1875-y
21. Shahbazi MN, Wang T, Tao X, Weatherbee BAT, Sun L, Zhan Y, Keller L, Smith GD, Pellicer A, Scott RT, Seli E, Zernicka-Goetz M (2020) Developmental potential of aneuploid human embryos cultured beyond imp lantation. Nat Commun 11(1):3987. https://doi.org/10.1038/s41467-020-17764-7
22. World Medical A (2013) World Medical Association Declaration of Helsinki: ethical principles for medical research involving human subjects. JAMA 310(20):2191–2194. https://doi.org/10.1001/jama.2013.281053
23. World Medical A (2024) World medical association declaration of Helsinki: ethical principles for medical research involving human participants. JAMA. https://doi.org/10.1001/jama.2024.21972
24. Gardner DK, Schoolcraft WB (1999) In vitro culture of human blastocysts. The Parthenon Publishing Group
25. Ethics Advisory Board DoH, Education, and Welfare (HEW) (1979) HEW support of research involving human in vitro fertilization and embryo transfer. Report and Conclusions. Washington, USA
26. Warnock committee DoHaSS (1984) Report of the committee of inquiry into human fertilisation and embryology. London, UK

Methods in Molecular Biology (2026) 3048: 143–162
DOI 10.1007/7651_2025_644

Published online: 22 May 2025

Generation of Human Naïve Pluripotent Stem Cell-Derived Blastoids in Thermoformed Microwell Platforms

Dorian G. Luijkx, Erik J. Vrij, Ge Guo, and Stefan Giselbrecht

Abstract

Due to the inaccessibility of early human embryos for large, robust studies, many questions regarding the mechanisms of early embryogenesis remain. To address these questions, multiple research groups have developed human stem cell-based models of the pre-implantation blastocyst stage. These models, known as blastoids, mimic several key processes that natural blastocysts undergo as they prepare to implant into the uterine wall. One of the main advantages of blastoids is their scalability, making them suitable for both screenings and molecular studies. To leverage this advantage, we established a protocol for the parallel formation and culture of blastoids in thermoformed microwell arrays. Thin-walled thermoformed microwell platforms allow for uniform, large-scale generation of blastoids and enable in situ high-resolution imaging through both widefield and confocal microscopy. Here we present a step-by-step protocol for the culture of blastoids in thermoformed microwell platforms.

Key words Blastocysts, Blastoids, Embryo models, Naïve pluripotent stem cells, Stem cell-based embryo models, Thermoformed microwells

1 Introduction

Within the first week after fertilization, human embryos undergo several developmental processes that are crucial for a successful implantation into the uterine wall and subsequent growth of a healthy fetus. Although a large portion of embryos, estimated around 40%, don't develop beyond the implantation stage, the reasons for this low fertility rate are not known [1]. For a long time, tools to address questions on early human embryogenesis were lacking. However, in recent years, stem cell biologists have discovered methods for culturing both maternal and embryonic tissues to investigate fertility and early embryogenesis. With endometrial organoids and endometrium monolayers to mimic the uterine environment, fertility on the maternal side can now be studied on a molecular level. To recapitulate the embryonic side, human stem cells can be induced into an early developmental state (referred to as "naïve pluripotent stem cells") [2–6]. These cells can in turn be induced in a suspension culture to mimic processes of early embryogenesis and give rise to 3D structures similar to embryos in the blastocyst stage (Fig. 1a) [7–12]. These structures called blastoids contain all three cell types also present in the blastocyst with the same organization of these cell types. This means blastoids consist of a dense clump of epiblast- and hypoblast-like cells that make up the inner cell mass, surrounded by a single cell layer cyst of trophectoderm-like cells. Some of the main advantages of blastoids are that during their formation, they recapitulate some of the processes also observed in human embryos and that they can be formed in high numbers. The high number of structures is necessary due to the variability that is inherent in current stem cell cultures. As a result of this variability, even embryo models cultured in the exact same conditions are overall similar, but not identical, differing in their exact cell numbers and undergoing formation processes asynchronously. Some of the structures don't recapitulate embryonic processes or morphology at all and remain uncavitated cell aggregates (Fig. 1b). Thus, reliably generating very large numbers of structures is crucial for statistically robust conclusions. The scalability of blastoids opens doors not only for hypothesis-driven studies on the mechanisms of early embryogenesis, but also for screening of compound libraries, which with natural embryos is infeasible [13].

To achieve the numbers required for these kind of studies, we have adapted and optimized the existing blastoid protocols to our thermoformed microwell platforms (Fig. 1c). Thermoformed microwells are fabricated as an array of smaller wells, typically with diameters ranging between 200 and 800 μm but adaptable to any size or shape that can fit, e.g., into any standard tissue culture format [14]. In these arrays, cells are distributed across the microwells and thus confined to a smaller space, allowing single cells to aggregate to cell clusters faster and more easily, ensuring most cells contribute to the aggregate. Thus, the platform gives control over the cell number per aggregate, through the defined number of microwells and the reduction of cell loss (Fig. 1d). Moreover, a unique feature of thermoformed microwells over some of the other microwell platforms is that they are formed from a single, very thin, and transparent polymer film. The microwells can be manufactured with different shapes, e.g., with flat or U-shaped bottom. The resulting thin and transparent bottom and walls (<50 μm) make them suitable for high-resolution imaging, also over long periods of time, thus allowing for tracking of any single blastoid as it forms. On a final note, the proprietary technology to manufacture thermoformed microstructures, the so-called SMART technology, is highly versatile [14]. Beyond the size and depth of the microwells, the polymer film material can be changed to optimize the optical, mechanical, or chemical properties, or the polymer film can be treated with various surface and bulk polymer modification techniques to create functionalized structures which have an overlaid 3D micro- or nanotopography or cylindrical pores of a defined size for

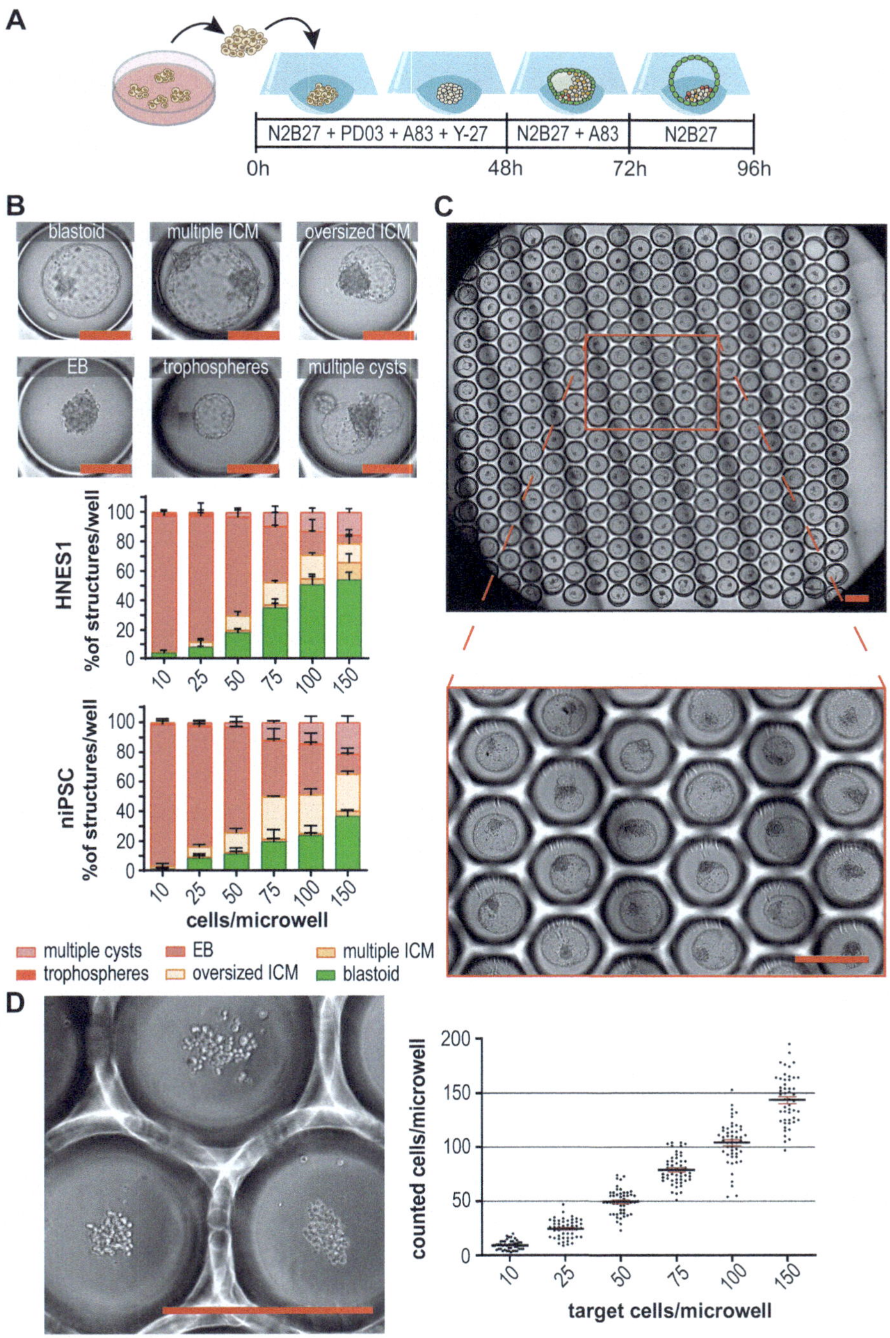

Fig. 1 Optimization of blastoid culture in microwells in the absence of LPA. (**a**) Schematic of the blastoid culture protocol without LPA. (**b**) Brightfield image with an inset of HNES1 blastoids in 500 μm microwell array in a 24-well plate, scale 500 μm. (**c**) Yields of structures formed per microwell array in a cell number titration experiment for both HNES1 and niPSC 75.1.c2 with example images of the categories, scale 200 μm. (**d**) Brightfield image of nPSCs 30 min after seeding, when single cells are still distinguishable, scale 500 μm, and quantification of targeted number of cells/microwell vs. cells/microwell counted after seeding

instance. Thermoformed microwell arrays can easily be incorporated into microfluidic chips and microbioreactors, for example, which opens possibilities for microwell-supported dynamic and controlled 3D cell culture applications [15].

In this protocol, we describe how thermoformed microwell arrays are prepared for cell culture, how naïve pluripotent stem cells are cultured, and how blastoids are formed in the microwells. For downstream analysis, we provide protocols for immunostaining and flow cytometry on blastoids, as well as a method for single structure handling. A functional application of blastoids, namely the assessment of the blastoids adhesion to the endometrium, is explored in the chapter on "Microfluidic on-chip assay for quantifying blastoid adhesion to endometrium epithelium," also found in this book.

2 Materials

2.1 Biological Materials

1. DR4-irradiated MEFs.
2. Naïve induced pluripotent stem cells.
3. Human embryo-derived naïve pluripotent stem cells.

2.2 Reagents

1. MEK inhibitor PD0325901 (PD03; Axon Med Chem cat# 1408).
2. Tankyrase inhibitor XAV-939 (XAV; MedChem Express cat# HY-15147).
3. aPKC inhibitor Gö-6983 (Gö; MedChem Express cat# HY-13689).
4. Recombinant human leukemia inhibitory factor (LIF; Sigma Aldrich cat# SRP3316).
5. Activin/nodal receptor inhibitor A83-01 (MedChem Express cat# HY-10432).
6. Rho associated kinase inhibitor Y-27632 (Y-27; Axon Med Chem cat# 1683).
7. Hippo inhibitor 1-Oleoyl lysophosphatidic acid sodium salt (LPA; BioTechne cat# 3854/1).
8. Dulbecco's phosphate buffered saline (1×), no calcium, no magnesium (DPBS; Sigma Aldrich cat# D8537).
9. DMEM/F12 (Sigma Aldrich cat# D8437).
10. Neurobasal (Gibco cat# 21103049).
11. B27 supplement (Gibco cat# 17504044).
12. N2 supplement (Made in-house).
13. DMEM/Hams F12 (Sigma Aldrich cat# D6421).

14. Culture-grade water (VWR cat# 392-0439).
15. Apotransferrin (R&D Systems cat# 3188-AT).
16. Insulin (Sigma Aldrich cat# I9278).
17. Sodium selenite (Sigma Aldrich cat# S5261-10G); Caution (*see* **Note 1**).
18. Progesterone (Sigma Aldrich cat# P8783-1G).
19. Putrescine (Sigma Aldrich cat# P5780-5G).
20. L-glutamine (Gibco cat# 25030024).
21. 2-Mercaptoethanol (Gibco cat# 11528926)
22. BSA V Fraction (Gibco cat# 15260037).
23. Fetal Bovine Serum (Gibco cat# 10270106).
24. Non-essential amino acids (Gibco cat# 11350912).
25. Gelatin (Sigma Aldrich cat# G1393).
26. Geltrex (Gibco cat# A1413302).
27. Pluronic F108 (Sigma Aldrich cat# 542342).
28. Accutase (STEMCELL Technologies cat# 07920).
29. TrypLE Express (Gibco cat# 12605036).
30. STEMCELL Banker (amsbio cat# 11924).
31. Trypan Blue (Sigma-Aldrich cat# T8154).
32. Ethanol, absolute (VWR cat# 8.18760.2500).
33. Ethanol, 70% (VWR cat# 84010059.5000).
34. Paraformaldehyde (ThermoFisher Scientific cat# 11400580); Caution (*see* **Note 2**).
35. Triton X-100 (Sigma Aldrich cat# T8787).
36. Donkey serum (VWR cat# AGRIAS101563).
37. Glycine (Sigma Aldrich cat# G5417).
38. Tween20 (Sigma Aldrich cat# P9416).
39. 4′,6-Diamidino-2-phenylindole dihydrochloride (DAPI; Sigma Aldrich cat# 32670).
40. Rabbit polyclonal anti-NANOG (1:200; Abcam cat# ab21624, RRID: AB_446437).
41. Goat polyclonal anti-OCT3/4 (1:300; R&D Systems cat# AF1759, RRID: AB_354975).
42. Goat polyclonal anti-SOX17 (1:100; R&D Systems cat# AF1924, RRID: AB_355060).
43. Rabbit monoclonal anti-GATA3 (1:300; Abcam cat# ab199428, RRID: AB_2819013).
44. Rabbit monoclonal anti-NR2F2 (1:100; Abcam cat# ab211776, RRID: AB_2893028).

45. Donkey anti-Rabbit IgG (H + L) Antibody, Alexa Fluor™ 647 (1:500; Invitrogen cat# A-31573, RRID: AB_2536183).
46. Donkey anti-Goat IgG (H + L) Antibody, Alexa Fluor™ 568 (1:500; Invitrogen cat# A-11057, RRID: AB_2534104).
47. Donkey anti-Goat IgG (H + L) Antibody, Alexa Fluor™ 488 (1:500; Invitrogen cat# A-11055, RRID: AB_2534102).
48. Goat anti-Rabbit IgG (H + L) Antibody, Alexa Fluor™ 750 (1: 500; Invitrogen cat# A-21039, RRID: AB_2535710).

2.3 Equipment

1. Hypoxic (5% O_2, 5% CO_2) incubator with humidity and gas control.
2. Normoxic (20% O_2, 5% CO_2) incubator with humidity and gas control.
3. Biological safety cabinet.
4. −20 °C freezer.
5. −80 °C freezer.
6. Centrifuge.
7. Water bath.
8. Inverted microscope.
9. Hemocytometer.
10. 500 mL bottle-top vacuum filtration system.
11. Cell strainers, 40 μm and 70 μm (ThermoFisher Scientific cat#11587522 and 11597522).
12. 6-Well Flat-Bottom Plate, Tissue Culture-Treated.
13. Microwell plate (300MICRONS cat# STATARRAYS MCA96–8.736-PS).
14. Microwell chips (in-house fabricated polycarbonate film).
15. O-rings (ERIKS cat# 10023241).
16. Pipettes (Eppendorf, P1000, P200, P20 and P10).
17. Conical tubes, 15 mL and 50 mL.
18. Common consumables (serological pipettes, pipette tips).

2.4 Stock Solutions and Medium Compositions

1. PD0325901

   Dissolve in DMSO (stock 10 mM). Aliquots can be stored at −20 °C for up to 6 months.
2. XAV939

   Dissolve in DMSO (stock 20 mM). Aliquots can be stored at −20 °C for up to 6 months.
3. Gö6983

   Dissolve in DMSO (stock 10 mM). Aliquots can be stored at −80 °C for up to 6 months. Protect from light.

4. Human LIF
   Dissolve in culture-grade water (stock 10 μg/mL). Aliquots can be stored at −20 °C for up to 12 months.
5. A83-01
   Dissolve in DMSO (stock 10 mM). Aliquots can be stored at −20 °C for up to 6 months.
6. Y-27632
   Dissolve in DMSO (stock 20 mM). Aliquots can be stored at −20 °C for up to 6 months.
7. 1-Oleoyl lysophosphatidic acid sodium salt (LPA)
   Dissolve in DMSO (stock 2.5 mM). Aliquots can be stored at −20 °C for up to 6 months.
8. N2 (home-made)
   Dissolve apotransferrin in culture-grade water (stock 100 mg/mL). Similarly dissolve sodium selenite in culture-grade water (stock 3 mM) and dissolve putrescine in culture-grade water (160 mg/mL). Dissolve progesterone in absolute ethanol (0.6 mg/mL). To prepare the N2, use DMEM/Hams F12 as the base medium and add apotransferrin (final concentration 10 mg/mL), BSA V Fraction (final concentration 0.75%), insulin (final concentration 2.5 mg/mL), progesterone (final concentration 1.98 μg/mL), putrescine (final concentration 1.6 mg/mL), and sodium selenite (final concentration 3 μM).
   Once aliquoted, the solution can be stored in −80 °C for up to 6 months.
9. MEF medium
   Prepare 500 mL MEF medium by mixing 420 mL of DMEM high glucose medium with 75 mL FBS (final concentration 15%). Add 5 mL NEAA solution (final concentration 1×) and 1 mL 2-mercaptoethanol (final concentration 0.1 mM).
   Medium can be stored at 4 °C for 4–5 weeks.
10. N2B27 medium
   Prepare 500 mL N2B27 basal medium by mixing 245 mL DMEM/F12 with 245 mL Neurobasal medium. Add 2.5 mL N2 supplement, 5 mL B27 supplement, 2.5 mL L-glutamine (final concentration 1 mM, due to L-glutamine present in DMEM/F12), and 1 mL 2-mercaptoethanol (final concentration 0.1 mM). Filter the N2B27 medium with a 500 mL bottle-top vacuum filtration system.
   Medium can be stored at 4 °C for 4–5 weeks.
11. PXGL medium
   Prepare PXGL medium by supplementing N2B27 basal medium with PD03 (final concentration 1 μM), XAV (final

concentration 2 μM), Gö (final concentration 2 μM) and LIF (final concentration 10 ng/mL).

Medium can be stored at 4 °C for 1 week.

12. PALY medium

Prepare PALY by supplementing N2B27 basal medium with PD03 (final concentration 2 μM), A83 (final concentration 2 μM), LPA (final concentration 0–2.5 μM depending on the cell line), and Y-27 (final concentration 10 μM) (*see* **Note 3**).

Medium can be stored at 4 °C for up to 2 days.

13. Gelatin solution

Prewarm the 2% stock solution in a water bath at 37 °C. Dilute 7.5 mL gelatin solution in 92.5 mL sterile DPBS (final concentration 0.15%).

The solution can be stored at 4 °C until the expiration date of the stock solution.

14. Pluronic F108 anti-adherence solution (*see* **Note 4**)

Weigh 500 mg of Pluronic F108 in a 50 mL tube. Dissolve in 50 mL sterile PBS (final concentration of Pluronic F108 will be 1% (*w/v*)). To help dissolve Pluronic F108, place the tube in a water bath at 37 °C and/or use a roller shaker at 50 rpm for several hours or overnight.

Once fully dissolved, the solution can be stored at 4 °C for up to 6 months.

15. Geltrex coating solution

Thaw undiluted Geltrex on ice or at 4 °C (*see* **Note 5**). Dissolve Geltrex in ice-cold DMEM (final concentration 1% (*v/v*)) and use immediately to coat tissue culture plates.

16. Fixation buffer

Dilute 37% PFA to 4% PFA in PBS (*v/v*).

Store at 4 °C and use within one week. Aliquots can also be frozen at −20 °C and thawed for immediate use.

17. Permeabilization buffer

Dilute Triton-X 100 to 0.25% in PBS (*v/v*).

Store at room temperature for 6 months.

18. Washing buffer

Dilute BSA V Fraction (7.5%) to 0.5% in PBS (*v/v*).

Store at 4 °C for up to 6 months.

19. Blocking buffer

Dilute BSA V Fraction (7.5%) to 2%, serum of host 2nd AB species to 5%, glycine to 0.5%, Triton-X100 to 0.1%, and Tween20 to 0.2% in PBS (*v/v*).

Store aliquots at −20 °C for up to 6 months. Avoid repeated freeze-and-thaw cycles.

20. Antibody buffer

Prepare by mixing 25% of blocking buffer with 25% Triton-X 0.1% solution and 50% PBS.

Store aliquots at −20 °C for up to 6 months. Avoid repeated freeze-and-thaw cycles.

3 Methods

3.1 Human Naïve Pluripotent Stem Cell Culture

3.1.1 MEF culture (see Notes 6 and 7)

1. Prepare MEF culture plate(s) by coating 6-well tissue culture plates with 1 mL gelatin solution (0.15% solution) per well.
2. Incubate at 37 °C for 30–60 min.
3. Take MEF medium from the fridge and prepare a 15 mL tube with 4 mL MEF medium.
4. Retrieve a vial of frozen MEFs from the liquid nitrogen and thaw in a water bath at 37 °C until only a small clump of ice remains.
5. Thoroughly clean the vial with 70% ethanol before opening it in a sterilized biosafety cabinet.
6. Rapidly transfer the cell suspension to the 15 mL tube with MEF medium the moment it is completely thawed. The cell suspension should be pipetted into the bottom of the tube and gently tapped to mix the cell suspension with the medium.
7. Centrifuge the tube at 200× *g* for 3 min.
8. Aspirate the supernatant, gently tap the tube to break up the cell pellet. and resuspend in 1 mL MEF medium.
9. Count cells using a hemocytometer and Trypan Blue.
10. Calculate how many cells you have and how many wells you can plate, aiming for 35,000 cells/cm^2 (350,000 cells per 6 wells plate well).
11. Resuspend MEF in MEF medium. The total number of mLs should be equivalent to the number of wells.
12. Prepare the culture plate(s) by aspirating the gelatin and adding 1 mL of MEF medium per well.
13. Distribute the cells across the wells by adding 1 mL of cell suspension per well.
14. For an even spread of cells, move the plate(s) gently in 8-shaped movements (back-and-forth and side-to-side motions).
15. Incubate the MEFs at 37 °C in a humidified incubator under normoxic conditions (20% O_2, 5% CO_2).

3.1.2 Thaw Naïve Pluripotent Stem Cells (nPSCs; see Note 8)

1. Prewarm wash buffer and PXGL medium at 37 °C.
2. Thaw an aliquot of Geltrex on ice. Dilute it 1:4 in DMEM/F12.

3. Prepare a 15 mL tube with 4.5 mL of wash buffer.
4. Thaw a cryovial with nPSCs in 37 °C water bath until small clump of ice remains.
5. Resuspend gently in warmed wash buffer.
6. Centrifuge the tube at 200× *g* for 3 min.
7. Aspirate the wash buffer, gently tap the tube to break up the cell pellet and resuspend cells in 2 mL PXGL medium supplemented with 10 μM Y-27.
8. Aspirate the MEF medium from a well coated with MEFs. Wash once with PBS.
9. Plate the nPSCs in PXGL medium supplemented with 10 μM Y-27 on the washed MEF plate. Add Geltrex at 2 μL/cm^2 of diluted solution from **step 2**.
10. Incubate the nPSCs at 37 °C in a humidified incubator under hypoxic conditions (5% O_2, 5% CO_2) (*see* **Note 9**).

3.1.3 Passage nPSCs

1. Prewarm wash buffer and PXGL medium at 37 °C.
2. Thaw an aliquot of Geltrex on ice. Dilute it 1:4 in DMEM/F12.
3. Prepare a 15 mL tube with 4.5 mL of wash buffer and prewarm the Accutase at 37 °C.
4. Aspirate PXGL medium from well with nPSCs on MEFs.
5. Add 500 μL of Accutase and incubate cells at 37 °C for 3–5 min, until individual cells can be distinguished within colonies and cells start detaching from the plate.
6. Firmly pipette the Accutase on the bottom of the well to dislodge the cells. Repeat up to 10 times until most cells are dissociated from the plate.
7. Collect the cell suspension and transfer to the tube with warmed wash buffer.
8. Centrifuge the tube at 200× *g* for 3 min.
9. Aspirate the wash buffer, gently tap the tube to break up the cell pellet, and resuspend cells in 200 μL to 1 mL PXGL medium, depending on the density of cells on the original plate (*see* **Note 10**).
10. Aspirate the MEF medium from a well coated with MEFs. Wash once with PBS.
11. Add 1.9 mL PXGL medium to the washed well with MEFs. Plate 100 μL of nPSC cell suspension in PXGL medium on this well (total volume 2 mL).
12. Supplement the medium with 10 μM Y-27 and the diluted Geltrex from **step 2** at 2 μL/cm^2 (20 μL per 6 wells plate well).

13. Incubate the nPSCs at 37 °C in a humidified incubator under hypoxic conditions (5% O_2, 5% CO_2).

3.1.4 Freeze nPSCs (see Note 11)

1. Aspirate the PXGL medium and dissociate cells by incubating with 500 μL Accutase per 6 wells plate well for 3–5 min at 37 °C.
2. Collect cell suspension in Accutase.
3. Transfer the cell suspension to a tube with 4.5 mL wash buffer per well. Depending on the number of wells dissociated, you may need either a 15 or a 50 mL tube.
4. Centrifuge the tube at 200× *g* for 3 min.
5. Aspirate the supernatant, gently tap the tube to break up the cell pellet, and resuspend cells in 1 mL of STEMCELL Banker per number of wells the cells were collected from (*see* **Note 12**).
6. Divide the cell suspension over cryovials, adding 0.5 mL per cryovial.
7. Store cryovials in a Mister Frosty or equivalent slow-rate cooling chamber in a −80 °C freezer for one day.
8. The next day, transfer the cryovials to a liquid nitrogen storage.

3.2 Microwell Preparation

3.2.1 Assembly in 24-Well Plate (Chips; Fig. 2)

1. Use a puncher with a 15 mm diameter to cut microwell array chips from the thermoformed film (*see* **Note 13**).
2. Sterilize O-rings by autoclaving at 120 °C for at least 20 min (*see* **Note 14**).
3. In a biosafety cabinet, add 500 μL sterile PBS per 24 wells plate well.
4. Using sterile forceps, dip microwell array chips in a 50 mL tube with 70% ethanol and place them in the 24-well plate with the bottoms of the microwells facing down (*see* **Note 15**).
5. Insert an O-ring on top of the microwell chip with forceps. Press the O-ring down to tightly lock the microwell chip between the bottom of the well and the O-ring.
6. Aspirate the PBS and continue immediately with the sterilization steps.

3.2.2 Sterilization and Anti-surfactant Treatment (300MICRONS Plates and Chips)

1. Add 50/200 μL 100% ethanol on top of the microwells to wet the surface. If air bubbles persist within microwells, remove these by pipetting up and down or by centrifuging the plate for a minute.
2. Remove the 100% ethanol and replace it with 200/500 μL of 70% ethanol.
3. Incubate at room temperature for 60 min.
4. Wash wells thrice with PBS to remove ethanol residues.

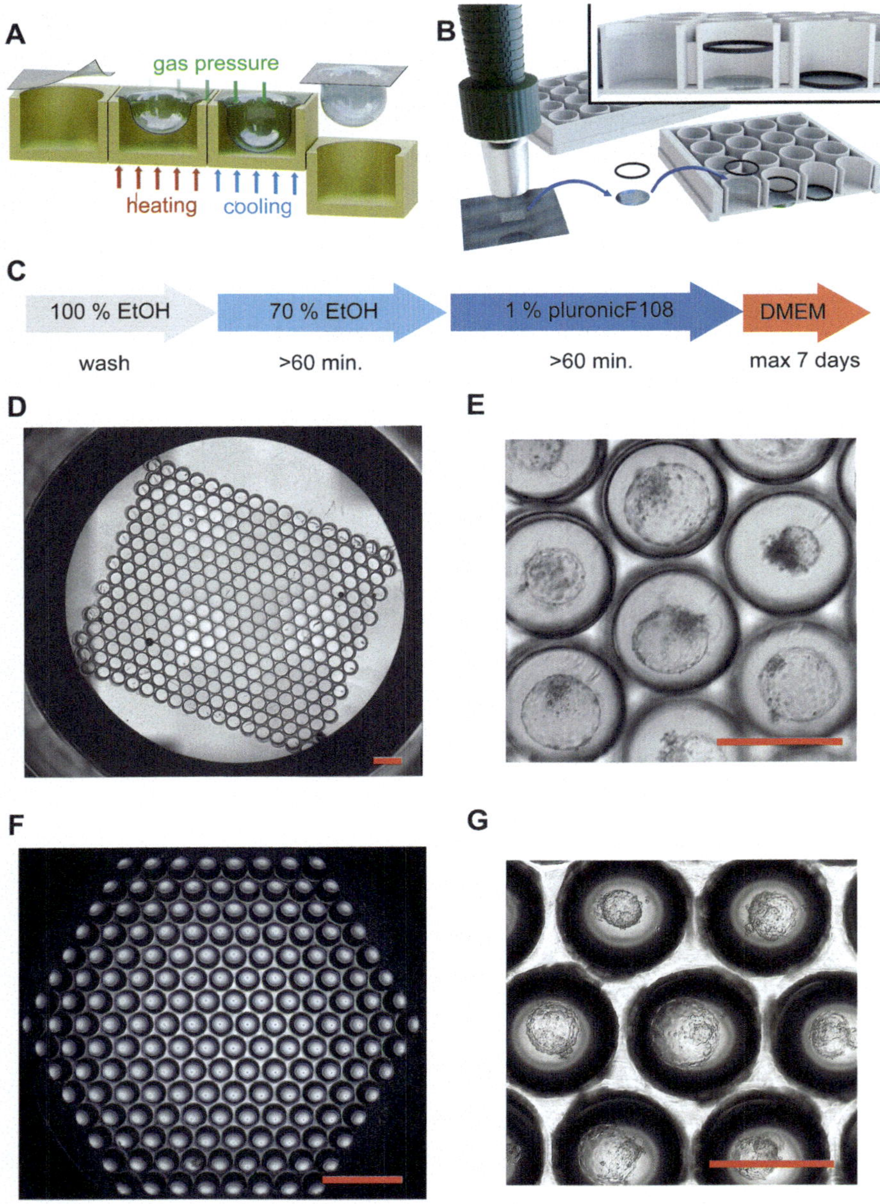

Fig. 2 Microwell fabrication and assembly. (**a**) Schematic of the microthermoforming process. (**b**) Schematic of microwell chip assembly. (**c**) Schematic of the sterilization and anti-adherence treatment prior to cell culture. (**d**) Brightfield image of a 500 μm microwell array chip assembled in a 24-well plate. (**e**) Brightfield image of HNES1 blastoids in a 500 μm microwell chip. (**f**) Brightfield image of a 300MICRONS 96-well plate. (**g**) Brightfield image of niPSC blastoids in a 500 μm microwell array of 300MICRONS. Scales for **d** and **f** are 1,000 μm and scales for **e** and **g** are 500 μm

5. Add 50/200 μL 1% (*w/v*) pluronic F108 anti-adherence solution.
6. Incubate at room temperature for 30 min (*see* **Note 16**).
7. Aspirate the pluronic F108 and add 100/200 μL DMEM.
8. Store plates at 4 °C until use (*see* **Note 17**).

3.3 Blastoid Formation

3.3.1 Geltrex Culture—Day −3 to Day 0 (Fig. 3)

1. Use 1 mL Geltrex coating solution per 6-well plate well and incubate at room temperature for 60 min. Alternatively, incubate for 30 min at 37 °C or store the plate at 4 °C for up to a week.
2. Prepare a 15 mL tube with 4.5 mL of wash buffer and prewarm the Accutase at 37 °C.
3. Aspirate PXGL medium from well with nPSCs on MEFs.
4. Add 500 μL of Accutase and incubate cells at 37 °C for 3–5 min, until individual cells can be distinguished within colonies and cells start detaching from the plate.
5. Firmly pipette the Accutase on the bottom of the well to dislodge the cells. Repeat up to 10 times until most cells are dissociated from the plate.
6. Collect the cell suspension and transfer to the tube with warmed wash buffer.
7. Centrifuge the tube at 200× *g* for 3 min.
8. Aspirate the wash buffer, gently tap the tube to break up the cell pellet and resuspend cells in 200 μL to 1 mL PXGL medium, depending on the density of cells on the original plate (*see* **Note 10**).
9. Aspirate Geltrex solution from the Geltrex-coated plate.
10. Add 1.9 mL PXGL medium to the Geltrex-coated plate. Plate 100 μL of nPSC cell suspension in PXGL medium on this well (total volume 2 mL).
11. Supplement the medium with 10 μM Y-27 and the diluted Geltrex from **step 2** at 2 μL/cm^2 (20 μL per 6 wells plate well).
12. Incubate the nPSCs at 37 °C in a humidified incubator under hypoxic conditions (5% O_2, 5% CO_2).
13. Refresh daily with 2 mL PXGL medium for the two following days.

3.3.2 Microwell Seeding—Day 0

1. On the third day of nPSC culture on a Geltrex-coated well, prewarm the prepared microwell plate at 37 °C.
2. Prewarm the PALY medium and wash buffer.
3. Remove the DMEM from the microwells and add 50 μL per well of a 96 well plate or 100 μL per well of a 24 well plate of PALY medium. Keep plate at 37 °C.
4. Prepare a 15 mL tube with 4.5 mL of wash buffer.

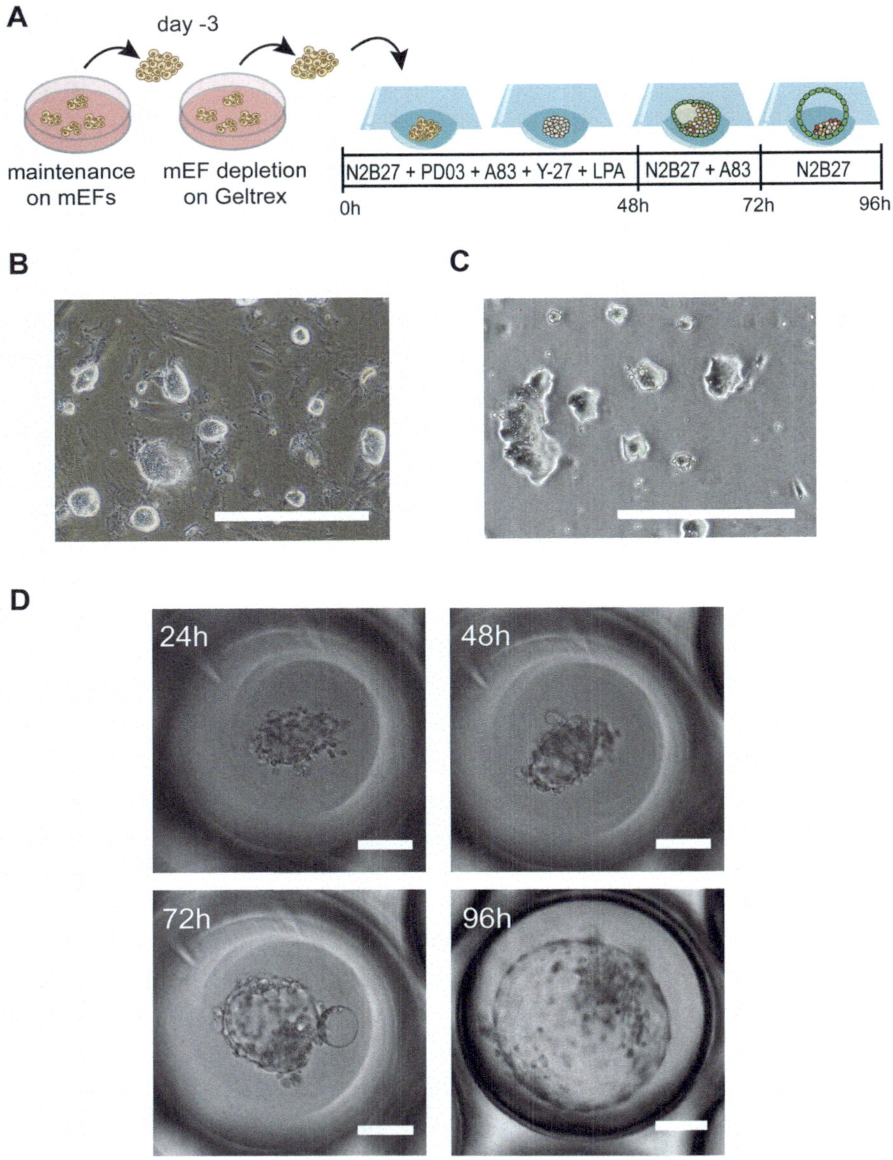

Fig. 3 Blastoid culture protocol. (**a**) Schematic of the blastoid culture protocol, from 2D nPSC culture to seeding cells and forming blastoids in suspension culture. (**b**) Brightfield image of niPSC colonies growing on DR4-irradiated MEFs (3 days after passaging). (**c**) Brightfield image of niPSC colonies growing on Geltrex, ready to be used for blastoid culture (3 days after passaging). (**d**) Images of a forming blastoid at 24, 48, 72, and 96 h after seeding in the microwells. All scales are 100 μm

5. Aspirate the PXGL medium from the Geltrex-coated well with nPSCs.
6. Add 500 μL TrypLE Express and incubate at 37 °C for 3–5 min, until individual cells can be distinguished within colonies and cells start detaching from the plate.

7. Firmly pipette the TrypLE Express on the bottom of the well to dislodge the cells. Repeat up to 10 times until most cells are dissociated from the plate.
8. Collect the cell suspension and transfer to the tube with warmed wash buffer.
9. Centrifuge the tube at 200× *g* for 3 min.
10. Aspirate the wash buffer, gently tap the tube to break up the cell pellet, and resuspend cells in 1 mL PALY medium (per well).
11. Count cells using a hemocytometer and Trypan Blue.
12. Calculate how many cells you have and how many microliters of cell suspension you need per well with microwells (*see* **Note 18**). If this volume is out of range (below 20 or above 200 μL for a 96 wells plate well, or below 100 μL or above 400 μL for a 24 wells plate well), resuspend cells in a higher or lower volume as needed and recount (*see* **Note 19**).
13. Seed the appropriate volume of cell suspension by pipetting the cell suspension dropwise on the surface of the medium in the well, with the pipette tip in the center of the well and perpendicular to the bottom of the well.
14. Incubate the plate at 37 °C for 10–15 min to allow cells to settle into the microwells (*see* **Note 20**).
15. Briefly check cells have settled down homogenously across the microwells (*see* **Note 21**).
16. Add PALY medium to make the total volume 250 μL in a 96 wells plate well and 500 μL in a 24 wells plate well.
17. Incubate the microwell plate at 37 °C in a humidified incubator under hypoxic conditions (5% O_2, 5% CO_2).

3.3.3 Medium Changes—Day 1 to Day 4

1. On day 1 of the blastoid culture, check that cells have aggregated within the microwells.
2. Prewarm PALY medium and refresh with 150 μL PALY medium in 96 wells plate wells, 300 μL in 24 wells plate wells. Continue incubation at 37 °C in a humidified incubator under hypoxic conditions (5% O_2, 5% CO_2).
3. On day 2, prewarm N2B27 medium and refresh wells with 175 μL N2B27 supplemented with 0.5 μM A83 in 96 wells plate wells, 350 μL in 24 wells plate wells. Continue incubation at 37 °C in a humidified incubator under hypoxic conditions (5% O_2, 5%CO_2).
4. On day 3, prewarm N2B27 medium and refresh wells with 175 μL N2B27 supplemented with 0.5 μM A83 in 96 wells plate wells, 350 μL in 24 wells plate wells.

5. On day 4, blastoids have reached a diameter of 200 μm, contain all three blastocyst lineages, and are ready for downstream analysis.

3.4 Downstream Analysis Options

3.4.1 Immunostaining—Day 0

1. To collect blastoids from the microwells, use a P1000 pipette with a cut-off tip to gently pipette up and down three times, perpendicular to the microwells. Then pipette up the medium with structures and transfer to a tube (*see* **Note 22**).
2. Centrifuge the tube at 60× *g* for 1 min.
3. Aspirate the supernatant and add 1 mL PBS.
4. Centrifuge the tube at 60× *g* for 1 min.
5. Aspirate the supernatant and add 500 μL of fixation buffer.
6. Incubate 15 min at room temperature.
7. Remove the fixation buffer and add 1 mL PBS (*see* **Note 23**).
8. Centrifuge the tube at 60× *g* for 1 min or leave structures to settle for 10–15 min.
9. Aspirate the supernatant and add 1 mL PBS.
10. Repeat **steps 8** and **9**, so that structures are washed with PBS three times.
11. Aspirate the PBS and add 500 μL permeabilization buffer.
12. Incubate 30 min at room temperature.
13. Remove the permeabilization buffer and add 500 μL blocking buffer.
14. Incubate for 60 min at room temperature.
15. Remove the blocking buffer and add 100–200 μL antibody buffer with the desired primary antibodies.
16. Shield from light and incubate overnight at 4 °C (*see* **Note 24**).

3.4.2 Immunostaining—Day 1

1. Check that structures are settled at the bottom of the tube, then aspirate the antibody buffer.
2. Add 500 μL washing buffer and incubate for 5 min.
3. Centrifuge the tube at 60× *g* for 1 min or leave structures to settle for 10–15 min.
4. Aspirate the washing buffer.
5. Repeat **steps 2–4** four times.
6. Add antibody buffer with the desired secondary antibodies.
7. Shield from light and incubate for 1–2 h at room temperature.
8. Check that structures are settled at the bottom of the tube, then aspirate the antibody buffer.
9. Add 500 μL washing buffer and incubate for 5 min.
10. Centrifuge the tube at 60× *g* for 1 min or leave structures to settle for 10–15 min.

11. Aspirate the washing buffer.
12. Repeat **steps 2–4** two times.
13. Store samples at 4 °C in 500 μL washing buffer until imaging.

3.4.3 Flow Cytometry

1. To collect blastoids from the microwells, use a P1000 pipette with a cut-off tip to gently pipette up and down three times, perpendicular to the microwells. Then pipette up the medium with structures and transfer to a tube (*see* **Note 22**).
2. Centrifuge the tube at 60× *g* for 1 min.
3. Remove supernatant and add 500 μL TrypLE Express.
4. Incubate at 37 °C for 15 min.
5. Mechanically disrupt cells with P200 and P1000 pipettes until structures are dissociated. If structures are not dissociated after pipetting up and down 10 times, continue incubation for another minute.
6. Dilute cell suspension with 4.5 mL PBS + 0.15% BSA.
7. Centrifuge the tube at 200× *g* for 3 min.
8. Aspirate the supernatant and resuspend in 1 mL PBS + 0.15% BSA.
9. Filter through a 40 μm cell strainer into a new tube.
10. Store samples on ice.
11. Set aside a sample as negative control.
12. Add the desired conjugated antibody/antibodies to the other samples.
13. Incubate for 30–45 min at 4 °C.
14. Aspirate the supernatant and add 1 mL cold PBS.
15. Centrifuge the tube at 200× *g* for 3 min.
16. Aspirate the supernatant and resuspend in 1 mL PBS.
17. Analyze samples with a flow cytometer.

3.4.4 Picking Single Structures

1. Set up an inverted microscope cleaned with 70% ethanol, preferably in a sterilized biosafety cabinet.
2. Prepare a tube or plate with prewarmed N2B27 or PBS, depending on your downstream analysis, to collect the blastoids in. Keep at 37 °C in the incubator until you are ready to start picking blastoids.
3. Collect blastoids from the microwells, use a P1000 pipette with a cut-off tip to gently pipette up and down three times, perpendicular to the microwells. Then pipette up the medium with structures and transfer to a petri dish (*see* **Notes 22 and 25**).

4. Put the petri dish under the microscope and take a P10 pipette. Take a P10 pipette tip and gently bend the end of the tip in a 90–120° from the pipette (*see* **Note 26**).
5. Take your collection vessel from the incubator and start looking for well-formed blastoids under the microscope.
6. Pick up blastoids by putting your pipette tip in the medium, gently sucking up 1 or 2 μL of medium before you approach your blastoid(s) of choice and finally slowly sucking up your blastoids (*see* **Note 27**).
7. Slowly pipette out in your collection vessel, leaving a bit of medium in your pipette tip to prevent air bubbles (*see* **Note 28**).
8. Repeat this as many times as you need to collect a sufficient number of blastoids for your analysis. Blastoids should not be kept out of the incubator for more than 15 min. Work in batches or take a break every 15 min or so, putting the petri dish and collected blastoids in the incubator.

4 Notes

1. *CRITICAL*: Sodium selenite is a toxic chemical and should be handled with care. Please take appropriate safety precautions when handling this chemical and dispose it responsibly.
2. *CRITICAL*: Paraformaldehyde is a toxic chemical and should be handled with care. Please take appropriate safety precautions when handling this chemical and dispose it responsibly.
3. For HNES1 cells, LPA is not strictly required, but for HDF75-based niPSCs, 0.5 μM LPA in the first 48 h of blastoid culture is recommended for more robust high yields.
4. Pluronic F108 1% (*w/v*) acts as an anti-surfactant and can thus be used as an alternative to anti-adherence solution (STEMCELL Technologies).
5. Undiluted Geltrex should be kept at 4 °C at all times to prevent gelation.
6. MEF need to be plated at least one day, preferably two days, prior to seeding of nPSCs.
7. Naïve pluripotent stem cells require good-quality inactivated MEFs. We recommend the DR4 strain.
8. This protocol assumes nPSCs are frozen in stocks of about 250,000 cells per vial. Smaller stocks may need to be cultured on 12-well plates or 24-well plates coated with MEFs for optimal cell expansion.

9. If density of colonies is very high, resuspend in 500 μL to 1 mL. If density of colonies is low or colonies are small, resuspend in 200–500 μL.
10. In case of poor cell survival (after thawing), we advise to dissociate the cells and transfer them to a smaller wells plate (e.g., 12- or 24-well plates).
11. Most protocols assume culture on 6-well plates. We advised against expansion culture in T flasks, as in our hands nPSCs have always grown poorly in those.
12. Depending on the cell density on the wells, this may differ. To store the full contents of a single well in one cryovial, resuspend the cells in 0.5 mL per well. To divide the contents of the well in 3, resuspend cells in 1.5 mL per well.
13. As an alternative to a puncher, any cylinder with the diameter of 15 mm and a scalpel to cut along the edge of this cylinder will serve.
14. O-rings can be stored at room temperature in 70% ethanol after autoclaving.
15. Be careful not to trap air bubbles under the film, as these will reduce visibility of the culture later on.
16. At this point in the procedure, plates can also be stored overnight at 4 °C.
17. Alternatively, incubate plates overnight at 37 °C and check plates for contamination to ensure clean material before starting an experiment.
18. To calculate the number of cells to seed per well, you multiply the number of cells you want per microwell with the number of microwells. However, this does not account for some cells that may settle on the edge of the well and thus don't go into the microwells. To compensate for this "loss," we commonly use a correction factor. In our hands, (9/7) works well, but it may vary based on equipment and individuals.
19. Cell suspension volumes are constrained in their maximum by the final volumes used in the culture. Minima are defined based on the minimal volume needed for homogenous seeding of cells across the microwell array within one well.
20. We advise against centrifuging at this step, as cells may end up unevenly distributed across the microwells.
21. Take pictures at this stage to retrospectively be able to determine the average number of cells per microwell.
22. Cut-off tip to reduce sheer stress and limit damage to the blastoids.

23. Fixation buffer contains paraformaldehyde and should therefore be disposed of with care. Check waste disposal regulations at your institution.
24. Overnight incubation with primary antibodies can optionally be done on a shaking platform. Make sure it is set at no more than 30 rpm to prevent damage to the blastoids.
25. Use a small dish, e.g., a 35 mm, and avoid transferring blastoids in a large medium volume to make finding your blastoids under the microscope.
26. By bending the pipette tip, it becomes easier to use the pipette to pick up blastoids from the petri dish.
27. Collecting some medium before picking up the blastoid (s) prevents them from potentially being exposed to air (bubbles) and keeps you from causing a flow near the blastoid that may cause it to move.
28. It is important to pipette up the blastoids slowly to prevent them from getting damaged by sheer stress.

References

1. Wilcox AJ, Harmon Q, Doody K, Wolf DP, Adashi EY (2020) Preimplantation loss of fertilized human ova: estimating the unobservable. Hum Reprod 35:743–750
2. Takashima Y et al (2014) Resetting transcription factor control circuitry toward groundstate pluripotency in human. Cell 158:1254–1269
3. Guo G et al (2016) Naive pluripotent stem cells derived directly from isolated cells of the human inner cell mass. Stem Cell Rep 6:437–446
4. Khan SA et al (2021) Probing the signaling requirements for naive human pluripotency by high-throughput chemical screening. Cell Rep 35:109233
5. Bayerl J et al (2021) Principles of signaling pathway modulation for enhancing human naive pluripotency induction. Cell Stem Cell 28:1549
6. Bredenkamp N et al (2019) Wnt inhibition facilitates RNA-mediated reprogramming of human somatic cells to naive pluripotency. Stem Cell Rep 13:1083–1098
7. Yanagida A, Spindlow D, Nichols J, Dattani A, Smith A, Guo G (2021) Naive stem cell blastocyst model captures human embryo lineage segregation. Cell Stem Cell 28:1016
8. Kagawa H et al (2021) Human blastoids model blastocyst development and implantation. Nature 601:600
9. Liu X et al (2021) Modelling human blastocysts by reprogramming fibroblasts into iBlastoids. Nature 591:627–632
10. Fan Y et al (2021) Generation of human blastocyst-like structures from pluripotent stem cells. Cell Discov 7:81
11. Karvas RM et al (2023) 3D-cultured blastoids model human embryogenesis from pre-implantation to early gastrulation stages. Cell Stem Cell 30:1148–1165.e1147
12. Luijkx DG, Ak A, Guo G, van Blitterswijk CA, Giselbrecht S, Vrij EJ (2024) Monochorionic twinning in bioengineered human embryo models. Adv Mater 36:2313306
13. Shankar V, van Blitterswijk C, Vrij E, Giselbrecht S (2024) Automated, high-throughput phenotypic screening and analysis platform to study pre- and post-implantation morphogenesis in stem cell-derived embryo-like structures. Adv Sci 11:2304987
14. Giselbrecht S et al (2006) 3D tissue culture substrates produced by microthermoforming of pre-processed polymer films. Biomed Microdevices 8:191–199
15. Carvalho DJ et al (2023) Thyroid-on-a-Chip: an organoid platform for in vitro assessment of endocrine disruption. Adv Healthc Mater 12: 2201555

Methods in Molecular Biology (2026) 3048: 163–179
DOI 10.1007/7651_2025_645

Published online: 17 May 2025

Microfluidic On-Chip Assay for Quantifying Blastoid Adhesion to Endometrial Epithelium

Aslı Ak, Dorian Luijkx, Daniel Carvalho, Stefan Giselbrecht, Ron van Golde, and Erik Vrij

Abstract

The ability of the endometrium to accept and support embryo implantation is crucial, but factors influencing this process remain elusive. This method aims to obtain precise quantitative information on factors causally affecting the initial stages of embryo implantation. We developed a personalized implantation-on-chip platform using in vitro models of the endometrium (organoids) and the embryo (blastoids) to quantify functional embryo attachment. Here, we describe a microfluidic platform for precisely assessing functional receptivity of endometrial epithelium through blastoid adhesion. Endometrial organoids were expanded and transformed into epithelial monolayers within custom-made microfluidic chips. These chips were then infused with large numbers of blastoids (>100) per chip. Followed after 48 h of co-culture, blastoids were exposed to a controlled stepwise increasing flow rate (50, 100 and 400 μL/min), while the rate of adhered blastoids was precisely measured from image-based readouts. Our method offers a robust platform for studying endometrial epithelial receptivity and testing therapeutic interventions with potential impact for infertile patients.

Key words Blastoids, Endometrium-on-chip, Human embryo implantation, Receptivity

1 Introduction

Embryo implantation in humans requires a precise synchronization between the blastocyst and a receptive endometrium, proceeding through three critical stages following hatching: apposition, adhesion, and invasion [1, 2]. Receptivity of the endometrium is coordinated by steroid hormones throughout the menstrual cycle and is specifically heightened during the window of implantation (WOI) [3]. During apposition, the blastocyst positions itself near the luminal side of the endometrium. Adhesion follows, marking the first physical contact between the blastocyst and the luminal endometrium, a process further strengthened by adhesion molecules and cytokines released by both the endometrium and the blastocyst [4, 5]. Finally, invasion occurs, with the blastocyst penetrating the stromal layer of the endometrium, a crucial step for subsequent post-implantation development [6].

Most of our understanding of implantation derives from studies on mammalian models, which often exhibit significant physiological differences from human implantation processes [7]. This discrepancy limits our comprehension of the mechanisms behind implantation and the reasons for its frequent failures in humans. Consequently, there is a critical need for human-specific in vitro models to better understand and delineate the factors involved in implantation.

The ethical and technical challenges associated with in vivo patient research further constrain our ability to study implantation, particularly implantation failure [8, 9]. This limitation has driven the development of advanced stem-cell-based models, including organoids and blastoids. In this chapter, we demonstrate the use of blastoids in conjunction with a novel endometrium-on-chip system to precisely quantify adhesion rates under various conditions, including hormonal induction for receptivity.

We transformed endometrial organoids (Fig. 1a) into a 2D endometrial epithelium (Fig. 1b) within a microfluidic chip. Concurrently, human naïve induced pluripotent stem cells were cultured to form blastoids (Fig. 1c, d). After a 48-h co-culture of blastoids on the chip (Fig. 1e), we quantified the adhesion rates through imaging after applying stepwise increasing flow rates (Fig. 1f). This platform can be used to precisely assess how introduced factors affect functional endometrial receptivity. As such, it has the potential to identify potential strategies to reinstate a receptive state in sub/infertile patients.

2 Materials

2.1 Biological Materials

1. Patient-derived endometrial organoids.
2. Human naïve pluripotent stem cell-derived blastoids (see Chapter "Generation of Human Naïve Pluripotent Stem Cellderived Blastoids in Thermoformed Microwell Platforms" by Luijkx et al. in this volume).

2.2 Reagents

1. 8-Bromo-cAMP, sodium salt (cAMP, Tocris Bioscience cat# 1140).
2. Acetic acid glacial (Sigma Aldrich cat# A6283).
3. Activin/nodal receptor inhibitor A83-01 (MedChem Express cat# HY-10432).
4. B27 supplement, minus Vitamin A (Gibco cat# 12587010).
5. BSA V Fraction (Gibco cat# 15260037).
6. Collagen type I, rat-tail (ibidi cat# 50205).
7. Collagenase, Type I (ThermoFisher Scientific cat# 11500536).
8. CMRL-1066 (Sigma Aldrich cat# C0422).

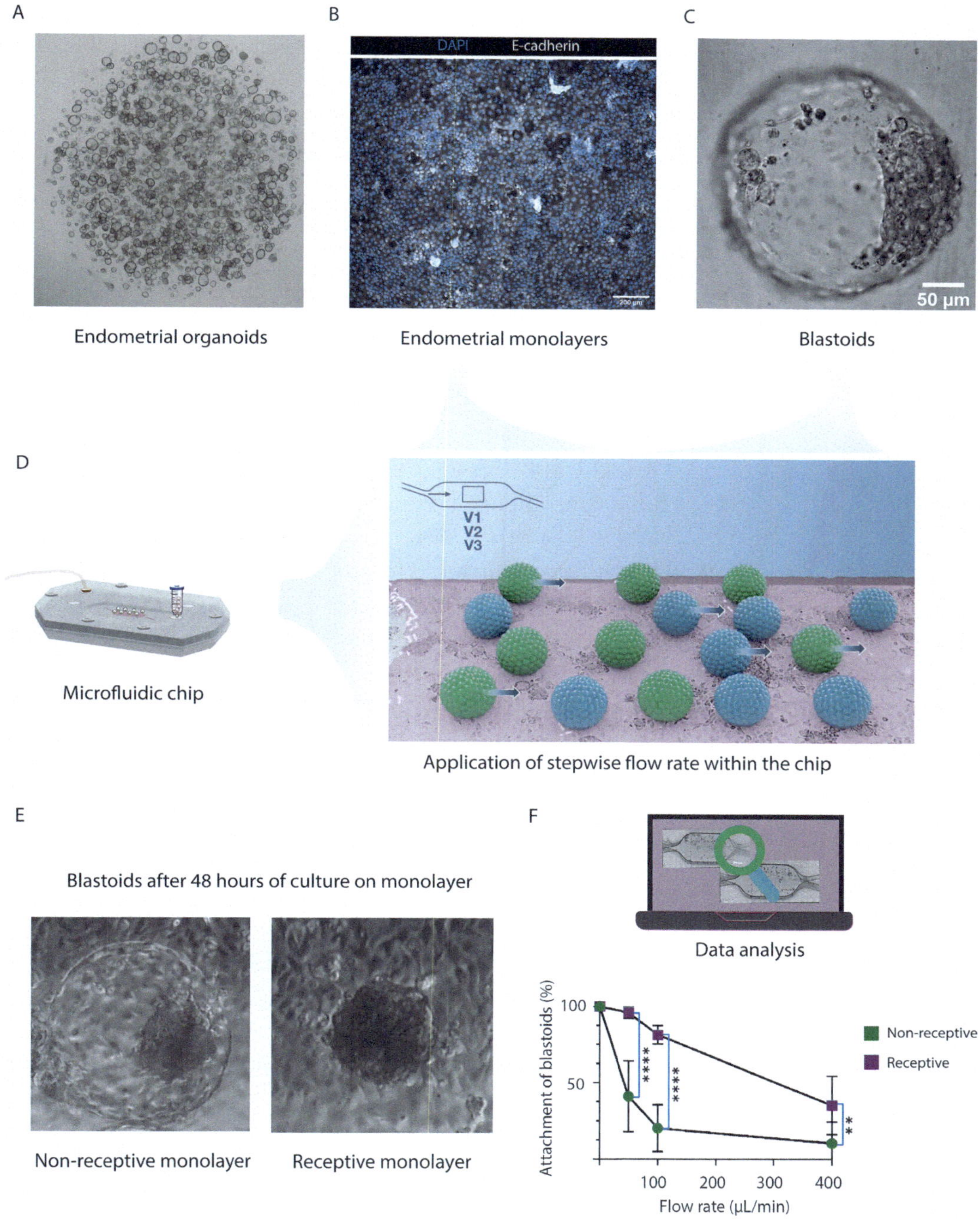

Fig. 1 (**a**) Brightfield image of endometrial organoids on day 7. (**b**) Immunofluorescence staining of endometrial monolayers with DAPI (blue) and E-cadherin (gray); scale bar: 200 μm. (**c**) Brightfield image of day 3 human naïve induced pluripotent stem cells-derived blastoid; scale bar 50 μm. (**d**) Graphical illustration of microfluidic chip and the co-culture with blastoids (and flow application) within the culture chamber. (**e**) Brightfield images of blastoids on non-receptive endometrium (on left) and on receptive endometrium (on right) after 48 h of co-culture; scale bar: 50 μm. (**f**) Graphical illustration of data analysis, representing an adhesion assay result comparing blastoids on non-receptive and receptive monolayers. Legend: green represents non-receptive endometrium and purple represents receptive endometrium

9. Cultrex Basement Membrane Extract (BME), Type 2 (R&D Systems cat# 3533-010-02).
10. DMEM/F12 (Sigma Aldrich cat# D8437).
11. Donkey serum (VWR cat# AGRIAS101563).
12. Dulbecco's phosphate buffered saline (1X), no calcium, no magnesium (DPBS; Sigma Aldrich cat# D8537).
13. Embryonic Stem Cell qualified FBS (ThermoFisher Scientific cat# 11500526).
14. FGF2; bFGF (ISOKine-bFGF, Neuromics, cat# PR80001).
15. Geltrex (Gibco cat# A1413302).
16. HEPES (N-2-hydroxyethylpiperazine-N-2-ethane sulfonic acid, cat# 15630080).
17. Human recombinant Epidermal Growth Factor (EGF, Stemcell Technologies, cat# 78006.1).
18. Human recombinant Fibroblast Growth Factor-10 (FGF-10 Peprotech cat# 100–26).
19. Human recombinant Hepatocyte Growth Factor (HGF, Peprotech cat# 100–39).
20. Human recombinant Noggin (Stemcell Technologies, cat# 78060).
21. Human recombinant R-spondin-1 (Stemcell Technologies, cat# 78213.1).
22. Hyaluronidase (Sigma Aldrich cat# H3884).
23. Insulin-Transferrin-Selenium (ITS-G; cat# 41400045).
24. L-glutamine (Gibco cat# 25030024).
25. N2 supplement-A (Stemcell Technologies, cat# 07152).
26. *N*-Acetyl-L-cysteine (NAC, Sigma Aldrich cat# A9165).
27. Nicotinamide (Sigma Aldrich cat# N3376).
28. p38 MAPK inhibitor SB202190 (Stemcell Technologies, cat# 72634),
29. Paraformaldehyde (PFA, ThermoFisher Scientific cat# 11400580); *Caution* (*see* **Note 2**).
30. Progesterone (Sigma Aldrich cat# P8783).
31. Rho associated kinase inhibitor Y-27632 (Y-27632, Axon Med Chem cat# 1683).
32. Sodium pyruvate (Sigma Aldrich cat# 11360070).
33. STEMCELL Banker (amsbio cat# 11924).
34. Tankyrase inhibitor XAV-939 (XAV, MedChem Express cat# HY-15147).
35. Triton X-100 (Sigma Aldrich cat# T8787).

36. TrypLE Express (Gibco cat# 12605036).
37. β-Estradiol (Sigma Aldrich cat# E2758).

2.3 Medium Compositions

2.3.1 Base Organoid Medium

Prepare 500 mL base organoid medium with DMEM/F12 supplemented with N2 (1X), B27 (1X), and 10 mM HEPES. Medium can be stored at 4 °C for 7–8 weeks.

2.3.2 Organoid Medium

Supplement the base organoid medium with insulin-transferrin-selenium (1X), nicotinamide (2 mM), N-acetyl-L-cysteine (1.25 mM), noggin (100 ng/mL), R-spondin-1 (100 ng/mL), EGF (50 ng/mL), HGF (50 ng/mL), SB202190 (10 μM), FGF-10 (10 ng/mL), A83-01 (500 nM), and bFGF (2.5 ng/mL). Medium can be stored at 4 °C for 1 week.

2.3.3 Collagen Type I Coating Solution

Thaw collagen in room temperature. Dilute the acetic acid solution to 17.5 mM in sterile cell culture-grade water (v/v) and filter-sterilize it. Measure the pH with pH strip (pH = 2–3). Thoroughly mix the acetic acid solution with collagen to reach the desired concentration.

2.3.4 IVC-1 Medium

Prepare IVC-1 medium by supplementing CMRL-1066 medium with fetal bovine serum (FBS)-qualified (10%), L-glutamine (1%), N2 (1%), B-27 (2%), sodium pyruvate (1%), NAC (25 μM), β-estradiol (E2, 10 nM), progesterone (P4, 1 μM), and Y-27632 (10 μM). Store at 4 °C and use it within 5 days.

2.3.5 Fixation Buffer

1. Dilute 37% PFA to 4% PFA in PBS (v/v).
2. Store at 4 °C and use it within one week. Aliquots can also be frozen at −20 °C and thawed for immediate use.

2.3.6 Permeabilization Buffer

1. Dilute Triton-X 100 to 1% in PBS (v/v).
2. Store at room temperature for 6 months.

2.3.7 Washing Buffer

1. Dilute BSA V Fraction (7.5%) to 0.5% in PBS (v/v).
2. Store at 4 °C for up to 6 months.

2.3.8 Antibody Buffer

1. Prepare by mixing 25% of blocking buffer with 25% Triton-X 0.1% solution and 50% PBS.
2. Store aliquots at −20 °C for up to 6 months. Avoid repeated freeze-and-thaw cycles.

2.4 Equipment

1. 1- and 3 mL Polypropylene (PP) soft syringes (without Luer-lock).

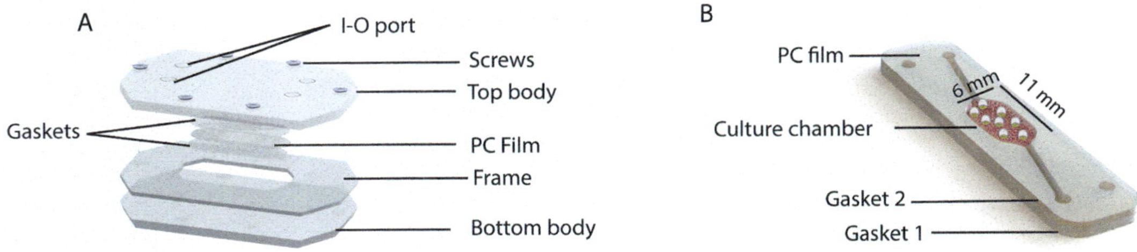

Fig. 2 (**a**) Graphical illustration of the housing components of the microfluidic chip, including gaskets, polycarbonate film, and screws. (**b**) Zoomed-in illustration of two gaskets (creating channels) placed on top of each other with a PC film in between. The culture chamber is visualized with illustrative blastoids on top of endometrial monolayers, showcasing the dimensions (6 × 11 mm)

2. 15–20 cm long silicone tubing (Tygon S3 E-lab, cat# E-3603, Darwin Microfluidics, ID 0.8 mm × OD 2.4 mm).
3. −20 °C freezer.
4. 25'G blunt needle (ID 0.25 mm × OD 0.52 mm).
5. 3 mL plastic syringes.
6. −80 °C freezer.
7. Barbed to male Luer slip adapter (Darwin Microfluidics, cat# ID-P-854, ID 1.5 mm).
8. Biological safety cabinet.
9. Cell strainers, 40 μm.
10. Centrifuge.
11. Common consumables (serological pipettes, pipette tips).
12. Conical tubes, 15 mL and 50 mL.
13. Electric screwdriver with hexagon key (Hazet, cat# 2570186 – 62, 2.5 mm).
14. The in-house fabricated housing components—comprising the bottom, frame, and top—are micromachined from translucent polycarbonate (PC) polymer plates with thicknesses of 4 mm, 1.5 mm, and 6 mm, respectively. Each component contains six holes with a pitch of 0.5 mm. The top component additionally features I/O ports reamed to a Luer geometry (Fig. 2a).
15. Hypoxic (5% O_2, 5% CO_2) incubator with humidity and gas control.
16. Inverted microscope.
17. M3 × 12 mm hexagon socket cap stainless steel screws.
18. Male Luer plugs (Microfluidic ChipShop, cat# CS-10000230).
19. Negative mold for gaskets (PC sheet, thickness: 4 mm) and flat PC lid.
20. Table clamps.

21. Non-porous PC film (11 × 45 mm (*W* × *L*), thickness: 50 μm).
22. Normoxic (20% O_2, 5% CO_2) incubator with humidity and gas control.
23. O-rings (ERIKS cat# 10023241).
24. Pipettes (Eppendorf, P1000, P200, P20 and P10).
25. SYLGARD™ 184 Silicone Elastomer Kit.
26. Polydimethylsiloxane (PDMS) gaskets (elastomer base and curing agents 10:1) (Fig. 2b).
27. PTFE Syringe Filter (pore size: 0.2 μm).
28. Single-Syringe Infusion Pump.
29. Thermoformed microwells (300MICRONS GmbH, STATARRAY MCA96-8.736-PS, ~450 μm inner microwell diameter).

3 Methods

3.1 Endometrial Organoid Culture and Passaging

1. Cut the endometrial tissue into approximately 1 mm^3 pieces with a surgical blade.
2. Place the endometrial tissue into a digestion medium containing collagenase type I (4.8 mg/mL) and hyaluronidase (0.1 mg/mL) and apply mechanical trituration.
3. Place Cultrex BME on ice to thaw at least 30 min prior to seeding.
4. Use a 40 μm cell strainer to collect the epithelial cells. Flip the strainer and wash the stainer into a 15 mL tube with DMEM/F12. Centrifuge at 300× *g* for 5 min.
5. Discard the supernatant. Plate the cells in a 12-well plate by mixing 100% Cultrex BME gel and seed four of 15 μL domes (per well) (*see* **Note 1**).
6. Incubate at 37 °C for 20 min for the solidification of the gel. Add 750 μL of pre-heated organoid medium supplemented with Y-27632 to each well (*see* **Note 2**). Refresh the medium every 2 days and passage every 7–10 days.
7. After 7–10 days of organoid culture, collect each well up to four domes in a 1.5 mL Eppendorf's tube with the existing medium. By using a 25'G needle and 1 mL soft syringe, break the organoid domes into small clumps (exactly 7 times). Centrifuge the tubes at 4 °C for 5 min at 800× *g*.
8. Using a 200-P pipette, carefully remove the gel layer on top of the pellet along with the supernatant (*see* **Note 3**). (Here, you can proceed to Subheading 3.5 for monolayer formation).

9. Pool all organoid pellets in different Eppendorf's tubes in one by using 1 mL ice-cold DMEM/F12. Centrifuge the tube at 4 °C for 4 min at 400× *g*.
10. Discard the supernatant and calculate the volume of Cultrex BME to be mixed with the pellet. Seed four 15 μL domes (per well) in a 12-well plate.
11. Repeat **step 5**.

3.2 Endometrial Organoid Freezing and Thawing

1. Prepare 10 mL of 10% FBS in DMEM/F12 in a 15 mL tube (v/v).
2. Thaw a cryovial with an organoid line in 37 °C water bath for 1–2 min.
3. Quickly bring the vial to the hood and gently resuspend in the medium.
4. Centrifuge the tube at room temperature at 400× *g* for 4 min.
5. Check for a visible pellet. Aspirate the supernatant without disturbing the pellet.
6. Repeat **step 5–8** (including) of Subheading 3.1.
7. Gently resuspend the pellet in each Eppendorf's tube in 1 mL of STEMCELL Banker solution and transfer to a pre-chilled cryovial (*see* **Note 4**).
8. Store cryovials in a Mister Frosty or equivalent slow-rate cooling chamber in a −80 °C freezer for one day.
9. The next day, transfer the cryovials to a liquid nitrogen storage.

3.3 Gasket Fabrication

1. Turn on the drying oven and set the temperature to 80 °C.
2. Cover the working station with paper for possible spilling of PDMS.
3. Mix the PDMS elastomer base and curing agent in a 10:1 ratio by weighing in a clean plastic cup.
4. Stir the mixture thoroughly for 2–3 min to ensure homogeneity with a spoon or Pasteur pipette.
5. Place the plastic cup in a vacuum degassing chamber for the removal of air bubbles (pressure: 0.2 bar).
6. Clean the gasket molds and flat PC lids by first removing the PDMS residue using a tape. Then using an air-gun clean the molds further. Finally, wipe with 70% ethanol.
7. Once the air bubbles are not visible anymore, use a 1 mL plastic syringe to collect the mixture and slowly pour into the negative mold. The mold should be filled out fully. In case of newly formed air bubbles, pinch them using a needle to remove them.

8. Overlay the flat PC lid on the molds in a gradual manner. This step ensures uniform thickness and proper gasket shape during curing.
9. Use table clamps to stabilize and compress the lid and mold together.
10. Transfer the clamped mold into a drying oven on top of an aluminum foil sheet to avoid potential spills on the oven. Cure for 2 h at 80 °C.
11. After 2 h, take out the clamps and place the mold in a clean hood. Carefully remove the lid from the mold. Use a blunt needle (not to damage the gasket, 20'G) to peel off the gasket from the mold. Place the gaskets on a new aluminum foil and transfer back to the oven. To minimize potential cytotoxicity, conduct a post-curing step by continuing the process at 120 °C for an additional 2 h.
12. Store the gaskets in a clean, sealed-off, and dry environment at room temperature until further use.

3.4 Chip Assembly and Preparation

1. Prior to assembly, clean the housing components with ethanol and air-dry. Clean the gaskets first with ethanol and later with PBS, and air-dry. Put the housing components, gaskets, and screws in glass beakers for autoclave (121 °C, 15 min).
2. Cut the non-porous PC film to dimensions of 11 × 45 mm using a mold and surgical scissors. Immerse the cut film in ethanol, then air-dry it by placing it against the wall of a well plate.
3. Dilute collagen type 1 to 1 mg/mL with 17.5 mM acetic acid (v/v) and mix thoroughly (*see* **Note 5**).
4. Assemble the chip starting with positioning the frame on top of the bottom body (Fig. 3: **Step 1**). Place the first gasket within the frame using a sterile tweezer, then the PC film and the second gasket. Carefully, place the top body in one movement. Place the screws in six holes and tighten with the screwdriver (Fig. 3: **Step 2**).
5. Cut a 3 mL soft syringe with a surgical blade from the middle and place it on the outlet port of the chip, close off with a PTFE syringe filter (Fig. 3: **Step 3**).
6. Take up 100–150 μL of 1 mg/mL collagen solution with a 1 mL syringe and place it on the inlet port of the chip. Slowly, and by tilting the chip for smooth flow, push the syringe and fill the chamber and channels with the collagen solution. Place the chip in the incubator at 37 °C for 24 h for efficient coating (*see* **Note 6**, Fig. 3: **Step 4**).
7. Proceed to Subheading 3.5 for seeding monolayer in the microfluidic chip.

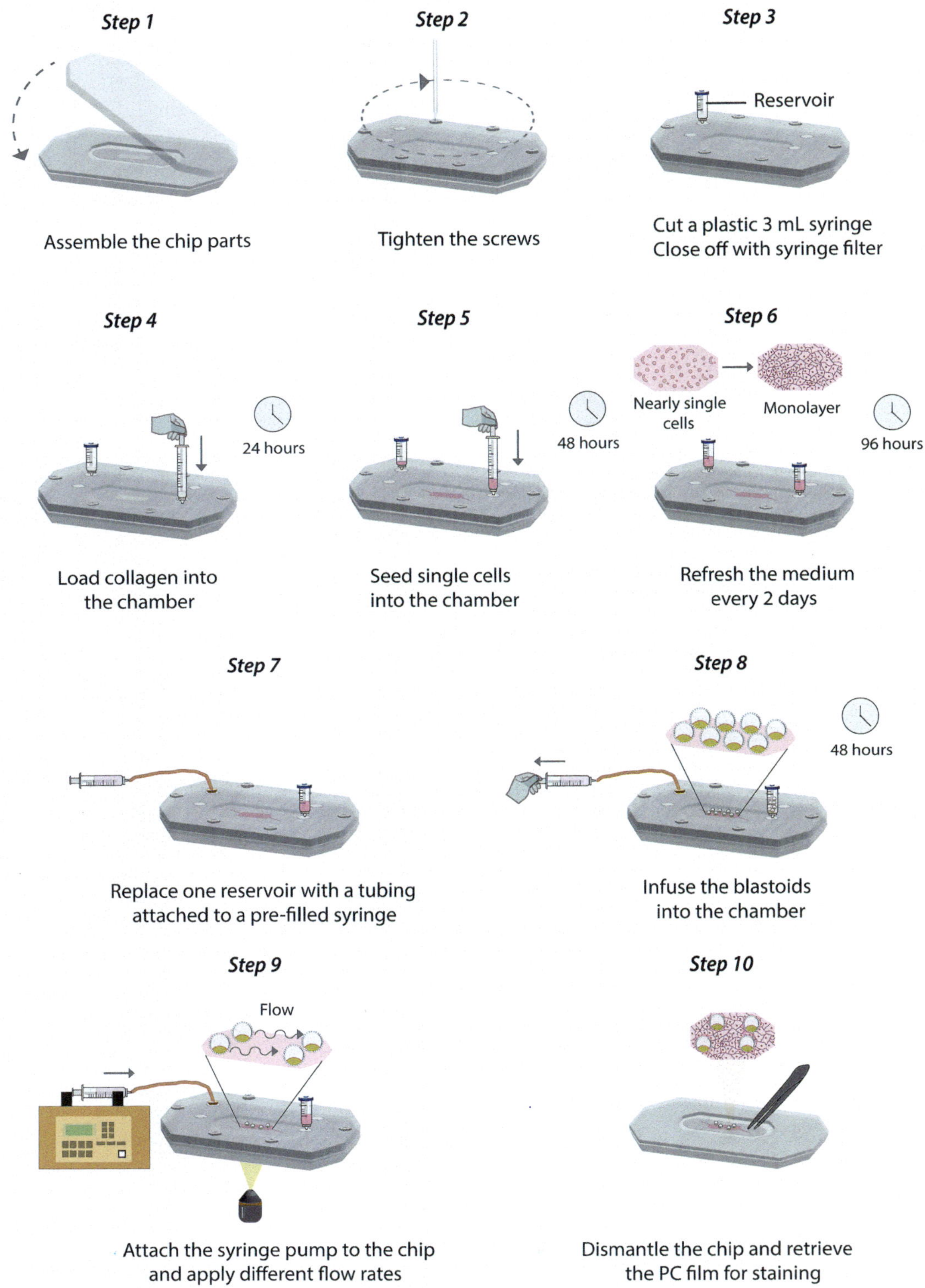

Fig. 3 Step-by-step protocol for endometrium-on-chip assembly and blastoid adhesion assay

3.5 Endometrial Monolayer Seeding in the Chip

1. Follow the organoid passaging protocol (Subheading 3.1) from **steps** 7 and **8**.
2. Resuspend the pellet in TrypLE Express (100 μL/dome) and place the tubes in 37 °C for 5–6 min (*see* **Note** 7). In the meantime, prepare 0.1% BSA in DMEM/F12 (v/v, washing buffer) in a 15 mL tube and keep it in room temperature. After incubation, insert a 200-P tip (without filter) into a 1000-P tip and thoroughly pipette the suspension 10 times. Then, mix the suspension with washing buffer in 1:10 volume in the 15 mL tube. For example, for 400-μL suspension, add 3.6 mL of washing buffer.
3. Centrifuge the tube at 400× g for 4 min at room temperature to obtain a pellet of nearly single endometrial cells.
4. In the meantime, take out the chips from the incubator and place them in a flow hood. Slowly pull the 1 mL syringe to take up the collagen coating solution. Discard the coating solution and by using the same syringe, take up 100–150 μL PBS. Apply the PBS to the chip's inlet port, tilting the chip as needed to fill the chamber and neutralize the low pH of the collagen. Quickly, using the syringe, aspirate the PBS from the chip and discard it.
5. Discard the supernatant medium and resuspend the pellet of endometrial cells in organoid medium supplemented with 10 μM Y-27632 (optionally add 10 nM β-estradiol as a first step toward induction of a status of receptive endometrium).
6. Count cells using a hemocytometer and Trypan Blue. Aim for 70,000 cells per chip to achieve the upper recommended cell density.
7. Thoroughly (but gently) mix the cell suspension and take up 100–150 μL suspension with the syringe. Place it on the inlet port of chip and by tilting the chip, push the syringe and fill the chamber with cells until reaching the reservoir on the other side (Fig. 3: **Step 5**).
8. 48 h later, cut another 3 mL syringe with a surgical blade in the middle (approximately at 1.5 mL point) and place it on the inlet port of the chip by swapping with the 1 mL syringe. By using a 1000-P, add 300–400 μL medium from the inlet reservoir and close off with a syringe filter. Slowly tilt the chip for equalizing the volume in both reservoirs (Fig. 3: **Step 6**).
9. For receptive endometrium, prepare the hormone cocktail, namely EPCX, with 10 nM β-estradiol (E2), 1 μM progesterone (P4), 250 μM 8-bromo-cAMP (C), and 10 μM XAV939 (X) in the organoid medium. After 48 h of E2 induction, switch to EPCX medium. At this point, the endometrial monolayer should be confluent. Refresh the medium within the

chamber statically every 2 days until day 6 of confluent monolayer culture (*see* **Note 8**).

10. At this point, the chamber can be visualized with a Brightfield microscope for assessing the confluency of the monolayer (*see* **Note 9**).

3.6 Blastoid Infusion in Endometrium-on-Chip

1. Autoclave the silicone tubing, barbed-to-male Luer slip adapter and blunt needle (121 °C, 15 min).
2. On day 6, replace one of the reservoirs with the barbed-to-male Luer slip adapter attached to a pre-filled (~1.5 mL) 3 mL syringe (with DMEM/F12) with needle and tubing (Fig. 3: **Step 7**). Fill the chamber with DMEM/F12 to avoid bubbles (*see* **Note 10**).
3. Collect the blastoids (*see* **Note 11**) from microwells (two wells for one Eppendorf's tube) using a pipette tip pre-coated with 1% BSA solution in PBS (v/v) and place in a 2 mL Eppendorf's tubes pre-coated with a 1% BSA solution in PBS (v/v) (*see* **Note 12**).
4. Centrifuge blastoids at 100× *g* for 1 min and gently resuspend in 100–150 μL IVC-1 medium (*see* **Note 13**). Collect the blastoids with a 1000-P tip and slowly load the suspension into the reservoir and close off the filter (Fig. 3: **Step 8**). Bring the chip under the microscope and by gently pulling the syringe, infuse the blastoids into the microfluidic chamber ensuring a homogenous distribution. Once the blastoids are positioned, bring the chip to the hood and add an additional 320 μL IVC-1 medium to the reservoir. Carefully place the chip with attached tubing to the normoxia incubator (*see* **Note 14**) and observe for the next 48 h (*see* **Note 15**).

3.7 Microfluidic Flow-Based Measurement of Blastoid Adhesion

1. At the end of 48 h, carefully bring the chip to a Brightfield microscope (*see* **Note 16**). Image the whole chamber to determine the precise number and location of blastoids before applying microfluidic flow.
2. Attach the syringe of the chip setup to the syringe pump while keeping all elements at the same height (to avoid any internal flow) (Fig. 3: **Step 9**).
3. Before applying any flow, image the whole chamber (*see* **Note 17**).
4. Apply the first flow rate; 50 μL/min for 90 s.
5. Stop the flow, and image the whole chamber.
6. Apply the second flow rate; 100 μL/min for 90 s.
7. Stop the flow, and image the whole chamber.
8. Apply the final flow rate; 400 μL/min for 90 s.

9. Stop the flow, and image the whole chamber (*see* **Note 18**).
10. At the end of the imaging, detach the syringe from the pump. Dismantle the chip parts and clean with 70% ethanol.

3.8 Immunofluorescent Staining

1. For immunofluorescent staining of blastoids and/or endometrial cells, collect the PC film from the chip using a tweezer (Fig. 3: **Step 10**).
2. Cut the film into 4 × 4 mm with a scissor. Place the PC film in a well of a 24-well plate and place an O-ring on top of it to stabilize the film.
3. Add 200 μL fixation buffer to the well and leave it at room temperature for 10–15 min (*see* **Note 19**).
4. Wash with PBS at least two times for 2 min, add Triton-X (1%) and leave it at 4 °C for 30 min.
5. Remove wash buffer and add blocking buffer (donkey or goat serum) and leave it at 4 °C for 30 min.
6. Prepare an antibody staining solution with the desired primary antibodies. Add the antibody solution to the samples and leave it at 4 °C overnight.
7. Wash the samples at least three times with PBS for 2 min.
8. Prepare an antibody buffer containing the desired secondary antibodies, add to samples, and leave it at 4 °C overnight.
9. Wash the samples two times and transfer the film with a tweezer to an imaging plate (e.g., ibidi μ-Dish 35 mm) and stabilize with an O-ring.
10. To avoid drying of the sample, add 100 μL of PBS on top of the film.

4 Notes

1. Passaging ratio of the organoids depends on the number of organoids per dome. If the number per dome is above 200, it is recommended to use a 1:3–4 ratio. If the number per dome is less than 100, use a 1:2 ratio. Optimal seeding densities are crucial, as both overcrowding and insufficient cell numbers can impede organoid proliferation.
2. During standard endometrial organoid culture, after each passage, organoid medium is supplemented with 10 μM Y-27632 for 48 h to promote cell survival and reduce stress-induced apoptosis. During subsequent medium refreshment steps, Y-27632 is omitted. The number of domes and the volume of organoid medium vary depending on the well plate format: (I) 6-well plate: each well contains approximately 10 domes

suspended in 1.5–2 mL of organoid medium; (II) 24-well plate: each well contains approximately 3 domes suspended in 500 μL of organoid medium; (III) 48-well plate: each well contains a single dome suspended in 200 μL of organoid medium.

3. After centrifuging at 4 °C, three distinct layers should form: the pellet at the bottom, a thin gel layer above it, and the remaining medium on top. Carefully use a 200 μL pipette to collect the gel layer directly and slowly. For better visibility of these layers, use a 1.5 mL Eppendorf tube rather than a 2 mL tube, as it provides clearer visualization. To distinguish the gel layer, illuminate the tube with a light source. Once the gel layer is collected, remove the remaining medium.
4. Keep the cells on ice until resuspending in STEMCELL Banker solution.
5. Allow the collagen to thaw at room temperature for at least 30 min before mixing. Collagen solutions at high concentrations (10 mg/mL) can be challenging to handle using pipettes. It is suggested to use positive displacement tips for viscous solutions (or cutting the tip), which allows pipetting them with an increased accuracy.
6. Start the assembly and coating of the chip one day prior to passaging of the organoids. On the passaging day, follow the monolayer protocol to seed in the chips.
7. At this stage, the cells are sensitive to all types of external stimuli, especially temperature fluctuations. After centrifuging at 4 °C, all the next steps should be performed at room temperature. Otherwise, the single cells clump together and adversely affect monolayer formation. Additionally, TrypLE Express enzymatically dissociates the organoid glands into single cells, but physical dissociation is as important at this point. Make sure to set the pipette volume slightly higher than the total volume of the Eppendorf's tube for an effective dissociation.
8. As this protocol contained multiple conditions, including hormone induction at various concentrations, the seeding to the chip differed in terms of medium composition. Here, using 2 mL Eppendorf's tubes is suggested as it is easier to collect with a syringe (to be attached to the chip) compared to 1.5 mL tubes. To seed cells in the chip non-induced and induced conditions at the same time, resuspend the pellet of endometrial cells in organoid medium and divide proportionally to chip numbers. For example, if there are 6 chips with 3 non-induced and 3 induced conditions, resuspend the pellet in 1.2 mL organoid medium (200 μL per chip). Then, add 10 μM Y-27632 to organoid medium. Seed half of the suspension in

three chips. To the remaining cell suspension, add 10 nM E2 and seed the cells in the remaining three chips. Leave the 1 mL syringe in each chip attached for 48 h and then switch to reservoir.

9. The height of the 1 mL syringe attached to the chip is approximately 10 cm and thus may not fit under the microscope. For this reason, we did visualize the chamber for the first 48 h. Despite this limitation, the 48-h period allowed sufficient time for the cells to settle and adhere within the chamber. After that, since reservoirs have half the height compared to 1 mL syringe, imaging was possible.
10. Fill a 3 mL syringe attached to a blunt needle with DMEM/F12. Replace one of the reservoirs with a barbed-to-male Luer and attach the syringe through the silicone tubing. Gently push the syringe to fill up the chamber until reservoir. This step is necessary to avoid air bubbles when attached to the syringe pump and start the flow. Remove the excessive DMEM/F12 in the remaining reservoir for the next step.
11. Blastoids should be on day 3 of culture in microwells. Blastoids were cultured in PALY medium for 48 h, and a consequent 24 h with N2B27 medium supplemented with A83-01. Blastoids should present a cavity with a visible inner cell mass (ICM) on day 3. Gently flush out the structures from the microwells and collect in Eppendorf's tube. For further details, see chapter "Generation of Human Naïve Pluripotent Stem Cellderived Blastoids in Thermoformed Microwell Platforms".
12. The collection of blastoids should be very gentle, as these structures are very delicate. Flush out the blastoids by setting the pipette volume to 250 μL. Coating the pipette tip and tube with a 1% BSA solution avoids the attachment of blastoids to these surfaces, which is strongly recommended.
13. 500 μm diameter wells contain 91 microwells. Collect two wells of blastoids into a 2 mL Eppendorf's tube for each chip, which equals a maximum of 182 blastoids. To prevent bubble formation during later steps, pre-equilibrate IVC-1 medium in the incubator for 30 min prior to use. Immediately after centrifugation, resuspend the blastoids in IVC-1 medium and keep in the incubator at 37 °C until infusing them into the chip.
14. Try to keep the syringe and the chips on the same plane while placing the chips in the incubator. Additionally, try to avoid the entanglement of the tubing.
15. During the critical 48-h period following infusion, regularly monitor the color of the medium. A shift toward yellow indicates increased metabolic activity, potentially due to a high

density of blastoids, which can disrupt the pH balance. If this is the case, apply a very low flow rate (e.g., 10 μL/min for 3–4 min). Observe the chamber during flow to avoid any movement of blastoids. Do not leave the chips outside of the incubator for longer than suggested, as this might interfere with their development.

16. Keep the chip and the syringe approximately on the same plane all the time, as this avoids the internal flow that may displace blastoids. Place the syringe part of the setup to a stable platform for precise imaging.
17. The chamber size for this chip is 6 × 11 mm ($W \times L$), which was not possible to image in one time with a standard Brightfield microscope (without an automated stage). To overcome this, image the chamber in several frames (we imaged in average 30 frames). Stitch all the frames together using Image Composite Editor Software (Microsoft corporation), which detects the frames and composes the final stitched image (without filling out automatically).
18. There should be four final images per chip at the end, being "before flow," "after 50 μL/min," "after 100 μL/min," and "after 400 μL/min." First, count the number of blastoids in the "before flow" image and detect the positions (and the number) of blastoids. Here, we used ImageJ Software. As the images are Brightfield and the chip is full of different size and shape of cells (including the monolayer), it was not possible to automate the counting. Select each blastoid with circling tool, count (Ctrl + M) and delete the circled blastoid to know which one was counted, as deleting leaves a black hole. Record the number of blastoids for "before flow" condition and set this as 100% per chip. Bring both "before flow" and "after 50 μL/min" images side by side and detect the blastoids moved by eye in the same manner. Using the pencil tool, annotate dislodged blastoids. Record the number of blastoids for after "50 μL/min" condition and calculate the percentage. Apply the same to "after 50 μL/min" and "after 100 μL/min" and then to "after 100 μL/min" and after "400 μL/min." This collective dataset (per condition in the chip) will generate an adhesion profile relative to flow rates. With a sufficient number of chips, statistical comparisons can be performed. We recommend using a two-way ANOVA followed by Šidák's post hoc test for analysis.
19. Dispose PFA with care.

References

1. Diedrich K, Fauser BCJM, Devroey P, Griesinger G (2007) The role of the endometrium and embryo in human implantation. Hum Reprod Update 13(4):365–377. https://doi.org/10.1093/humupd/dmm011
2. Aplin JD, Ruane PT (2017) Embryo–epithelium interactions during implantation at a glance. J Cell Sci 130(1):15–22
3. Bergh PA, Navot D (1992) The impact of embryonic development and endometrial maturity on the timing of implantation. Fertil Steril 58(3):537–542
4. Lessey BA, Castelbaum AJ, Sawin SW, Sun J (1995) Integrins as markers of uterine receptivity in women with primary unexplained infertility. Fertil Steril 63(3):535–542
5. Lipari CW, Zhao Y, King JA, Alexander CJ, Bankowski BB, Lai T-H, Garcia J, Vlahos NP (2008) Variations in L-selectin ligand expression in the luteal-phase endometrium after final oocyte maturation with human chorionic gonadotropin compared with a gonadotropin-releasing hormone agonist. Fertil Steril 90(3): 876–878
6. Kim SM, Kim JS (2017) A review of mechanisms of implantation. Dev Reprod 21(4):351–359. https://doi.org/10.12717/dr.2017.21.4.351
7. Lee KY, DeMayo FJ (2004) Animal models of implantation. Reproduction 128(6):679–695. https://doi.org/10.1530/rep.1.00340
8. Clark AT, Brivanlou A, Fu J, Kato K, Mathews D, Niakan KK, Rivron N, Saitou M, Surani A, Tang F (2021) Human embryo research, stem cell-derived embryo models and in vitro gametogenesis: considerations leading to the revised ISSCR guidelines. Stem Cell Rep 16(6):1416–1424
9. Rivron N, Pera M, Rossant J, Martinez Arias A, Zernicka-Goetz M, Fu J, van den Brink S, Bredenoord A, Dondorp W, de Wert G (2018) Debate ethics of embryo models from stem cells. Nature 564(7735):183–185

Methods in Molecular Biology (2026) 3048: 181–210
DOI 10.1007/7651_2025_638

Published online: 22 May 2025

Linking Single-Cell Dynamics to Cell Fate in Differentiating hPSCs

Seth Teague, Zhiyuan Yu, and Idse Heemskerk

Abstract

Protocols for human pluripotent stem cell differentiation commonly yield a heterogeneous mix of cell types. To understand the source of heterogeneity at the single-cell level, it may be necessary to link final cell state to the cell's history and initial state, for example to determine gene expression or morphogen signaling over the course of differentiation. Here we present methods to quantify and track single cells in time-lapse fluorescence microscopy during stem cell differentiation and link single-cell dynamics to the final resulting state in the same cells.

Key words Cell tracking, Differentiation, Human pluripotent stem cells, Iterative immunofluorescence, Signaling dynamics

1 Introduction

Stem cell systems have a remarkable ability to self-organize into morphological structures and cell fate patterns in response to uniform exogenous differentiation cues. This has enabled powerful stem cell-based embryo and organ models that recapitulate aspects of in vivo development. At the same time, it poses a problem in stem cell differentiation protocols that aim to produce a single cell type, for example for applications to regenerative medicine, which often yield a heterogeneous mixture instead. Thus, a better understanding of heterogeneous differentiation in vitro can both provide insight into embryonic development and improve stem cell technology. Final state heterogeneity may reflect initial state heterogeneity, i.e., cells may be biased toward a particular fate from the start, or it may reflect heterogeneous response to exogenous differentiation cues. One approach to discern these possibilities at the single-cell level is to track cells over time in order to relate earlier cell state or signaling activity to cell fate.

Here we present a method to track single-cell dynamics in live cells over the course of differentiation linked to the final state of the same cells measured after fixation. In particular, this allows relating signaling activity over time to final gene expression at the single-cell

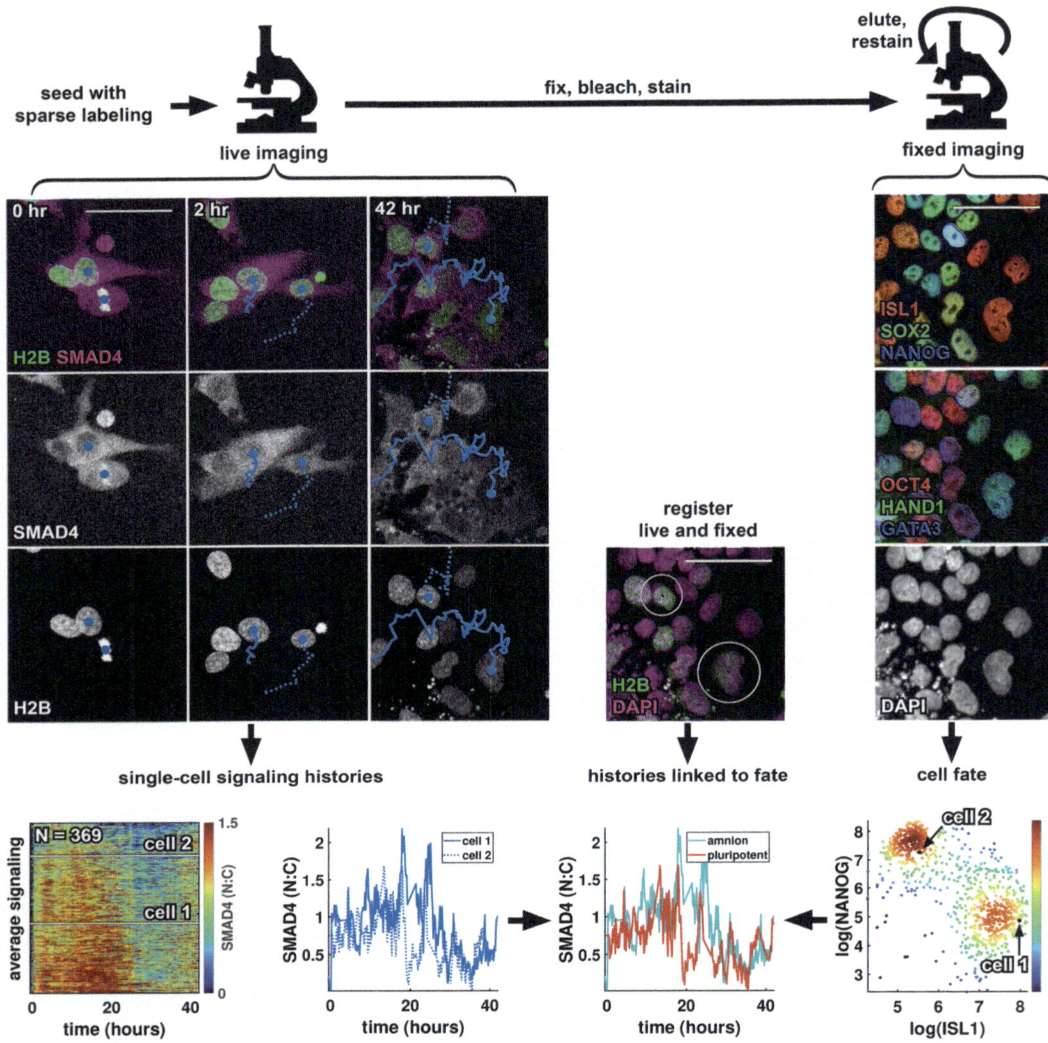

Fig. 1 Graphical overview of the method, demonstrating measurement of single-cell dynamics of GFP::SMAD4 nuclear localization linked to expression of markers of pluripotency and amnion fate as an example application. (Reproduced from Teague et al., 2024, distributed by Nature Communications under a Creative Commons CC BY license)

level. Our protocol uses time-lapse live-cell fluorescence microscopy to measure dynamics, followed by fixation of the sample, bleaching of the live fluorophores [1], and subsequent (iterative [2–4]) immunofluorescence or fluorescence in situ hybridization to measure gene expression in the same single cells (Fig. 1). This approach requires a cell line with a fluorescent readout of interest. For example, we developed this approach to measure nuclear localization of fluorescently tagged SMAD4 protein as a readout of BMP signaling dynamics and linked this to amnion-like differentiation measured by iterative immunofluorescence staining for seven cell fate markers [5]. The live readout need not be cell signaling activity—the same approach can link dynamics of gene expression, cell cycle, displacement over time, etc. to cell state. Several studies have taken a similar approach to link single-cell dynamics of pluripotency marker expression to cell fate after differentiation [6, 7]. In addition to the experimental protocol, we describe a computational pipeline to extract single-cell histories from live data using cell tracking and link these to the quantitative readouts of the final state from fixed data in the same cells. We describe how to do this for either standard adherent culture or micropatterned hPSC colonies treated with BMP4 (2D gastruloids) using three different microscopes (Nikon and Andor spinning disk confocal, Olympus widefield epifluorescence), but a similar approach could be followed for other experimental setups and different microscopes.

2 Materials

2.1 Wet Lab Materials

1. Matrigel.
2. mTeSR1 (StemCell Technologies).
3. Accutase.
4. ROCK inhibitor Y-27632 (RI).
5. DMEM/F12 medium.
6. Micropatterned culture dish or multi-well slide (*see* **Note 1**).
7. Laminin 521 (Biolamina).
8. Phosphate-buffered saline with calcium chloride and magnesium chloride (PBS+/+).
9. Phosphate-buffered saline without calcium chloride or magnesium chloride (PBS−/−).
10. PBS−/− with 0.01% sodium azide by weight.
11. Paraformaldehyde 4% solution in PBS−/−.
12. Bleaching buffer: 3% H_2O_2 and 20 mM HCl in PBS−/−.
13. Permeabilization buffer: 0.1% Triton-X 100 in PBS−/−.
14. Blocking buffer: 3% donkey normal serum (if working with donkey secondary antibodies), 1% Penicillin/Streptomycin (100×), 0.1% Triton-X 100 diluted in PBS−/−.
15. PBST: 0.1% Tween-20 diluted in PBS−/−.
16. Imaging buffer (25 mL volume): 6 mL distilled water, 1.76 g N-acetyl-L-cysteine (Sigma-Aldrich), 17 mL 1 N NaOH, adjusted to a pH of 7.4.

17. Elution buffer (20 mL volume): 12 mL distilled water, 0.750 g L-glycine, 3.604 g urea, 5.772 g guanidine hydrochloride, 0.400 g Tris(2-carboxyethyl) phosphine hydrochloride (TCEP), adjusted to a pH of 2.5.

2.2 Software

1. MATLAB version 2020b or newer with the Image Processing Toolbox, Computer Vision Toolbox, and Parallel Computing Toolbox.
2. Cellpose 1.0 or higher [8].
3. Fiji distribution of ImageJ [9] (https://imagej.net/).
4. Anaconda distribution of Python with the Conda package manager (anaconda.com/download).
5. Github Desktop (desktop.github.com/download/).
6. Code for our custom image-processing pipeline, found at https://github.com/seth414/HeemskerkLabMethods.
7. An example dataset on which the image-processing and cell tracking pipeline can be tested is provided in the University of Michigan Deep Blue Data repository, at https://doi.org/10.7302/720m-v383.

3 Methods

3.1 Human Pluripotent Stem Cell (hPSC) Culture

1. Prepare tissue culture plates for adherent cell culture by mixing cold DMEM/F12 with 1% matrigel, transferring the mixture to the plates, and incubating for at least 1 h at 37 °C or overnight at 4 °C (*see* **Note 2**).
2. Culture the cells in mTeSR1 pluripotency maintenance medium and change the medium every 24 h (*see* **Note 3**). Incubate the cells at 37 °C with 5% CO_2.
3. Passage the cells at 70–80% confluency (*see* **Note 4**).
4. After seeding from a single-cell suspension following Accutase passaging or thawing, culture the cells in medium supplemented with 10 μM RI for 24 h.

3.2 Differentiation Protocol

3.2.1 hPSC Differentiation in Standard Adherent Culture

1. Coat a multi-well slide or tissue culture plate with 1% matrigel in DMEM/F12 as for routine maintenance.
2. Follow the steps for accutase passaging to obtain a single-cell suspension and seed at the desired cell density (*see* **Note 5**). To facilitate single-cell tracking, consider seeding cells with sparse labeling (*see* **Note 6**).
3. Culture the cells overnight in mTeSR1 with 10 μM RI following seeding.

4. Continue culture to reach the desired colony morphology and cell density before beginning differentiation (*see* **Note 7**).
5. See Subheading 3.3 "Live-cell imaging" for steps to set up acquisition on the microscope before adding treatment to induce differentiation.
6. Treat cells with desired differentiation medium (e.g., mTeSR1 with 50 ng/mL BMP4 and 5 μM IWP2 for amnion-like differentiation).

3.2.2 2D Gastruloid Differentiation

1. Incubate micropatterned plates with laminin 521 diluted 1:20 in PBS+/+ at 37 °C for 2 h.
2. Add enough room temperature (RT) PBS+/+ to the well in addition to the laminin solution to nearly fill the well and then remove an equal amount (i.e., wash without completely emptying the well) three times.
3. Fully remove the solution in each well, then add and remove room temperature PBS+/+ once more. Use a pipette, do not aspirate with a vacuum to avoid damaging the coating.
4. Add RT PBS+/+ to each well and store the plate at 4 °C until ready for seeding. Each time the well is fully emptied, quickly add the next solution to avoid the laminin drying out.
5. Create a suspension of single cells in mTeSR1 with RI by accutase passaging and seed at the desired cell density (*see* **Note 5**). For 2D gastruloids, use a density of about 200,000 cells/cm^2.
6. Allow cells to settle for 5 min, then move the plate to the incubator.
7. 30 min after seeding, wash with PBS−/−, then add mTeSR1 with RI.
8. Visually inspect the colonies on a microscope to confirm that cells are present in micropatterned discs on the plate with few or no cells outside of these discs.
9. Incubate cells in mTeSR1 with 10 μM RI for 24 h before removing RI and adding 50 ng/mL BMP4 to induce differentiation.

3.3 Live-Cell Imaging

Instructions are provided for three microscope control and acquisition programs: Nikon NIS Elements, Andor Fusion, and Olympus cellSens Dimension.

1. If possible, ensure that cells are in phenol red-free medium during imaging.
2. Turn on the microscope incubation system at least an hour before image acquisition to allow the temperature to equilibrate and prevent thermal drift.

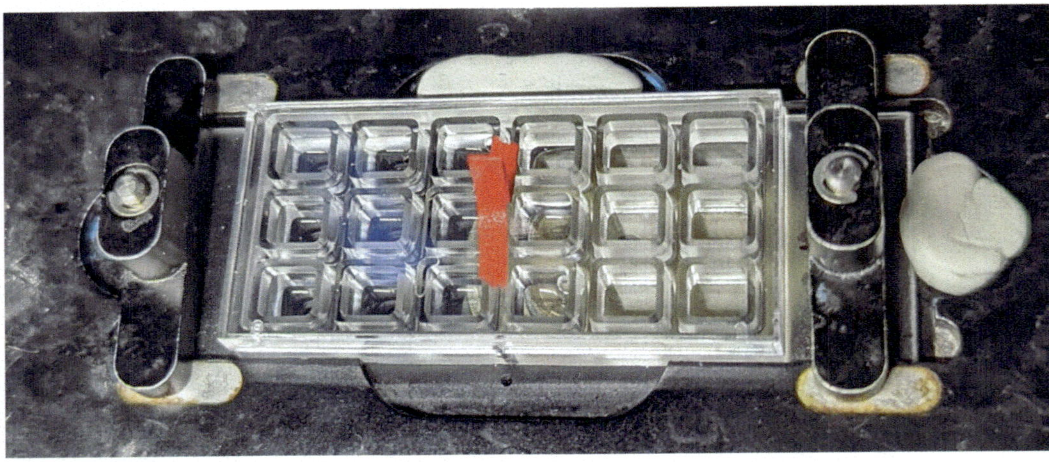

Fig. 2 Image of an Ibidi μ-Slide 18-well plate fixed in place with putty in a microscope sample holder. A small handle is added using a small piece of red label tape to allow the plate lid to be removed without disturbing the plate's position

3. Turn on the microscope and start the microscope software (Fusion/NIS Elements/Olympus cellSens Dimension).
4. Mount the sample in the plate holder. Use of an intermediate magnification (20–40×), high NA objective is recommended to ensure sufficient data quality for analysis of single-cell signaling or gene expression histories (*see* **Notes 8** and **9**).
5. (Optional) Fix the sample in place with putty (*see* **Note 10**) and make a "handle" out of tape stuck to the plate lid to allow the lid to be lifted more easily without disturbing the plate (*see* **Note 11** and Fig. 2).
6. Find the panel or tab used for multi-position time-lapse protocol definition in the microscope software.
   (a) In Andor Fusion, this is the "Protocol Manager" tab (Fig. 3). In the Protocol Manager, in the "Type" dropdown menu, select "Multi-position."
   (b) In NIS Elements, it is the "ND Acquisition" panel (Fig. 4).
   (c) In cellSens, this is the "Process Manager" panel (Fig. 5).
7. Bring the sample into focus. In NIS Elements, then check on the "Use PFS" box in the "XY" tab of the ND Acquisition panel. If the box is already checked, uncheck and recheck it to ensure that the PFS is switched on. After turning on PFS, the focus may have automatically shifted below the cells. Bring them back into focus.
8. Select predefined channels according to the live fluorophores to be imaged.

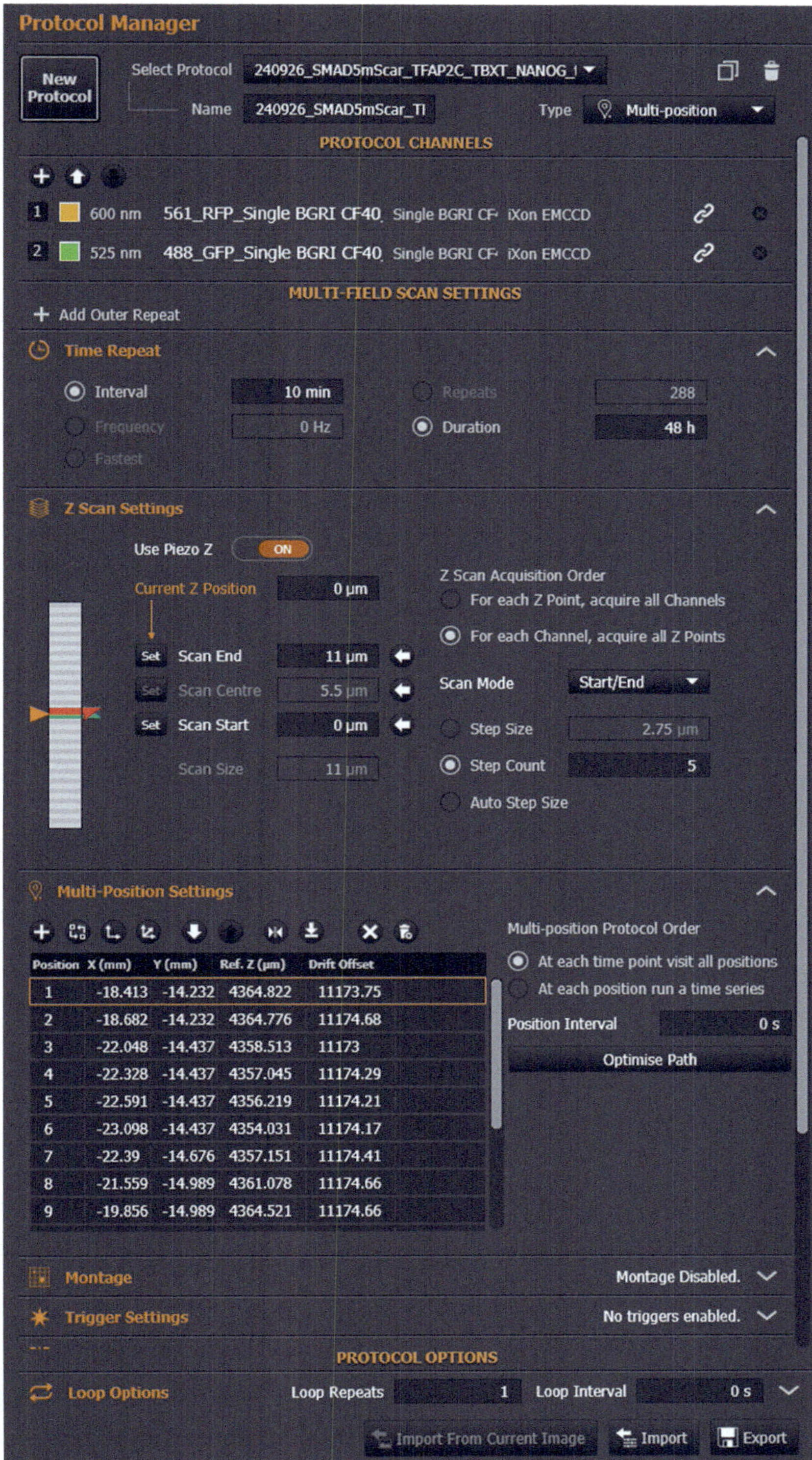

Fig. 3 A screenshot of the Protocol Manager tab in the Andor Fusion software

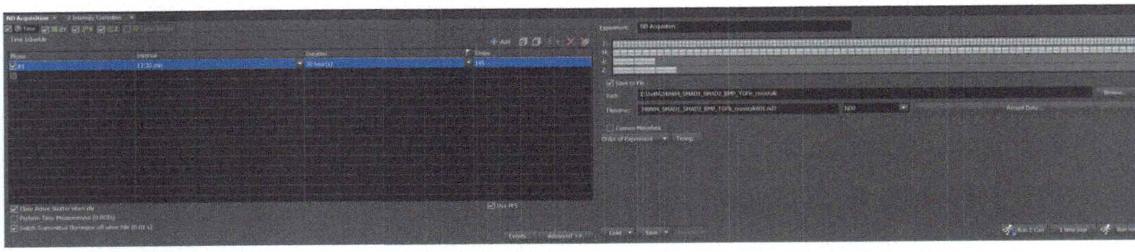

Fig. 4 A screenshot of the ND Acquisition panel in the NIS Elements software with the "Time" tab selected

(a) In Andor Fusion, choose channels from the dropdown menu in the "Protocol Channels" section of the Protocol Manager tab.

(b) In NIS Elements, select channels by checking them in the "λ" tab of the ND Acquisition panel.

(c) In cellSens, select channels in the "C" tab of the Process Manager by checking the "Acquire" box in that channel's settings.

9. Set the laser intensity and exposure time for each channel (*see* **Note 12**).

   (a) In Andor Fusion, this is done in the "Acquisition Control" panel to the left of the Protocol Manager tab.

   (b) In NIS Elements, select the channel in the "Acquisition" (not ND Acquisition) tab to set the laser power and exposure time.

   (c) For an Olympus epifluorescence microscope with an LED for excitation instead of a set of laser lines, the intensity cannot be adjusted, only the exposure time. In cellSens, select the channel in the "C" tab of the Process Manager, then in the "Camera Control" panel, set the exposure time to a value appropriate for the fluorophore being imaged. In the "C" tab of the Process Manager, click "Read Settings" to update the value used in the protocol to the current setting in the Camera Control panel.

10. Set the duration of the acquisition to the length of the differentiation/patterning protocol and set the interval between subsequent time points (*see* **Notes 13** and **14**).

    (a) In Andor Fusion, this is set in the "Time Repeat" section of the Protocol Manager.

    (b) In NIS Elements, set this in the "Time" tab of the ND Acquisition panel.

    (c) In cellSens, set the recording time and interval in the "t" tab of the Process Manager.

11. Set the z stack mode, size, and spacing (*see* **Notes 15** and **16**).

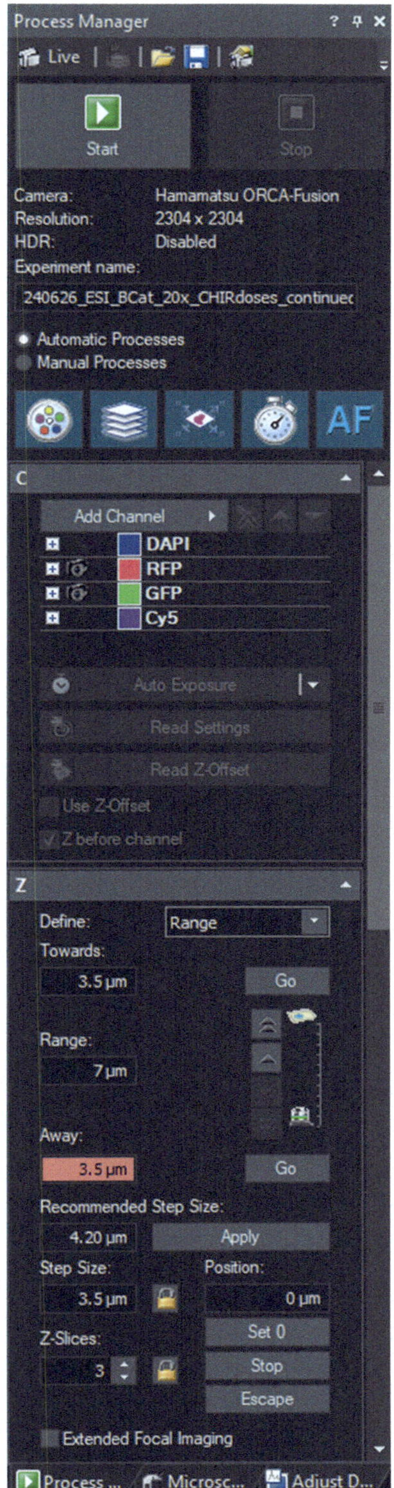

Fig. 5 A screenshot of the Process Manager panel in the Olympus cellSens software

(a) In Andor Fusion in the "Z Scan Settings" section of the Protocol Manager, ensure that "Use piezo z" is toggled on and set the current position to 0. Set the z stack mode to "Start-End," set the start position to zero, and the end position to the maximum expected height of the sample by the end of differentiation. Set the number of z slices to the desired spatial resolution in z.

(b) In the "Z" tab of the ND Acquisition panel, set the z stack mode to "Asymmetric scan defined by range," and set the "Below" position to zero and the "Above" position to the maximum expected height of the sample by the end of differentiation. Set the number of z slices to the desired spatial resolution in z.

(c) In cellSens, go to the "Z" tab of the Process Manager, set the "Away" position to 0 and the "Towards" position to the maximum expected height of the sample by the end of differentiation. Set the step size to the desired spatial resolution in z.

12. Set the locations of fields of view for multi-position acquisition. If acquiring tiled montages to stitch into a larger field of view, ensure that individual positions are chosen sufficiently far apart to prevent overlap between different montages (*see* **Note 17**).

13. For each field of view, move the stage to the location of interest, find the focal plane at which the lowest cells or nuclei are most in focus, go down another ~1–3 μm, and add the position to the multi-position list (*see* **Note 18**).

(a) In Andor Fusion, toggle on "Drift Stabilization Active" in the "Drift Stabilization Settings" (*see* **Note 19**) section of the Protocol Manager and add the position to the table by clicking the "+" button in the "Multi-Position Settings" section.

(b) In NIS Elements, add the position to the multi-position table by clicking the "+" button in the "XY" tab of the ND Acquisition panel.

(c) In cellSens, in the "XY" tab of the Process Manager, check the "Use current Z-position" box, and toggle on the button to save ZDC offset when each new position is added. Add the position to the multi-position list by clicking the "+" button. To view the multi-position list and edit or delete positions, click the "Open Position List Dialog" button.

14. (Optional) Define tiled montages.

(a) In Fusion, in the "Protocol Options" section of the Protocol Manager, export the protocol file as a .bkp file. Open the "multiMontageFusion.m" script in MATLAB, set "protocolFile" to the path to the .bkp file, "resolution" to the width of a pixel in μm for the currently used objective and camera, "gridSize" to the size of the tiled montage (e.g., 2 × 3 for a montage 2 fields of view tall by 2 fields of view wide), "montageOverlap" to the percent overlap between adjacent fields of view. Run the script, then return to Fusion and import the newly generated .bkp file with an updated multiposition list.

(b) In NIS Elements, in the "XY" tab of the ND Acquisition panel, save the positions list as a .xml file. Open the "multiMontageElements.m" script in MATLAB, set "protocolFile" to the path to the .xml file, "resolution" to the width of a pixel in μm for the currently used camera and objective, "gridSize" to the size of the tiled montage, and "montageOverlap" to the percent overlap between adjacent fields of view. Run the script. Return to NIS Elements and load the newly generated xml file in the "XY" tab of the ND Acquisition panel.

15. Once all positions are defined, save a file with the information to reload them into the software to a file location that can be accessed again at the end of the protocol (*see* **Note 20**).
    (a) In Andor Fusion in the "Protocol Options" section of the Protocol Manager, export the protocol file as a .bkp file.
    (b) In NIS Elements, in the "XY" tab of the ND Acquisition panel, save the positions list as a .xml file.
    (c) In cellSens, save the process definition with the save icon at the top of the Process Manager.
16. Set the image file name format.
    (a) In Andor Fusion, this is the "Image Name" field of the "Navigation" tab of the Acquisition panel.
    (b) In NIS Elements, on the right side of the ND Acquisition panel, check "Save to File" and set the path and filename for the image file.
    (c) In cellSens, set this in the "Saving" tab in the "Acquisition settings" window near the top of the Process Manager.
17. Start acquisition.
    (a) In Andor Fusion, make sure that "Drift Stabilization Active" is toggled on and press "Acquire" near the bottom of the window to begin the time lapse.
    (b) In NIS Elements, press "Run protocol" in the bottom right of the ND Acquisition panel.
    (c) In cellSens, check the box beside "Z-drift compensation" and select "Single shot ZDC mode" from the accompanying dropdown menu. Then press the "Start" button at the top of the Process Manager.
18. (Optional) Add treatment to the cells to begin differentiation (*see* **Note 21**).

3.4 Fixation and Immunostaining

1. Fix cells immediately after live-cell imaging to minimize cell movement between the end of live imaging and fixation.
2. To fix the sample, remove media, add 4% paraformaldehyde, and incubate for 20 min at room temperature (*see* **Note 2**).
3. Wash twice with PBS−/−; i.e., remove the 4% PFA solution and add PBS−/−, then remove and replace with PBS−/− again.
4. (Optional) Take the sample to the microscope and load in the same position in the plate holder with putty, load the protocol for live-cell imaging, change the time-lapse settings to acquire a single time point, and run (*see* **Note 22**).
5. If immunostaining will not be done immediately, replace the PBS−/− with PBS−/− + 0.01% sodium azide and store at 4 ° C until ready for staining (*see* **Note 23**).
6. Prepare permeabilization and blocking buffers. These should be prepared fresh each time they are to be used.
7. Permeabilize the cells by adding permeabilization buffer to each well.
8. Wash twice with PBS−/−.
9. Add bleaching buffer to each well and incubate at room temperature for 1 h under a bright lamp (*see* **Note 24**).
10. Wash twice with PBS−/−.
11. Incubate for 30 min at room temperature in blocking buffer (see Note 25).
12. Prepare the primary antibody solution: dilute primary antibodies in blocking buffer at concentrations indicated by their vendors.
13. Remove blocking buffer, add the solution of primary antibodies, and incubate overnight at 4 °C or for 2 h at room temperature (*see* **Note 25**).
14. Remove the primary antibody solution, add PBST to each well, and incubate 20 min at room temperature.
15. Repeat the previous step twice, for a total of 3 PBST washes.
16. Dilute secondary antibodies and DAPI in blocking buffer. For Alexa Fluor secondary antibodies, dilute each secondary antibody 1:500 to a final concentration of 4 μg/mL. DAPI should be diluted to 2 μg/mL in the same solution as the secondary antibodies.
17. Remove PBST and add the secondary antibody solution to each well. Incubate at room temperature for 1 h (*see* **Note 25**).
18. Wash twice with PBST with no incubation step.

19. Wash twice with PBST with 20 min of incubation at room temperature after each wash.
20. Remove PBST. If immunofluorescence imaging will be done immediately, add imaging buffer to each well, otherwise store cells in PBS−/− + 0.01% sodium azide. Remove the PBS−/− with sodium azide and replace with imaging buffer immediately before imaging.

3.5 Immunofluorescence Imaging

1. Turn on the microscope and open control software prior to bringing the plate to the microscope as before. [Do not turn on the incubator].
2. Bring the plate back to the microscope on which live-cell imaging was performed and use the same objective.
3. Mount the sample in the plate holder and use putty to push the plate to the same corner of the sample holder as for live imaging.
4. Load the multi-position list used for live imaging:
   (a) In the Andor Fusion software, click "Import protocol" near the bottom right of the Protocol definition panel, and choose the .bkp file saved from the live-cell imaging protocol.
   (b) In NIS Elements, in the "XY" tab of the ND Acquisition panel, click "Load" and choose the .xml file saved from the live-cell imaging protocol.
   (c) In cellSens, click the "Load" button at the top of the Process Manager and load the process definition used for live acquisition.
5. Bring the cells into focus and toggle on the Autofocus (Andor Fusion) or check "Use PFS" (NIS Elements).
6. Set laser intensity and exposure time according to the brightness of the stain in each channel (*see* **Note 26**).
7. (optional) Update the focal position for each field of view in the multi-position list:
   (a) In NIS Elements, in the "XY" tab check on "Move Stage to Selected Point," select the point in the position list, then adjust the focus as for the live imaging protocol. Once the focus is set correctly, click the arrow to the right of the XYZ coordinates in the position list to update the focus for that position.
   (b) In Andor Fusion, select the position from the multi-position list, click the "Move stage to selected position" button and adjust the focus as for the live imaging protocol. Toggle the Autofocus back on and select "Replace/update with the current stage position."

   (c) In cellSens, open the Position List Dialog from the "XY" tab of the Process Manager. For each position in the table, select the position, press "Go to Position," adjust the focus, and press "Update Z" under "Stage coordinate" and "Read Offset" under "ZDC offset."

8. Update the image file name format.
9. Start acquisition.

3.6 Iterative Immunostaining

If measurement of the expression or localization of additional proteins is desired, perform iterative immunofluorescence following the approach from Gut et al., 2018 [2], as modified in Freeburne et al., 2023 [4].

1. Elute the antibodies by removing the imaging buffer, adding elution buffer, and incubating for 10 min on an orbital shaker at 50 rotations per minute.
2. Repeat **step 1** two more times, then wash the sample ×3 with PBS−/−.
3. The sample may then be stained again following the protocol from Subheading 3.4 with the following changes:
   (a) Permeabilization and bleaching are not required after the first round of immunostaining.
   (b) Triton-X 100 need not be included in the blocking buffer after the first round (*see* **Note 27**).
4. Repeat Subheading 3.5.

3.7 Live-Cell Image Analysis

The instructions for computational image analysis presented here follow our custom image-processing pipeline, but other methods may be substituted if they segment individual cells in each frame of the time lapse and quantify the relevant data to be tracked.

1. To access our custom image-processing code, install GitHub Desktop and clone the repository github.com/seth414/HeemskerkLabMethods (File - > Clone repository, choose the "URL" tab, and enter the URL of the repository). This will copy the code to a folder on your device. You can choose the location of this folder; in Windows the default location is in a folder called "GitHub" in the default "Documents" folder.
2. Install Cellpose v1 in a conda virtual environment following the instructions at https://github.com/MouseLand/cellpose or in Freeburne et al., 2023 [4].
3. Before beginning processing, run the setup.m script in the GitHub repository. This adds the directory containing the script and all subfolders of that directory to the MATLAB path.
4. Make a folder containing the live-cell image data.

5. Copy the "preprocess_images.m" script from the "template scripts" folder of the GitHub repository into the folder with the live-cell image data and open the script in MATLAB.
6. If tiled montages of overlapping images were acquired to stitch into larger fields of view, set "makeStitchedImages" to true and both "makeMIPs" and "writeZslices" to false. Otherwise, set "makeStitchedImages" to false and both "makeMIPs" and "writeZslices" to true.
7. Set "nucChannel" to the channel used as a nuclear marker, e.g., a fluorescently tagged H2B protein. Channels are indexed from 0, so if the nuclear marker is in the second imaged channel, set nucChannel = 1.
8. Enter a label for each imaged condition or well in a cell array for "conditions."
9. In the cell array for the "channelLabel" variable, enter a label for each channel in the order in which they were imaged.
10. If doing stitching in the preprocessing step, set "gridsize" to the size of the grid of overlapping images and "montageOverlap" to the percent overlap between adjacent images in the grid (as a percentage, i.e., for 15% overlap, set montageOverlap = 15, not 0.15).
11. If not doing stitching and if any channels need to be segmented in addition to the nuclear channel, set the field "writeChannels" of "zSliceOptions" to include those channels, again indexing channels from 0. For example, if the nuclear channel was imaged in the first channel but cell membranes will also be segmented using the signal from the third channel, nucChannel will be set to 0 and zSliceOptions.writeChannels should be set to 2.
12. After setting the metadata and processing options, run the "preprocess_images" script.
13. Segment cell nuclei using Cellpose (*see* **Note 28**).
14. Copy the "segmentation_cellpose.ipynb" iPython notebook from the "template_scripts" folder of the GitHub repository to the data folder.
15. Open an "Anaconda Prompt" window, or, if anaconda is added to the machine's path, optionally instead open a Command Prompt (Windows) or Terminal (macOS or Linux) window.
16. In this window, navigate to the data folder containing the "segmentation_cellpose.ipynb" notebook.
17. Activate the Cellpose environment (type "conda activate cellpose" and enter).
18. Open a Jupyter Notebook ("jupyter notebook") or JupyterLab window ("jupyter lab"). If one of these packages is not installed

in the cellpose environment, Jupyter Notebook can be added by typing "pip install notebook" or JupyterLab can be installed with "pip install jupyterlab". A new tab will open in the default browser with the interface for JupyterLab or JupyterNotebook.

19. Open the segmentation_cellpose.ipynb notebook in Jupyter.
20. Run the notebook cell-by-cell until reaching the "run the model once" cell.
21. Set the diam parameter to the approximate diameter of a single nucleus in the data. For the sample data at 20× in which each pixel is ~0.6 × 0.6 μm, a diam value of 20 pixels works well.
22. In the model definition, set "gpu" to True or False (*see* **Note 29**).
23. Run the "run the model once" and "DISPLAY RESULTS" cells for a preview of the segmentation results on a single image. If nuclei are accurately segmented, run the next cell ("RUN CELLPOSE") to run the model on the entire dataset. Otherwise, try tuning the diam parameter, or see the Cellpose documentation (https://cellpose.readthedocs.io/en/v1.0.2/) for additional settings and troubleshooting tips.
24. Copy the combineSegmentations.m script from the template_-scripts folder in the GitHub repository into the data folder.
25. (Optional) Edit the parameters for postprocessing nuclear segmentations like minarea, maxarea, maxZsize, etc.
26. Run the script. As an output the script will save a TIFF file with the suffix "_FinalSegmentation" and a .mat file with the suffix "_masks" for every time point in every field of view.
27. Copy singleCellQuantificationLive.m from the GitHub repository into the data folder.
28. Run the "clear; close all; clc," "setup," and "iterate over positions, time points, and channels, and read out intensities" sections. This will save a data structure of class "Position" with cell data and metadata for every field of view (position) in the file "positions.mat" and write a CSV file for every field of view with data to load into TrackMate later for single-cell tracking.

3.8 Single-Cell Tracking

Single-cell tracking is described here using Trackmate [10, 11], as it is a fast, popular, and widely available package, but the broad approach does not depend on the specific cell tracking software used (*see* **Note 30** for instructions to run our single-cell tracking code integrated with the image-processing pipeline in MATLAB).

1. Open Fiji.
2. Install the plugin to load precomputed cell features into TrackMate by adding its update site in Fiji software (Help → Update… → Manage Update Sites → search and add TrackMateCSVImporter).
3. Import time-lapsed MIPs (2D tracking) or z-stacks (3D tracking) into Fiji (*see* **Note 31**).
4. Open TrackMate in Fiji (Plugins - > Tracking - > TrackMate). You may be prompted to swap the z and t axes for your data. Choose yes if appropriate.
5. Open the TrackMate CSV importer (Plugins - > Tracking - > TrackMate CSV importer).
6. In the new window, under "CSV file" enter the path to the CSV file with data for the position of interest (e.g., for the first position this will be data_table_p0000.csv), or click "Browse" and navigate to the location of the CSV file of interest and select it.
7. After selecting the CSV file, uncheck the boxes next to "Compute all features" and "Import tracks."
8. Ensure in the fields below that "X column" uses "POSITION_X," "Y column" uses "POSITION_Y," "Z column" uses "POSITION_Z," "Frame column" has "FRAME" selected, "Quality column" uses "MEAN_INTENSITY_CH," and "ID Column" uses "ID." Click "Import."
9. In the new window that opens, advance to the "Select a tracker" page and choose "LAP Tracker" from the dropdown menu.
10. Advance to the next page ("Settings for tracker") and set parameters for tracking (*see* **Note 32**). We recommend checking "Allow gap closing" and "Allow track segment splitting" but not "Allow track segment merging." Under each selected header, under "Feature penalties" click the green "+" button and select "Quality" from the dropdown menu.
11. Press "Next" and the tracks will be computed and displayed over the loaded image sequence.
12. Advance until reaching the "Display options" page. At the bottom of the page, select the "Tracks" button (between "TrackScheme" and "Spots"), go to the "Edges" tab in the new window, and click "Export to CSV." Navigate to the main folder with the live image data, name the CSV in the format "trackmate_p####_edges.csv" (e.g., for the first position use "trackmate_p0000_edges.csv") and press save.
13. If desired, continue analyzing features of single-cell tracking results in the TrackMate graphical user interface.

14. Once tracking has been done for all desired positions/fields of view, add tracking results to the existing data structure containing quantified cell data. To do so, open "singleCellQuantificationLive" in the folder with live-cell image data and run the "clear; close all; clc," "setup," and "load data if processing has already been done" sections, then run the section titled "load trackmate results, parse into a digraph, and add to the Position object."

3.8.1 Manual Track Validation and Correction

Instructions are provided below to correct/validate single-cell tracking results with our custom code (*see* **Note 33** for alternative approaches).

1. Copy the "correctTracks.m" script into the directory with live-cell data.
2. To correct/validate tracking results for a given field of view (position), run the "load positions" section, set pidx to the index of the desired field of view in the "choose a position index for which to start/continue validation" section and run that section and the "load time lapse of adjusted MIPs" section.
3. Run the "Manual correction" section. A window will open showing the maximal intensity projection of the nuclear channel in the final frame of the live-cell image data overlaid with a scatterplot of cell centroid positions, tracking tails for single-cell tracking results and a number (index) for each cell. Use keyboard key presses to move through the time lapse, zoom and pan on the images, and edit and validate tracking results. Key mappings are as follows:
   (a) To pan forward and backward in time and examine results, press the "a" key to move back one time point or "s" to go back 10 time points, the "d" key to move forward one time point or the "w" key to go forward 10 time points.
   (b) Zoom in with the "+" key and out with the "−" key.
   (c) Pan up with "8," down with "2," left with "4," and right with "6." Reset the field of view and zoom with the "5" key.
   (d) Press "v" to choose a cell index from the final frame to validate (*see* **Note 34**). If the wrong cell index is chosen by mistake, press "u" and select the cell to "un-validate."
   (e) To edit the single-cell tracking results, press the "m" key and select an option from the list dialog—"add link," "remove link before," or "remove link after." If adding a new link, first use the mouse to click the desired cell centroid (just click nearby, the nearest point to the mouse click will be selected). Then pan forward in time with the "d" key to a frame with the same cell or its

progeny, press the "n" key, and select the cell to which to make the link (*see* **Note 35**). To remove a link, press "m," choose "remove link before" or "remove link after," and choose the cell from which to remove the link (*see* **Note 36**).

(f) After completing a session of validating and correcting tracking results, press "e" to close track overlay window and save results. Both the updated digraph and the Boolean array indicating which tracks have been validated are saved and can be loaded to continue validation later.

3.9 Immunofluorescence Image Analysis

1. Copy immunofluorescence image data to the desired location for processing. If there are multiple rounds of staining from iterative immunofluorescence, it is simplest to have a single parent directory for all the fixed data with the image data from each round of staining and imaging in a subfolder of that repository named "round1," "round2," etc., or similar.
2. Copy the "preprocess_images.m" script from the "template_scripts" folder of the GitHub repository into the folder with fixed image data. For multiple rounds of staining, preprocess_images should be added to and run in the individual folder for every round of staining. Edit the settings in "preprocess_images" and run following the instructions for live-cell image data.
3. Copy segmentation_cellpose.ipynb from the template_scripts folder into the folder containing the first round of immunofluorescence image data and run the Cellpose segmentation following the instructions for live-cell image analysis in Subheading 3.7.
4. Copy the combineSegmentations.m script from the template_scripts folder into the data folder for the first round of immunofluorescence image data. Run the script following the instructions from Subheading 3.7.
5. For datasets with multiple rounds of immunofluorescence staining, copy the "findRoundAlignment.m" script from the "template_scripts" folder in the GitHub repository into the parent folder containing all the immunofluorescence image data (*see* **Note 37**).
6. The "dirs" variable in this script is a cell array. For each entry of the cell, give the name of the subfolder containing image data for one round of immunofluorescence. E.g., "dirs = {"round1","round2","round3"};".
7. Run the script.
8. Copy "singleCellQuantificationFixed.m" into the parent folder containing all the immunofluorescence image data and set the

"dirs" variable the same way as in the findRoundAlignment script. If there is only one round of immunofluorescence, the script can be copied into the single fixed-cell image data folder and dirs set to {""}.

9. Set the "imageType" variable to "disordered" for data in standard adherent culture or "MP" for micropatterned colonies. If imageType is set to "MP", specify the colony radius in each condition, i.e., radii = [300, 300, 500, 700].
10. Run the "clear; close all; clc," "setup," and "iterate over colonies, folders, and channels, and read out intensities" code sections to quantify the fixed-cell data.

3.10 Alignment and Linking of Live-Cell Data to Immunofluorescence Data

1. Once single-cell quantification has been done for both live-cell data and fixed data, and single-cell tracking has been done, open the singleCellQuantifcationFixed.m script in the immunofluorescence data folder. If linking live-cell data immediately after doing quantification of the immunofluorescence data, and the immunofluorescence data is still in the MATLAB workspace from singleCellQuantificationFixed, skip **step 2** below.
2. Run the "clear; close all; clc" and "setup" code sections.
3. In the "Load live and fixed data" section, set the "liveDir" variable to the path to the folder containing live-cell image data. The default location in the script is in a folder called "live" in the parent directory of the folder with the immunofluorescence data. Run the code section.
4. In the "initialize lineage trace & map live to fixed" code section, set the "maxDist" variable to the largest distance in pixels expected between a live nucleus and the same nucleus in the immunofluorescence data after aligning the images (*see* **Note 38**). Run the code section. This will create a variable "lt" defined by the "LineageTrace" class which stores the live-cell and fixed-cell position data, metadata, and links from live to fixed cells.
5. A loop will run allowing visual inspection of links from live to fixed nuclei. A figure will show the live and fixed immunofluorescence data overlaid in green and magenta with an arrow plot showing links between live and fixed cells, and MATLAB will pause at each iteration. After inspecting the links visually for a given field of view, press any key to advance to the next position. Press any key after the last position to leave the loop.
6. In the same code block, single-cell tracking results that have already been computed are loaded and used to construct a tracking history for each live cell that is matched to a fixed cell. Histories are saved in the "histories" field of lt.

7. Run the block "make matrices of signaling histories and fixed-cell data" to construct matrices with single-cell dynamics for fluorescence readouts and nuclear morphology and corresponding matrices with immunofluorescence intensity in the fixed data and a vector indicating the position index and whether the cell has been manually validated for each history.

4 Notes

1. Micropatterned dishes may be purchased from a vendor such as CYTOO or custom-made as in [4, 12, 13]. We make micropatterned multi-well slides following the protocol in [4].
2. For the volume of cell culture medium, follow instructions for the specific tissue culture plates or multi-well slides used. Volumes for coating plates with Matrigel or Laminin and for immunofluorescence buffers should be at least enough to cover the bottom of the plate, usually about half the volume of medium used for culture.
3. Similar results may be obtained with other maintenance medium like StemFlex, Essential 8, or mouse embryonic fibroblast-conditioned HUESM medium with FGF2.
4. Whole-colony passage with L7 [14], EDTA, or STEMCELL ReLeSR is recommended for routine cell culture to maintain pluripotent morphology and reduce the accumulation of chromosomal abnormalities over many passages. Dissociation into single cells for passaging is required when seeding for differentiation to count cells and seed at the correct density. We use Accutase to dissociate cells, but similar results may be obtained using Trypsin, TryPLE, or ReLeSR.
5. Control of seeding density is critical for both standard and micropatterned differentiation and may have to be optimized; for linking dynamics to final state in standard culture, conditions with maximal heterogeneity are most informative. For 2D gastruloids, the presence of extraembryonic-like cells in the interior of the colony rather than restricted to the colony edge indicates that the cell density is too low. Few or no primitive streak-like cells and a very large region of ectoderm-like or pluripotent cells indicate that the cell density is too high.
6. Single-cell tracking is very difficult in densely packed tissues where any segmentation error may result in a cell in one frame being linked to the wrong cell in the next frame. To make tracking feasible in conditions with the same overall cell density or tissue organization, mix cells with a fluorescent marker with wild-type cells with no fluorescence and track only the fluorescently labeled cells. In Teague et al., 2024 [5], 10–20% of cells

were fluorescently labeled in experiments with single-cell tracking. The highest fraction of labeled cells for which tracking is reliable should be chosen to maximize the number of high-confidence single-cell histories obtained per experiment.

7. Optimal culture duration and conditions before initiating differentiation depend on the experiment. In Teague et al., 2024 [5], we seeded 16–20 h prior to adding BMP4 treatment and removed RI from the media 2.5 h before treatment. In other experiments, it may be desired to allow fully epithelialized colonies to form before differentiation, which would entail one or more days of culture after removal of RI before adding differentiation-inducing cues.
8. For applications in which careful quantification of fluorescence is desired, particularly in different cellular compartments (e.g., nuclear to cytoplasmic ratio of GFP::SMAD4 intensity), an intermediate-magnification immersion objective should be used, and we find 30× silicon oil the optimal tradeoff between field of view and resolution, as well as image quality and live imaging stability. If the tracked quantity is less sensitive, for example average velocity or displacement over time, a lower magnification air objective will generally enable easier acquisition of more cell trajectories.
9. Maintaining contact between an immersion objective and the plate for a long time-lapse can be difficult. A water objective may be used if the objective is equipped with a pump to automatically replenish the water between the sample and objective during acquisition, otherwise oil immersion is more stable. If losing the droplet during acquisition is a persistent problem, consider reducing the imaging area or number of fields of view to be imaged and optimizing the layout of fields of view to minimize the distance of travel between any two fields of view.
10. Generally, there is some room for movement of the cell culture dish or multi-well slide in the microscope sample holder. Placing putty in the sample holder corners fixes the plate securely in place, preventing motion during acquisition to facilitate single-cell tracking and enabling better realignment of the sample after fixation to measure fixed-cell data linked to the live-cell trajectories in the same cells. The plate should be pushed into one corner of the plate holder (we use the bottom left corner) before applying putty.
11. A tape handle allows the plate lid to be removed while the plate is in the sample holder on the microscope with minimal disturbance to its position during acquisition (Fig. 2).
12. hPSCs are highly sensitive to phototoxicity. To minimize phototoxicity, laser power and exposure time should be the

minimum to produce sufficiently high-quality data for downstream analysis. Conventional wisdom holds that increasing exposure time while holding laser intensity fixed is less phototoxic than increasing laser power while holding exposure time fixed. However, increasing exposure time increases the total acquisition time, reducing the number of fields of view that can be captured in a set time interval.

13. Naïvely, the duration of the time-lapse acquisition should be the duration of the differentiation/patterning process of interest. However, if the relevant heterogeneity in dynamics only occurs at later times, it is reasonable to restrict imaging to those times, which will reduce difficulty in tracking, reduce dataset size, make it easier to maintain focus and immersion oil/water contact, as well as reducing the risk of phototoxicity. For example, in micropatterned 2D human gastruloids, endogenous Nodal signaling begins about 24 h after initial treatment with BMP4 [15], so imaging the last 24 h of differentiation would likely be adequate for a study linking single-cell heterogeneity in Nodal signaling to resulting cell fate in this system.
14. The time resolution of acquisition is a critical parameter in determining the success of single-cell tracking later in the analysis—minimizing this interval can substantially improve the tracking results and making it too large can make effective tracking impossible. The needed time resolution depends on cell density, the degree of sparse labeling, and the amount of cell movement during differentiation. There are additional constraints on the possible time resolution. Shorter time intervals increase phototoxicity; this may be balanced by reducing the laser intensity, exposure time, or number of z slices if the signal quality is sufficient. Reducing the time interval also reduces the number of fields of view that can be acquired per interval, so a balance must be struck between the ease of tracking and the number of cells and conditions that can be acquired in a single experiment. High time resolution may also yield very large datasets.
15. An initial experiment should determine the maximal extent of the sample in z. A live sample at the end of the experiment is best, as some colony deformation may occur during fixation. For a micropatterned 2D human gastruloid differentiated for about 2 days, the height of a single colony is generally under 25 μm. For cells in sparse adherent culture, a colony height of 10–15 μm is typical.
16. There is a trade-off between the size and spacing of the z stack and phototoxicity—increasing the number of z slices increases total light exposure to the cells. As discussed in earlier notes, the number of z slices may be balanced with laser intensity,

exposure time, and time interval to avoiding phototoxicity. The needed z resolution depends strongly on context; in densely packed multilayered colonies, a high resolution is needed to accurately segment each cell, which is a prerequisite for successful cell tracking. In sparsely labeled monolayers of cells, even a single z slice may be adequate. This also depends on the tracked quantity—for example, if it is important to accurately separate nucleus and cytoplasm, then finer spacing will ensure that each cell is captured in the correct focal plane (s), which will improve downstream analysis.

17. Tiling overlapping fields of view and stitching the resulting images into a larger single field of view are useful to reduce the number of cells that cannot be tracked because they leave the field of view, depending on how motile cells are. For larger stem cell models like 700 μm diameter 2D human gastruloids, such tiling is also necessary to image the entire colony at high resolution.
18. We recommend setting the bottom focal plane slightly below the best focal plane for the lowest cells in each field of view because cell morphology may change over the course of differentiation. In particular, if cells differentiate to a more squamous morphology, they will flatten and the best focal plane for those cells will move down slightly.
19. Autofocus must be toggled on immediately before adding each new position in the Andor Fusion software, as it automatically disengages when the focus is manually adjusted.
20. The .bkp protocol file from Andor Fusion and the .xml multi-position file from NIS elements save the positions list, which can be loaded again to image the plate in the same fields of view after fixation. Andor Fusion can also import protocol information from a loaded image file, but it is easier to have a small protocol file to import than to keep an accessible copy of large time-lapse image files accessible on the microscope computer.
21. To measure the quantity of interest over the full course of differentiation, the inductive signal(s) for differentiation should not be added until the acquisition has already begun. A simple way to accomplish this with minimal disruption of the plate during acquisition is to change medium in each well to 2/3 of the standard before placing the plate on the microscope, and then after several imaging intervals, add 1/3 volume of treatment at 3× concentration. For example, micropatterned 2D human gastruloids are generally treated with 50 ng/mL BMP4 in base media to induce differentiation and patterning. In a micropatterned 18-well Ibidi u-Slide chamber slide, each well can be changed to 100 uL base media before setting up the live-cell acquisition and then 50 uL of treatment consisting of

150 ng/mL BMP4 can be added to each well, resulting in a volume of 150 uL media with 50 ng/mL BMP4 in each well. Alternatively, medium can be pipetted out and replaced with a full volume of 1× treatment; this may be the only option if differentiation is carried out in a different base media than cell maintenance or if some component, e.g., ROCK inhibitor, is removed at the time of treatment to induce differentiation. However, this poses a larger risk of moving the plate through contact with the pipette tip during image acquisition. Lifting the lid of the dish may also lead to unwanted movement, although a tape handle can alleviate this issue, as mentioned in **Note 11**.

22. Our approach to linking live-cell dynamics to fixed-cell immunofluorescence measurements relies on the ability to match individual cells from the live-cell time lapse to cells in the immunofluorescence images. If there was any cell movement between the last frame of the live time lapse and when the cells were fixed, it may be difficult to match live and fixed cells in dense cultures like 2D gastruloids with sparse labeling because only some cells are visible in the live data, but every cell is labeled with DAPI after immunostaining. This can be solved by taking one more image of the fluorophores after fixation but before bleaching—these can be matched to the live data as they are still sparsely labeled, and then these cells can be matched to the immunofluorescence images after staining. Bleaching could be omitted, reducing the number of channels that can be stained but enabling imaging of the nuclear marker that was used live together with the immunofluorescence stains to aid in alignment and matching. To improve live-to-fixed linking without compromising the number of markers to be measured with immunofluorescence, this strategy can be combined with iterative immunofluorescence.
23. We suggest storing the sample in PBS−/− with sodium azide added as an antimicrobial agent to prevent sample degradation.
24. We have noticed that for a small number of primary antibodies, the quality of the immunostaining is dramatically reduced on samples that have been bleached compared to those which have not. We recommend testing primary antibodies of interest with and without the bleaching step included to ensure that stain quality is not impaired.
25. Stain quality may be improved by performing the blocking and incubation of the sample with primary antibodies and secondary antibodies on an orbital shaker set to 50 rotations per minute. This is especially recommended in rounds 2 and higher of staining with iterative immunofluorescence.

26. When setting laser intensity and exposure time for immunofluorescence imaging after fixation, phototoxicity is no longer an issue and in general much more intense excitation can be used. Similarly, a more finely sampled z stack may result in more precise quantification of immunofluorescence data. However, with very bright settings and many z slices, photobleaching of the sample may be a concern. It is best to have positive and negative controls for the expression of each marker stained to use when setting the laser intensity and exposure time. Try to use settings that result in intensity values in a wide range of the dynamic range of the camera sensor without saturating it. For example, for a 16-bit sensor, intensity values should span a wide range between 0 and 65,535 for most precise quantification.
27. We have found that the inclusion of Triton-X 100 in the blocking buffer in every round of immunostaining often eventually results in deformation of the cells, with the effect becoming noticeable around round 7–8 of immunostaining and imaging.
28. For simplicity we use only Cellpose for segmentation here. In the past, we found that Cellpose performed very well at segmenting the nuclei that it does identify but had a high false negative rate. To fill in the gaps in the Cellpose segmentation, we took the approach of segmenting nuclei with both Cellpose and with the Pixel Classification Workflow in Ilastik [16], and using the binary Ilastik mask to fill in the gaps missed by Cellpose. Because Ilastik only segments foreground vs background rather than detecting individual nuclei, we use an algorithm for approximate convex decomposition of polygons to split connected components in the Ilastik segmentation into discrete nuclei. In the GitHub repository associated with this publication, we include code to use Cellpose, Ilastik, or both together in the "combineSegmentations.m" script if adding Ilastik is desired. Alternatively, the segmentation results using Cellpose may be substantially improved without the need for additional segmentation software by using Cellpose 2.0 [17] and training a custom model instead of using a pretrained generalist model.
29. If the machine used for image processing has a discrete NVIDIA graphics card and the instructions at https://github.com/MouseLand/cellpose were followed to ensure that the correct GPU drivers are installed and the cellpose environment is set up with the GPU version of torch, using the GPU for processing should result in a substantial speedup compared to running Cellpose on the CPU. Otherwise, the CPU may be used by setting gpu = False in the model definition.

30. The approach we present should be compatible with any accurate single-cell tracking algorithm. We provide instructions for using TrackMate as it is a popular algorithm with integration into ImageJ, but code to use our lab's own custom single-cell tracking algorithm is also included with this chapter. Our tracking approach, like TrackMate, is based on an older paper describing an algorithm for particle tracking [18] and is described in Supplemental Note 1 of [5]. To run our tracking code, open the same "singleCellQuantification.m" script used to process the live-cell image data, run the "load data if processing has already been done" block, move on to the "do automated single-cell tracking & save results" section, set the max_linking_distance, max_gap_distance, and max_gap_time variables, and run the section. This will save a directed graph for each field of view containing the tracking links.
31. To import a sequence of images, click File- > Import- > Image Sequence, select the folder containing images to open, and in the "Filter" box, type in a file pattern to identify images only for the channel and position of interest. For example, navigate to the MIP folder and enter "p0000_w0000" to select all maximal intensity projections for the first channel of the first position. If images are very large or there are many time points, checking "Use virtual stack" can reduce the memory requirements to load the data. Alternatively drag images of interest from the file explorer to Fiji and use Images - > Stack - > Images to Stack to make a stack of images.
32. The parameters to set are the "Max distance" for frame to frame linking, gap closing, and splitting, and "Max frame gap" for gap closing. For our test data, we recommend starting with a max distance for frame-frame linking of ~22 pixels (~13 μm), max distance for gap closing and splitting of ~35 pixels (~21 μm), and max frame gap of 2–5 frames. When visually inspecting or correcting tracking results, seeing many erroneous links in relatively dense cell populations may indicate that a lower "Max distance" parameter is more appropriate, while seeing that many tracks are lost because fast-moving cells are not linked between adjacent frames may indicate that a higher "Max distance" is needed.
33. While Trackmate has its own manual correction module called TrackScheme, it is not very scalable and is difficult to use. We recommend deepcell-label, an open-source application with its source code available at https://github.com/vanvalenlab/deepcell-label.git and a web application at https://label.deepcell.org/, for track correction. Note that, like TrackScheme, our code in MATLAB does not scale well to very large datasets and deepcell-label may be preferred for these datasets. The deepcell-label application contains several

useful features including the ability to add/delete/edit existing cell masks, break/combine cell tracks, define cell division, and annotate cells. However, the deepcell-label application does not currently support 3D tracking correction, so to manually correct 3D tracking results, TrackScheme should still be used. The steps below describe how to use deepcell-label for track correction.

(a) Export the tracking result from Trackmate as a label image. Combine the raw image of nuclei (TYXC) and its labeled image (TYXC) into a single npz file using the savez function implemented in Python's numpy package. If those images do not contain the fourth axis, a dummy fourth axis must be added using the expand_dims function from numpy.

(b) The web version of the deepcell-label application limits the file size to 128 MB. If the size of the npz file exceeds this limit, the image files must either be downsized (e.g., 16 bit to 8 bit) or the application locally deployed, after which the value of the DROPZONE_MAX_FILE_SIZE field in config.py should be manually increased. Follow the "runs locally" instruction in the GitHub repository for local deployment. For machines without properly configured development environments, it is recommended to deploy the application using the official Docker image.

(c) Drag and drop the npz file to the deepcell-label application. Dropdown instructions are available while working on a project in the application.

(d) Once finished, download the project as a zip file and parse the JSON files containing tracks and division events to other desired formats.

34. Track validation is done based on cells in the final frame because we are interested in linking live cells to fixed cells in the immunofluorescence data. For a given cell in the final frame, move backward through the time lapse to determine manually if links into the track associated with that cell are all accurate. If they are, press "v" and then select the index of that cell from the list dialog that opens to validate the track associated with that cell (recall that indices of individual cells are displayed on the final frame of the time lapse).

35. Note that when adding links between cells, the link will sometimes not be added. This is because there are some constraints on the tracking digraph—there can be no more than 2 links out of a given cell, and there can be no more than 1 link into any cell. If one of these criteria is not met, remove extraneous links before adding the new link.

36. In general, it is better to choose "remove link before" because each cell can have no more than 1 link into it, whereas there could be ambiguity if "remove link after" is chosen for a cell with more than one link out of it due to track splitting. After removing or adding a link, the tracks are recomputed and redisplayed. For small datasets this is very fast, but it may take several seconds for a data set with many hundreds or thousands of cells per frame or with many hundreds of time points.
37. To quantify multiple rounds of immunofluorescence, we only segment the first round, then read and quantify images from each round based on that segmentation after aligning subsequent rounds to the first round based on the DAPI stain. Sometimes subsequent rounds of staining result in shrinking or deformation of the sample, particularly in dense samples with layered cells. If the deformation is too great to determine a cell-cell correspondence between rounds, another experiment may be necessary. If, however, the data can be aligned closely enough to determine which cells in the later round correspond to cells in the first round, but not well enough to read out the intensity in the same mask, then an alternative to reading out every round using a single segmentation is to segment each round independently and link cells between rounds the same way fixed cells are linked to live cells.
38. Because some time passes between the final frame and imaging and fixation of the sample, there will likely be some cell movement between the final live-cell image and the immunofluorescence image. The maxDist parameter accounts for that movement and small tissue deformation and can be adjusted based on a preview of the link from live to fixed data.

References

1. Lin J-R, Fallahi-Sichani M, Sorger PK (2015) Highly multiplexed imaging of single cells using a high-throughput cyclic immunofluorescence method. Nat Commun 6:8390
2. Gut G, Herrmann MD, Pelkmans L (2018) Multiplexed protein maps link subcellular organization to cellular states. Science 361: eaar7042
3. Kramer B, Castillo J, Pelkmans L, et al (2023) Iterative indirect immunofluorescence imaging (4i) on adherent cells and tissue sections. BIO-Protoc 13
4. Freeburne E, Teague S, Khan H et al (2023) Spatial single cell analysis of proteins in 2D human gastruloids using iterative immunofluorescence. Curr Protoc 3:e915
5. Teague S, Primavera G, Chen B et al (2024) Time-integrated BMP signaling determines fate in a stem cell model for early human development. Nat Commun 15:1471
6. Valcourt JR, Huang RE, Kundu S et al (2021) Modulating mesendoderm competence during human germ layer differentiation. Cell Rep 37: 109990
7. Wolff SC, Kedziora KM, Dumitru R et al (2018) Inheritance of OCT 4 predetermines fate choice in human embryonic stem cells. Mol Syst Biol 14:e8140
8. Stringer C, Wang T, Michaelos M et al (2021) Cellpose: a generalist algorithm for cellular segmentation. Nat Methods 18:100–106
9. Schindelin J, Arganda-Carreras I, Frise E et al (2012) Fiji: an open-source platform for biological-image analysis. Nat Methods 9: 676–682

10. Tinevez J-Y, Perry N, Schindelin J et al (2017) TrackMate: an open and extensible platform for single-particle tracking. Methods 115:80–90
11. Ershov D, Phan M-S, Pylvänäinen JW et al (2022) TrackMate 7: integrating state-of-the-art segmentation algorithms into tracking pipelines. Nat Methods 19:829–832
12. Tewary M, Ostblom J, Prochazka L et al (2017) A stepwise model of Reaction-Diffusion and Positional-Information governs self-organized human peri-gastrulation-like patterning. Development dev.149658
13. Karzbrun E, Khankhel AH, Megale HC et al (2021) Human neural tube morphogenesis in vitro by geometric constraints. Nature 599: 268–272
14. Nie Y, Walsh P, Clarke DL et al (2014) Scalable passaging of adherent human pluripotent stem cells. PLoS One 9:e88012
15. Heemskerk I, Burt K, Miller M et al (2019) Rapid changes in morphogen concentration control self-organized patterning in human embryonic stem cells. elife 8:e40526
16. Berg S, Kutra D, Kroeger T et al (2019) Ilastik: interactive machine learning for (bio)image analysis. Nat Methods 16:1226–1232
17. Pachitariu M, Stringer C (2022) Cellpose 2.0: how to train your own model. Nat Methods 19:1634–1641
18. Jaqaman K, Loerke D, Mettlen M et al (2008) Robust single-particle tracking in live-cell time-lapse sequences. Nat Methods 5:695–702

Methods in Molecular Biology (2026) 3048: 211–238
DOI 10.1007/7651_2025_637

Published online: 22 May 2025

Application of CRISPR-Based Epigenome Editing Tools for Engineering Programmable Embryo Models

Gerrald A. Lodewijk, Sayaka Kozuki, Carly Guiltinan, Benjamin R. Topacio, and S. Ali Shariati

Abstract

Stem cell-based embryo models (SEMs) have the potential to transform our understanding of early human embryogenesis. A critical step in engineering SEMs is the generation of the major cell types that compose preimplantation embryos including two primary extraembryonic lineages: (i) trophoblast cells, which are crucial for implantation and the establishment of maternal-fetal exchange, and (ii) hypoblast cells, which contribute to yolk sac formation. In addition, both cell types provide key signaling cues necessary for embryonic development. CRISPR-based epigenome editors are programmable devices that allow for efficient and precise activation (CRISPRa) or repression (CRISPRi) of cell fate-determining factors by modulating endogenous regulatory elements. Here, we present a step-by-step method to implement CRISPRa for controlling cell fate in embryonic stem cells based on our work in generation of CRISPR-programmed mouse embryo models.

Key words CRISPR activation, Embryo models, Embryoids, Epigenome, Mammalian embryo, Pluripotent stem cells

1 Introduction

The mammalian embryo starts as a unicellular, totipotent zygote that undergoes successive rounds of division coupled with cell fate decisions to give rise to all cell types of the body and extraembryonic structures. The first cell fate decision is the segregation of the inner cell mass (ICM) from the outer trophoblast cells, which will become the placenta. Soon after, the ICM generates the hypoblast, which will form the yolk sac, and the epiblast, which remains as the pluripotent entity that will become the fetus. Together, these three cell types form a spatially ordered, hollow spherical structure known as a blastocyst that goes on to produce complex structures that comprise and support a pregnancy. The emergence of distinct cellular lineages from a common population is driven by external signals from the surrounding environment (i.e., extrinsic), which

Gerrald Lodewijk and Sayaka Kozuki contributed equally.

cells can selectively respond to by changing the epigenetic status of chromatin (i.e., intrinsic) to induce transcriptional programs required for differentiation.

Recent advances have harnessed the capacity of human embryonic stem cells (ESCs) to self-organize into human embryo-like structures [1–25]. Stem cell-based embryo models (SEMs) represent a promising alternative to trace features of embryogenesis that are otherwise inaccessible for direct study, especially in humans. Most current methods for the generation of human SEMs rely on some or all of the following: (i) stochastic differentiation of ESCs, (ii) biochemical optimization of media with extrinsic cell-specific morphogens and/or kinase inhibitors, and (iii) exogenous expression of transcription factors. Although representing a major technical breakthrough, the first-generation of human SEMs often have limited programmability capacity and may show variability in cellular compositions. One potential solution to these challenges is the use of programmable devices to drive cell fate decisions with greater precision and control [26–29]. CRISPR activation (CRISPRa) is an epigenome editing approach that allows for intrinsic control of cell fate by activating specific regulatory elements of cell-fate determining factors to initiate differentiation [30, 31]. CRISPRa requires only two components: deactivated Cas9 (dCas9) fused to a transcriptional activator, which binds specifically to target DNA without inducing double strand break, and a single guide RNA (sgRNA), which is complementary to the promoter of a gene of interest. These factors cooperate to drive expression of a specific gene by replicating natural chromatin changes that occur during cell fate decisions.

Previous studies have revealed that overexpression of transcription factors can differentiate ESCs into extraembryonic lineages [16, 32–38]. Recently, we have shown that CRISPRa can be used to efficiently activate cell fate-determining transcription factors in mouse ESCs to control differentiation toward the extraembryonic lineages. We have shown that the resulting cell types readily self-organize into pre-gastrulation embryo models with consistent cellular composition and spatial organization [20]. Here, we describe methods for application of CRISPRa to induce extraembryonic endoderm fate, highlighting the potential of CRISPR-based devices to engineer programmable embryo models. Collectively, this progress will contribute to the establishment of programmable systems that faithfully replicates the intrinsic epigenetic signals that drive cell fate determination during human embryogenesis while allowing for precise lineage-specific gene expression manipulation.

An overview of a typical steps in a CRISPRa experiment is shown in Fig. 1. Throughout our chapter, we have separated the sections where methodologies are different between mouse and human ESCs.

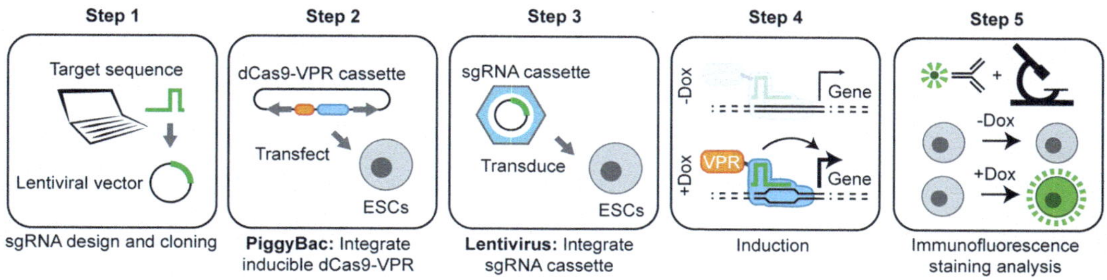

Fig. 1 Overview of steps to generate and validate inducible epigenome CRISPR systems in embryonic stem cells (ESCs)

2 Materials

2.1 Cell Lines

1. Mouse embryonic stem cells (mouse ESCs) (v6.5, RRID: CVCL_C865).
2. Human embryonic stem cells (human ESCs) (H9, RRID: CVCL_9773).
3. HEK293T cells (RRID:CVCL_0063).

2.2 Cell Culture Reagents for Mouse ESCs

1. ESGRO 2i medium (Millipore SF016-200).
2. Gelatin (0.1% solution in 1× PBS or MQ H_2O) (Sigma #G1890).
3. DMEM/F12 GlutaMax (ThermoFisher Scientific #10565018).
4. DMEM GlutaMax (ThermoFisher Scientific #10566016).
5. ES cell qualified FBS (Millipore #ES-009-B).
6. 100 mM Sodium pyruvate (ThermoFisher Scientific #11360070).
7. 100× MEM Non-essential amino acids (ThermoFisher Scientific #11140050).
8. 10,000 u/mL Pen-Strep (ThermoFisher Scientific #15140122).
9. 55 mM 2-Mercaptoethanol (ThermoFisher Scientific #21985023).
10. 1× PBS (ThermoFisher Scientific #10010023).
11. Accutase (Innovative Cell Technologies, #AT104).
12. Doxycycline (1 mg/mL solution) (MilliporeSigma #D5207-1G).
13. Hygromycin B (50 mg/mL solution) (ThermoFisher Scientific #10687010).
14. Puromycin (1 mg/mL solution) (STEMCELL Technologies #73342).

2.2.1 Standard Mouse ESC Differentiation Medium

1. DMEM GlutaMax.
2. 15% ES cell qualified FBS.

3. 1 mM Sodium pyruvate.
4. 1× Non-essential amino acids.
5. 100 u/mL Pen-Strep.
6. 0.1 mM 2-Mercaptoethanol.
7. Optional: 0.5 μg/mL Doxycycline (for induction of dCas9 transgenes).

2.3 Cell Culture Reagents for Human ESCs

1. mTeSR1 medium (STEMCELL Technologies #85850).
2. Geltrex solution (ThermoFisher Scientific #A1569601).
3. DMEM/F12 GlutaMax (ThermoFisher Scientific #10565018).
4. KnockOut Serum replacement (ThermoFisher Scientific #10828028).
5. 100 mM Sodium pyruvate (ThermoFisher Scientific #11360070).
6. 100× MEM Non-essential amino acids (ThermoFisher Scientific #11140050).
7. 10,000 u/mL Pen-Strep (ThermoFisher Scientific #15140122).
8. 55 mM 2-Mercaptoethanol (ThermoFisher Scientific #21985023).
9. 1× PBS (ThermoFisher Scientific #10010023).
10. Accutase (Innovative Cell Technologies, #AT104).
11. ROCK inhibitor Y-27632 (20 mM solution) (STEMCELL Technologies #72304).
12. Doxycycline (1 mg/mL solution) (MilliporeSigma #D5207-1G).
13. Hygromycin B (50 mg/mL solution) (ThermoFisher Scientific #10687010).
14. Puromycin (1 mg/mL solution) (STEMCELL Technologies #73342).
15. PEG Virus Precipitation Kit (ApexBio Technology #10187-096).
16. Polybrene (10 mg/mL solution; MilliporeSigma #TR-1003-G).

2.3.1 Standard Human ESC Differentiation Medium

1. DMEM/F12 GlutaMax.
2. 20% KOSR.
3. 1 mM Sodium pyruvate.
4. 1× Non-essential amino acids.
5. 100 u/mL Pen-Strep.

6. 0.1 mM 2-Mercaptoethanol.
7. Optional: 0.5 μg/mL Doxycycline (for induction of dCas9 transgenes).

2.4 Cell Culture Reagents for HEK293T Cells

1. DMEM GlutaMax (ThermoFisher Scientific #10566016).
2. HIFBS Premium Plus (ThermoFisher Scientific #A5669801).
3. Trypsin–EDTA (ThermoFisher Scientific #25200056).
4. 1× PBS (ThermoFisher Scientific #10010023).
5. 0.45 μm syringe filters (PES or PVDF) (Millipore Sigma, #SLHPR33RS, #SLHVR33RS).

2.4.1 HEK293T Cell Medium

1. DMEM GlutaMax.
2. 10% HIFBS Premium Plus.
3. 100 u/mL Pen-Strep.

2.5 Plasmids

1. PB-TRE3G-dCas9-VPR for CRISPRa experiments (Addgene #63800) [39].
2. PB-TRE3G-dCas9-KM for CRISPRi experiments (Addgene #122267) [40].
3. pSLQ1371 for sgRNA expression (Addgene #63710) [41].
4. pSLQ1653 for sgRNA expression in human ESCs (Addgene #63711) [41].
5. PiggyBac Transposase (System Biosciences #PB210PA-1)
6. pUC19 (Addgene #50005) [42].
7. DR8.74 for lentiviral production (Addgene #22036)
8. pVSVG for lentiviral production (Addgene #8454) [43].

2.6 Transfection Materials

1. Opti-MEM (ThermoFisher Scientific #31985062).
2. Lipofectamine Stem (ThermoFisher Scientific #STEM00003).
3. TurboFect (ThermoFisher Scientific #R0531).

2.7 sgRNA Cloning Materials

1. Thermocycler PCR machine (e.g., Bio-Rad #1861096 or similar).
2. Spectrophotometer (e.g., ThermoFisher Scientific #NDLPLUSGL or similar).
3. Heat block for microcentrifuge tubes.
4. Temperature controlled water bath.
5. Phusion High-Fidelity PCR kit (ThermoFisher Scientific #F553L).
6. DNA Clean & Concentrator 5 kit (Zymo Research #D4014).
7. Zymoclean Gel DNA Recovery kit (Zymo Research #D4008).

8. ZymoPURE Plasmid Miniprep kit (Zymo Research #D4210).
9. BstXI enzyme (New England Biolabs #R0113S).
10. XhoI enzyme (New England Biolabs #R0146S).
11. DpnI enzyme (New England Biolabs #R0176S).
12. NEBuffer r3.1 (New England Biolabs #B6003S).
13. 10 mM Tris-HCl, 1 mM EDTA solution, pH 8.0 – $T_{10}E_1$ (ThermoFisher Scientific #AM9858).
14. 1 mM Tris-HCl, 0.1 mM EDTA solution, pH 8.0 – $T_1E_{0.1}$ (diluted $T_{10}E_1$ 1:10 in H_2O).
15. DNA Agarose Gel Electrophoresis equipment (e.g., Bio-Rad #1640301 or similar).
16. Agarose (Bio-Rad #1613101).
17. TAE buffer (Bio-Rad #1610743).
18. 10,000× GelRed solution (Biotium #41008-500uL).
19. GeneRuler 100 bp DNA Ladder (ThermoFisher Scientific #SM0241).
20. GeneRuler 1 kb DNA Ladder (ThermoFisher Scientific #SM0311).
21. Quick CIP kit (New England Biolabs #M0525S).
22. Rapid DNA Ligation kit (ThermoFisher Scientific #K1423).
23. LB (Fisher Scientific #BP1425-500).
24. Terrific Broth (ThermoFisher Scientific #22711022).
25. Glycerol (Sigma #G2025-500ML).
26. LB-Agar (Fisher Scientific #BP1425-2).
27. Ampicillin (100 mg/mL solution) (Fisher Scientific #BP1760-25).
28. NEB Stable *E. coli* competent cells (New England Biolabs #C3040I).

2.7.1 Bacterial Growth Medium

1. 25 mL of 1× LB.
2. 25 mL of 1× Terrific Broth.
3. 50 μL of 100 mg/mL Ampicillin.

2.7.2 Preparing LB/Agar/Amp Plates

1. 500 mL H_2O.
2. 20 g LB-Agar.
   (a) Autoclave and let cool down to 60 °C.
   (b) Add 500 μL of 100 mg/mL ampicillin solution.
   (c) Mix and divide over approximately 20 standard 100 mm petri dishes.

(d) Let dishes fully cool down to set at room temperature.
(e) Store at 4 °C for later use if needed.

2.8 Guide RNA Design Guidelines

A typical SpCas9 single-guide RNA consists of a ~20-nucleotide long CRISPR RNA (crRNA) that specifies the protospacer sequence for the target DNA sequence. This is fused to a trans-activating CRISPR RNA (tracrRNA), which mediates sgRNA binding to SpCas9 protein. The full sgRNA sequence is expressed from a type III RNA-polymerase promoter (e.g., U6 promoter). Guide RNA directs CRISPR-based devices to their complementary protospacer genomic site for cleavage or epigenome editing. For working with SpCas9 and sgRNAs expressed from the U6 promoter, follow these general guidelines:

1. The protospacer sequence has to be 20–24 bp long (most design tools use 20 bp protospacer length).
2. The protospacer sequence should start with a G (if using U6 promoter).
3. The protospacer sequence should not contain the PAM sequence (NGG).
4. The protospacer sequence should not contain more than 3 Us in a stretch (if using U6 promoter).

Several online tools, such as CRISPOR [44] or CRISPick [45, 46], are available to design sgRNAs:

1. Note that CRISPick has the option to specify the tracrRNA sequence, and pSLQ1371 and pSLQ1653 use the Chen et al. 2013 sequence. It was found that the tracrRNA sequence has an effect on sgRNA protospacer efficiency; including this into the sgRNA prediction model leads to slightly better efficiency predictions when using RS3 scores to pick gRNA protospacer sequences [47].
2. It is recommended to test 3–4 sgRNAs per target region and validate the sgRNAs with optimal activity for further experiments.

For targeting promoter region using CRISPRa:

1. Target the region 50–600 bp upstream of the gene TSS.
2. Use chromatin accessibility such as ATAC-Seq to target regulatory regions that are either closer or further away from the TSS and show increased accessibility in your desired cell type.

For targeting promoter region using CRISPRi:

1. Target the region 400 bp upstream and 200 bp downstream of the gene TSS.

Cloning the sgRNA into pSLQ1371 or pSLQ1653 plasmids:

1. Design and order forward primer for each sgRNA:
   (a) The forward primer follows this template (60 nt): GAGAA*CCACCTTGTTG*NNNNNNNNNNNNNNNNNNNNGTTTAAGAGCTATGCTGGAAACAG.
      (i) The Italic part is the BstXI site.
      (ii) The stretch of N's is the protospacer for the target of interest.
      (iii) Order the universal reverse primer for pSLQ1371 and pSLQ1653:
      (iv) GCCGCCTAATGGATCCTAGTA.
2. Order the Sanger sequencing primer for pSLQ1371 and pSLQ1653:
   (a) GAGATCCAGTTTGGTTAGTACCGGG.

3 Methods

3.1 Guide RNA Cloning

1. Use pSLQ1371 or pSLQ1653 plasmid as a template to amplify the new sgRNA using the unique forward primer and the universal reverse primer (primer sequences can be found in the sgRNA design section).
   (a) pSLQ1371 is recommended for working with mouse ESC lines.
   (b) pSLQ1653 is recommended for working with human ESC lines.
2. Create a mix for each PCR reaction using a Phusion High-Fidelity PCR kit: (*see* **Note 1**).

Components	Amount (μL)
MQ H_2O	15.75
5× HF buffer	5
dNTPs	0.5
10 μM forward primer	1.25
10 μM reverse primer	1.25
1 ng/μL template plasmid	1
Phusion DNA polymerase	0.25
Total	*25*

3. Set up the thermocycler with the following cycling conditions:

Step	Temp (°C)	Time	# of cycles
1	98	3 min	1
2 3 4	98 59 72	15 s 20 s 30 s	32
5	72	5 min	1
6	8	hold	1

4. After the PCR is completed, analyze amplified DNA by agarose gel electrophoresis.
   - (a) For PCR products, load a 5 μL sample to verify the amplification worked.
   - (b) The agarose gel is prepared including 1× GelRed to visualize DNA bands in the gel using UV light.
   - (c) The expected amplicon size is 154 bp.
   - (d) Load a sample of GeneRuler 100 bp for accurate assessment of amplicon size.
5. Purify the successful PCR samples using the Zymo DNA Clean & Concentrator 5 kit (*see* **Note 2**).
   - (a) Pre-warm elution buffer: $T_1E_{0.1}$ at 60 °C.
   - (b) Elute twice in 10 μL $T_1E_{0.1}$ (20 μL in total eluate volume).
   - (c) When adding $T_1E_{0.1}$, incubate for 1 min before centrifuging.
6. Digest the PCR product using BstXI, XhoI, and DpnI in a 50 μL reaction (*see* **Notes 3–5**).

Components	Amount (μL)
Purified PCR product	20
Buffer 3.1 (10×)	5
MQ H_2O	23.75
BstXI enzyme	0.5
XhoI enzyme	0.5
DpnI enzyme	0.25
Total	*50*

Incubate at 37 °C for 1 h, followed by 80 °C for 20 min.

7. Clean up the PCR amplicon digestion reaction using the Zymo DNA Clean and Concentrator 5 kit.
   (a) Pre-warm elution buffer: $T_1E_{0.1}$ at 60 °C.
   (b) Elute twice in 10 μL $T_1E_{0.1}$ (20 μL in total eluate volume).
   (c) When adding $T_1E_{0.1}$, incubate for 1 min before centrifuging.
   (d) Store at −20 °C, or at 4 °C for short term (<1 day).
8. Digest the pSLQ1371 or pSLQ1653 plasmid with BstXI and XhoI.

Components	Amount (μL)
MQ H_2O	33
10× buffer 3.1	5
300 ng/μL pSLQ1371 or pSLQ1653	10
BstXI enzyme	1
XhoI enzyme	1
Total	*50*

   Incubate at 37 °C for 1 h, followed by 80 °C for 20 min.
9. After plasmid digestion is completed, analyze samples by agarose gel electrophoresis.
   (a) For plasmid digestion, load the complete sample (divide over 2–3 wells) for gel extraction.
   (b) The linearized and double digested plasmid runs at about 8,200 bp (pSLQ1371) or about 9,500 bp (pSLQ1653).
   (c) Add a sample of GeneRuler 1 kb to assess the approximate length.
   (d) Visualize the DNA bands in the agarose gel (containing 1× Gel Red) on a UV tray (wear appropriate personal protective equipment).
   (e) Use a scalpel or a small straight edge spatula to cut out the plasmid DNA from the gel and transfer to a microcentrifuge tube.
10. If the plasmid digestion looks complete, perform gel extraction according to the Zymo Gel DNA Recovery kit protocol.
    (a) Pre-warm elution buffer: $T_1E_{0.1}$ at 60 °C.
    (b) Elute twice in 10 μL $T_1E_{0.1}$ (20 μL in total eluate volume).

(c) When adding $T_1E_{0.1}$, incubate for 1 min before centrifuging.

11. Treat purified plasmid with phosphatase to remove 5′ phosphate groups (*see* Note 6).

Components	Amount (μL)
Purified digested plasmid	20
MQ H_2O	24
10× Cutsmart buffer	5
Quick CIP enzyme	1
Total	*50*

Incubate for 15 min at 37 °C, followed by 2 min at 80 °C.

12. Clean up the digested, dephosphorylated plasmid backbone DNA using the Zymo DNA Clean and Concentrator 5 kit.
    (a) Pre-warm elution buffer: $T_1E_{0.1}$ at 60 °C.
    (b) Elute twice in 10 μL T1E0.1 (20 μL in total eluate volume).
    (c) When adding T1E0.1, incubate for 1 min before centrifuging.
    (d) Store at −20 °C or at 4 °C for short term (<1 day).
13. Set up ligation reactions, and include a no-insert (no PCR digest added, use H_2O instead) control reaction to assess background ligation rate (*see* **Notes** 7 and **8**).

Components	Amount (μL)
14 ng/μL PCR digest	1
50 ng/μL plasmid digestion	1
5× rapid ligation buffer	4
MQ H_2O	13
T4 DNA ligase	1
Total	*20*

Incubate the ligation mix for 10 min at 22 °C or room temperature, then keep on ice.

14. Thaw NEB Stable *E. coli* competent cells per ligation reaction in an ice box and add 50 μL to microcentrifuge tubes (kept in ice) for each ligation reaction.

15. Mix 2.5 μL of the ligation reaction into 50 μL NEB Stable *E. coli* competent cells for the transformation.
    (a) Mix gently.
    (b) Keep the remaining ligation mix at −20 °C.
    (c) Incubate in ice for 30 min.
    (d) Heat shock for 30 s in a water bath set to 42 °C.
    (e) Incubate in ice for 2 min.
    (f) For recovery, add 200 μL NEBStable outgrowth medium, or bacterial growth medium (without Amp), and incubate at 30 °C, shaking 250 rpm for 1 h.
    (g) Plate out 125 μL per transformation on an LB/Agar/Amp plate for each sample.
    (h) Keep the remaining suspension at 4 °C if needed.
    (i) Incubate plates for 20–24 h at 30 °C.
16. Assess growth on LB/Agar/Amp plates and, if successful, pick individual colonies and add them to separate 14 mL round bottom tubes with 3 mL of bacterial growth medium (*see* **Note 9**).
    (a) Pick 2–4 colonies per plate.
17. Grow 20–24 h at 30 °C, shaking at 250 rpm.
18. Use ~2 mL of bacteria suspension to extract plasmid DNA using the ZymoPURE plasmid miniprep kit (*see* **Note 10**).
    (a) Pre-warm elution buffer, $T_1E_{0.1}$ at 60 °C.
    (b) Elute twice in 25 μL $T_{10}E_1$ (50 μL in total eluate volume).
    (c) Store at −20 °C or at 4 °C for short term (< 1 day).
    (d) Remaining bacteria suspension can be used to prepare a glycerol stock.
        (i) Mix 0.5 mL of bacteria suspension with 0.5 mL of 50% glycerol. Add to a cryovial and store at −70 °C to −80 °C.
19. Verify the presence of correct sgRNA sequence via Sanger sequencing, using the pSLQ Sanger sequencing primer.

3.2 Preparing Mouse ESCs for Inducible CRISPRa/CRISPRi Experiments

3.2.1 Maintenance of Mouse ESCs in a Six-Well Plate

1. Add 1.5 mL of 0.1% Gelatin solution to one well of a new six-well plate. Check that solution covers the whole well and spread onto uncovered areas by gently tapping the plate.
    (a) Incubate at least for 30 min in 37 °C incubator.
    (b) Remove Gelatin solution and add 1 mL 1× PBS.
    (c) Remove 1× PBS just before use (when adding culture medium or cells).

2. Remove medium from mouse ESCs that are ready to passage.
3. Wash with 1 mL 1× PBS.
4. Add 0.5 mL Accutase.
5. Incubate 3 min at room temperature, or until cells detach.
6. Add 5 mL DMEM/F12, pipette up and down 2–3 times to break up cell larger aggregates.
7. Transfer the complete volume of cell suspension from 6 to a 15 mL conical tube.
8. Centrifuge for 5 min at 200 × g.
9. Remove supernatant and resuspend the cell pellet in 1 mL DMEM/F12.
10. Remove 1× PBS from Gelatin-coated well in the new six-well plate and add 2 mL of 2i medium.
11. Seed 50,000 cells (approximately 1:20 split) into the new well.
12. Refresh 2i medium daily, cells are ready to passage again after 2–3 days.
    (a) Adjust volumes and cell number accordingly when using multi-well plates or dishes with different surface areas.

3.2.2 Transfection of PiggyBac Plasmids into Mouse ESCs

1. Day 1: Plate ~200,000 mouse ESCs per well in a 0.1% gelatin coated six-well plate.
2. Day 2: Replace 2i medium (optional: without Pen/Strep).
    (a) Prepare transfection mixes:
        (i) Tube 1: (*see* **Note 11**)

Opti-MEM	100 μL
PiggyBac Transposase plasmid	200 ng
PB-TRE3G-dCas9 (VPR or KM)	800 ng
pUC19	1000 ng

        (ii) Tube 2:

Opti-MEM	100 μL
PiggyBac Transposase plasmid	200 ng
PB-TRE3G-dCas9 (VPR or KM)	800 ng
pUC19	1000 ng

   (b) Briefly vortex, spin, and incubate for 5 min.
   (c) Add 100 μL of Tube 2 to Tube 1.
   (d) Briefly vortex, spin, and incubate for 10 min.
   (e) Add 200 μL to the prepared well.
3. Day 3: Replace 2i medium—split cells at high ratio (~250,000 cells per well, or about a 1:4 ratio) if high cell density is observed.
4. Day 4: Start selection by adding fresh 2i medium containing 100 ug/mL Hygromycin.
   (a) Keep Hygromycin B in culture medium until cells in a control well are dead.
      (i) For optimizing transfections, 50 ng of a (non-integrating) GFP expression plasmid can be added to assess transfection efficiency (*see* **Note 12**).
      (ii) Amount and ratio of PiggyBac Transposase and PB plasmid can be optimized to decrease or increase integration efficiency.
      (iii) A suitable control is cells transfected without PiggyBac Transposase to account for any cells that did not have integrated the PB plasmid transgene, but do have temporary resistance from HygR gene expressed from the PB plasmid.
      (iv) After selection is complete, cells can be cultured using routine maintenance procedures described above. Optional: a low concentration of Hygromycin B (e.g., 25 μg/mL) can be added to maintenance media to keep transgenic resistant ESCs (*see* **Note 13**).

3.2.3 Lentivirus Preparation

1. Day 1: Plate 500,000 HEK293T cells per well in a six-well plate.
   (a) Prepare 1 well for each sgRNA plasmid used.
2. Day 2: Replace medium with 2 mL DMEM +10% HIFBS approximately 1 h before the transfection.
   (a) Prepare transfection mixes.
      (i) Tube 1:

Opti-MEM	100 μL
VSVG	400 ng
DR8.74	600 ng
pSLQ1371 or pSLQ1653	1000 ng

(ii) Tube 2:

Opti-MEM	100 μL
TurboFect	8 μL

(iii) Each sgRNA plasmid (pSLQ1371 or pSLQ1653) is prepared in a separate tube. Scale the transfection mix (Tube 2) accordingly for all sgRNA lentivirus preparations (multiple reactions of Tube 1 mix).

(b) Mix by vortexing briefly 3 times and collect in the bottom of the tube by centrifugation shortly.

(c) Add 100 μL of tube 2 into tube 1, repeat the vortex and centrifuge steps.

(d) Incubate for 20 min at room temperature.

(e) Add the mix into the well of HEK239T cells (200 μL into one well of the six-well plate).

(f) After 6–24 h, add 1 mL DMEM +10% HIFBS +300 u/mL Pen-Strep.

3. Day 5: Collect the medium containing lentivirus into a 15 mL tube (*see* **Notes 14** and**15**).

(a) Centrifuge for 5 min at 300 × g.

(b) Filter the supernatant medium using a 0.45 μm filter and collect flow-through in a new 15 mL tube.

(c) The final volume should be about 2.5–3 mL of virus-containing medium.

3.2.4 Lentivirus Transduction of Mouse ESCs

1. Day 1: Plate ~50,000 mouse ESCs per well in a 0.1% gelatin coated six-well plate.
2. Day 2: Remove medium and replace with:

(a) 1 mL 2i medium

(b) 1 mL filtered lentivirus medium

3. Day 3: Replace medium with 2i medium.
4. Day 4: Split cells at high density (e.g., 1:4) into a new 0.1% gelatin-coated six-well plate.

(a) Seed cells in 2i medium containing 0.8 μg / mL Puromycin and keep cells in 2i medium with Puromycin for at least 2 passages (4–6 days) (*see* Note 13).

3.3 Preparing Human ESCs for Inducible CRISPRa/CRISPRi Experiments

3.3.1 Maintenance of Human ESCs in a Six-Well Plate

1. Add 1 mL of Geltrex solution to one well of a new six-well plate. Check that solution covers the whole well and spread onto uncovered areas by gently tapping the plate.
   (a) Incubate for 60 min in a 37 °C incubator, then incubate 60 minutes at room temperature
   (b) Remove Geltrex solution and add 1 mL DMEM/F12.
   (c) Remove DMEM/F12 just before use (when adding culture medium or cells).
2. Remove medium from human ESCs that are ready to passage (>70% confluency).
3. Wash with 1 mL 1× PBS.
4. Add 0.5 mL Accutase.
5. Incubate for 5 min at room temperature, or until colonies detach.
   (a) Plate can be gently tapped to release colonies.
6. Add 5 mL DMEM/F12.
7. Transfer the complete volume of cell suspension from 6 to a 15 mL conical tube.
8. Centrifuge for 5 min at 200 × g.
9. Remove supernatant, leave a small amount (~ 50 μL), and tap the tube to resuspend the pellet.
10. Add 1 mL mTeSR1 + 10 μM Y-27632.
    (a) Aim to keep small cell clusters intact rather than a single cell solution to promote cell survival and adherence.
11. Remove DMEM/F12 from Geltrex-coated well in the new six-well plate and add 2 mL of mTeSR1 + 10 μM Y-27632.
12. Seed about 80–120 μL of cell suspension into the well (1:8–1:12 ratio split).
13. Refresh mTeSR1 medium daily, and cells are ready to passage again after 4–6 days.
    (a) Adjust volumes and cell number accordingly when using multi-well plates or dishes with different surface areas.
    (b) Add Y-27632 only for the first day.

3.3.2 Transfection of PiggyBac Plasmids into Human ESCs

1. Day 1: Plate ~250,000 human ESCs per well in a Geltrex-coated six-well plate in mTeSR1 medium with 10 μM Y-27632.
2. Day 2: Replace mTeSR1 medium (optional: without Pen/-Strep, with 10 μM Y-27632).
   (a) Prepare transfection mixes:
      (i) Tube 1: (*see* **Note 11**).

Opti-MEM	100 μL
PiggyBac Transposase plasmid	200 ng
PB-TRE3G-dCas9 (VPR or KM)	800 ng
pUC19	1000 ng

(ii) Tube 2:

Opti-MEM	100 μL
Lipofectamine Stem	8 μL

(b) Briefly vortex, spin, and incubate for 5 min.

(c) Add 100 μL of Tube 2 to Tube 1.

(d) Briefly vortex, spin, and incubate for 10 min.

(e) Add 200 μL to the prepared well.

3. Day 3: Replace mTeSR1 medium—split cells at high ratio (~250,000 cells / well) if needed.
4. Day 4: Replace mTeSR1 medium containing 5 μg/mL Hygromycin.

   (a) Keep Hygromycin B in culture medium until cells in a control well are dead.

   (i) For optimizing transfections, 50 ng of a (non-integrating) GFP expression plasmid can be added to assess transfection efficiency (*see* **Note 12**).

   (ii) Amount and ratio of PiggyBac Transposase and PB plasmid can be optimized to decrease or increase integration efficiency.

   (iii) A suitable control is cells transfected without PiggyBac Transposase to account for any cells that did not have integrated the PB plasmid transgene, but do have temporary resistance from HygR gene expressed from the PB plasmid.

   (iv) After selection is complete, cells can be cultured using routine maintenance procedures described above. Optional: a low concentration of Hygromycin B (e.g., 1 μg/mL) can be added to maintenance media to keep transgenic resistant ESCs (*see* **Note 13**).

3.3.3 Lentivirus Preparation

1. Day 1: Plate 500,000 HEK293T cells per well in a six-well plate.

   (a) Prepare 1 well for each sgRNA plasmid used.

2. Day 2: Replace medium with 2 mL DMEM +10% HIFBS approximately 1 h before the transfection.

(a) Prepare transfection mixes.

(i) Tube 1:

Opti-MEM	100 μL
VSVG	400 ng
DR8.74	600 ng
pSLQ1371 or pSLQ1653	1000 ng

(ii) Tube 2:

Opti-MEM	100 μL
TurboFect	8 μL

(iii) Each sgRNA plasmid (pSLQ1371 or pSLQ1653) is prepared in a separate tube. Scale the transfection mix (Tube 2) accordingly for all sgRNA lentivirus preparations (multiple reactions of Tube 1 mix).

(b) Mix by vortexing briefly 3 times and collect in the bottom of the tube by centrifugation shortly.

(c) Add 100 μL of Tube 2 into Tube 1, repeat the vortex and centrifuge steps.

(d) Incubate for 20 min at room temperature.

(e) Add the master mix into the well of HEK239T cells (200 μL into one well of the six-well plate).

(f) After 6–24 h, add 1 mL DMEM +10% HIFBS +300 u/mL Pen-Strep.

3. Day 5: Collect the medium containing lentivirus into a 15 mL tube (*see* **Notes 14** and **15**).

(a) Centrifuge for 5 min at 300 × g.

(b) Filter the supernatant medium using a 0.45 μm filter and collect flow-through in a new 15 mL tube.

(c) The final volume should be about 2.5–3 mL of virus-containing medium.

3.3.4 Virus Precipitation and Concentration for Human ESCs Transduction

1. Prepare lentivirus as described above.
2. Centrifuge lentivirus-containing medium at 3200 × g for 15 min at 4 °C to remove cell debris.
3. Collect supernatant and add 5× PEG solution (e.g., 0.75 mL to 3 mL of lentivirus-containing medium). Invert to mix.

(a) 5× PEG stock can also be prepared by mixing in a 1:1 ratio PEG-6000 or PEG-8000 and 1.5 M NaCl. [48]

4. Refrigerate overnight at 4 °C (or up to 2 days at 4 °C).
5. Centrifuge at 3200 × g for 30 min at 4 °C. A white pellet (the virus) should be clearly visible.
6. Remove supernatant by aspiration.
7. Resuspend the lentivirus pellet in 100 μL Virus Resuspension Solution. Alternatively, lentivirus may be resuspended in 1× PBS or growth medium.
8. Aliquot the virus and store at −80 °C.
9. Optional: calculate virus titer [49].

3.3.5 Lentiviral Transduction of Human ESCs

1. Grow human ESCs in wells of a six-well plate until 70–80% confluent.
2. Passage human ESCs as described before using a 1:4 splitting ratio, maintaining small aggregates, in a new Geltrex-coated six-well plate.
3. The next day, feed human ESCs with mTeSR1 with 6 μg/mL of polybrene.
4. Return cells to the incubator for 15 min.
5. Add 20 μL concentrated lentivirus particles to the culture medium and gently mix by slowly swirling the plate.
   (a) To start, a concentrated lentivirus sample of 3 mL medium can be used as described previously.
   (b) To increase transduction rate, larger volumes of lentivirus containing medium can be precipitated instead (e.g., from a 10 cm dish of lentivirus transfected HEK293T cells), or transduction of human ESCs can be done in smaller wells (12- or 24-well plates) with the same amounts of virus.
6. Return cells to the incubator overnight.
7. The next day, repeat **steps 3–6**, but add three times more (60 μL) lentivirus than initially used.
8. Continue regular daily feeding and passage cells if necessary.
9. Start 0.25 μg/mL Puromycin selection 2 days after completing transduction.

3.4 Validating Guide RNA Efficiency Using Immunofluorescence Staining

The following sgRNA protospacer sequence induces *Gata6* expression in mouse ESCs and can be used as a positive control:

Mouse Gata6 sgRNA: GTAGCTCTTATTGTCCGCTA

It is also recommended to use a non-targeting sgRNA as an additional negative control, e.g., with the following protospacer sequence:

NT-sgRNA: GCGAGGTATTCGGCTCCGCG

Additional sgRNAs targeting other genes validated in mouse ESCs can be included if needed [20].

1. Seed ~25,000 cells per well in 24-well plates with 0.1% gelatin coating (for mouse ESCs) or Geltrex coating (for human ESCs).
   (a) Each cell line is seeded in duplicate to perform doxycycline and no doxycycline control treatments.
   (b) Doxycycline is used at a final concentration of 0.5 μg/mL in the culture medium to induce expression of dCas9-VPR or dCas9-KM.
   (c) Cells can be cultured in pluripotency medium (2i or mTeSR).
   (d) More efficient activation of genes may be observed by releasing cells in a differentiation medium (mouse ESC standard differentiation medium or human ESC standard differentiation medium, see materials) (*see* **Note 16**).
2. Replace medium daily, including doxycycline for experiment wells.
   (a) Gene activation of Gata6 is detectable at 24 h of doxycycline treatment.
   (b) Downstream targets, such as Foxa1 or Sox17 should be detectable 48–72 h after doxycycline treatment.
3. Validate induction of the CRISPRa target gene (e.g., Gata6) by immunofluorescence staining (Fig. 2), for example using the following antibodies (*see* **Note 17**):
   (a) Gata6: Cell Signaling Technology #5851S.
   (b) Foxa1: Abcam #ab55178.
   (c) Sox17: R&D Systems #AF1924.
4. Use DAPI to stain all nuclei.
   (a) Calculate the ratio of CRISPRa target (e.g., Gata6)-positive/ DAPI-positive cells in both no doxycycline control and in doxycycline treatment conditions to calculate activation efficiency compared to background (*see* **Notes 18** and **19**).

4 Notes

1. Prepare single-use aliquots of the 5X HF buffer provided in the Phusion PCR kit. The buffer is sensitive to freeze-thawing, which leads to reduced PCR efficiency. Always keep Phusion

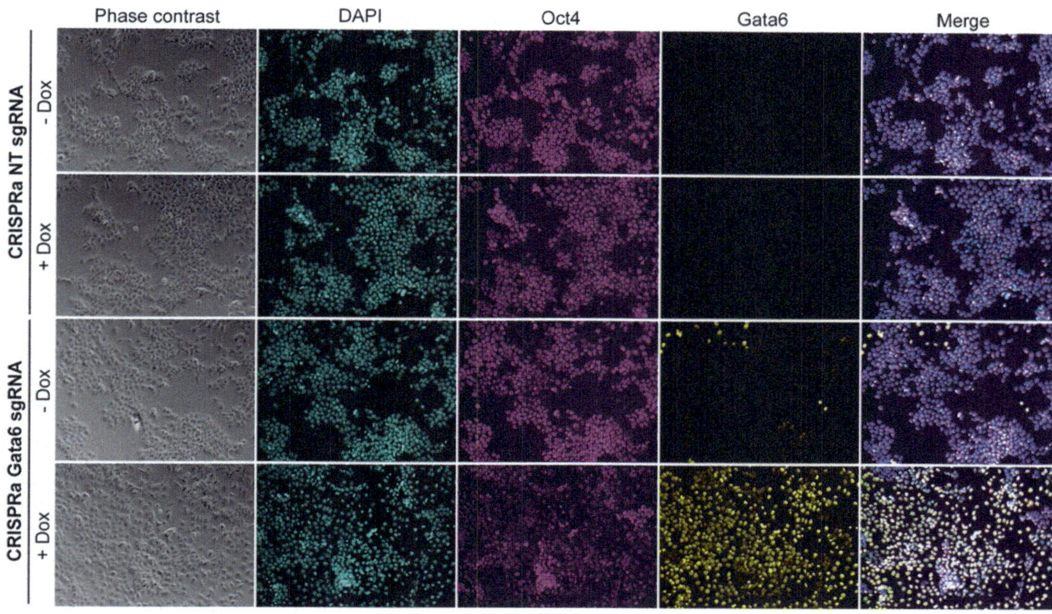

Fig. 2 Example of CRISPRa experiment immunofluorescence staining. Images shown are ESCs with integration of inducible dCas9-VPR and NT sgRNA or Gata6-targeting sgRNA, after doxycycline induction for 3 days in standard differentiation medium [20]

Polymerase tube in a −20 °C mini cooler when taken out of the freezer to maintain optimal activity.

2. NanoDrop measurements can be done after each purification step to ensure the protocols were properly followed.
3. To reduce background in cloning experiments, we recommend adding DpnI when digesting the sgRNA amplicon to fragment the template (plasmid) DNA. DpnI only digests methylated DNA (e.g., plasmid DNA extracted from bacteria). Not adding DpnI may lead to high numbers of background colonies after transformation due to remaining intact plasmid DNA in the sample. DpnI is only added to the sgRNA PCR amplicon digest.
4. Use single restriction enzyme and no enzyme controls when digesting the pSLQ1371 or pSLQ1653 plasmid DNA to ensure both restriction enzymes are working efficiently. Analyze the digested fragments by agarose gel electrophoresis. If one or both enzymes do not cut, they may have to be replaced. Full digestion and dephosphorylation of double cut plasmid DNA are important to minimize background or self-ligation of plasmid DNA and to prevent high number of background colonies during transformation. A no-insert DNA control reaction, that includes the pSLQ1371 or pSLQ1653 digested and dephosporylated DNA backbone, ligation buffer, ligation

enzyme, but without sgRNA DNA insert, is recommended during ligation to measure background ligation.

5. Prepare single-use aliquots of the 10× r3.1 buffer. The buffer may be sensitive to freeze-thawing, which can lead to reduced efficiency. Always keep enzyme tubes in a −20 °C mini cooler when taken out of the freezer to maintain optimal activity.
6. Prepare single-use aliquots of the 10× rCutsmart buffer. The buffers may be sensitive to freeze-thawing, which can lead to reduced efficiency. Always keep enzyme tubes in a −20 °C mini cooler when taken out of the freezer to maintain optimal activity.
7. Prepare single-use aliquots of the 5× Rapid Ligation buffer. The buffer may be sensitive to freeze-thawing, which can lead to reduced efficiency. Always keep enzyme tubes in a −20 °C mini cooler when taken out of the freezer to maintain optimal activity.
8. Purity of the prepared sgRNA amplicon and plasmid DNA is important to maximize ligation efficiency. Measure purity and concentration of final DNA products, e.g., using Nanodrop.
9. If few or no colonies are observed after transformation, check the following:
   (a) sgRNA PCR amplicons and plasmid backbones are properly prepared and show good concentrations and purity.
      (i) Restriction-digested, purified sgRNA PCR amplicons should not be treated with Quick CIP.
   (b) Ligation reagents are working.
   (c) Competent cells have appropriate colony forming capacity (cfu).
   (d) Restriction enzymes are working.
10. In general, use plasmid preparation with high purity and high concentrations. It is suggested to use plasmid purification kits that ensure low endotoxin concentrations (e.g., ZymoPURE kits) to achieve higher transfection efficiency and lower toxicity. When preparing the general plasmids for transfections (all plasmids listed in materials, and optionally pSLQ1731 and pSLQ1653), the ZymoPURE II Plasmid Midiprep Kit (Zymo Research #D4200) or Maxiprep Kit (Zymo Research #D4202) or equivalent endotoxin removing kits are recommended. Transfection efficiency of ESC lines can vary, and working with high purity plasmids can similarly improve transfections; however, using the antibiotic or fluorescent protein selection methods are typically required to select against non-transfected and non-integrated cells.

11. For transfection of PiggyBac constructs to integrate inducible dCas9 cassettes, the pUC19 DNA is solely used to increase the amount of DNA to create more efficient transfection particles. The pUC19 DNA can be replaced with more of the PiggyBac plasmid DNA or increased (reducing PiggyBac plasmid DNA) to alter the integration rate. The number and efficiency of PiggyBac-mediated transgene integration can be controlled by increasing or decreasing the amount of PiggyBac plasmid DNA transfected.
12. Transfection efficiency in various cell lines can be assessed by using a small amount of GFP-expressing plasmid (e.g., 5% of total plasmid transfected) with a constitutively active promoter. Avoid using the CMV promoter in human ESCs as it can be silenced rapidly through epigenetic repression. EF1a, PGK, or CAG promoters are recommended for human ESCs. Transfection conditions can be optimized by changing the concentration of transfected DNA and the amount of Lipofectamine Stem or TurboFect to either increase efficiency or reduce cell toxicity. The transfection duration can also be shortened (6 h) to reduce cell toxicity. We recommend adding ROCK inhibitor (Y-27632, 10 μM) during transfection of human ESCs to promote cell survival. Following this, use only pSLQ1653 for sgRNA expression in human ESCs. pSLQ1371 is suitable for use in mouse ESCs or other cells that efficiently use the CMV promoter.
13. It is recommended to use kill curves to assess the optimal Hygromycin B and Puromycin selection concentrations, especially when using other mouse ESC or human ESC lines than described here (mouse V6.5 ESCs and human H9 ESCs). A kill curve is done by using a range of selection antibiotic concentrations to find the optimal concentration for a particular cell line or condition. Too high concentrations may also kill resistant cells, and too low concentration may result in incomplete selection. Choose the lowest concentration that kills non-resistant cells in 2–3 days based on the kill curve results.
14. Transfection efficiency for HEK293T cells should be between 60 and 90% for efficient lentivirus generation. If observing low transfection efficiency, start with a healthy stock of HEK293T cells.
15. pSLQ1371 and pSLQ1653 both co-express mCherry and Puromycin resistance from the EF1a promoter. Transfection efficiency of HEK293T cells for lentivirus generation can be assessed using mCherry signal. Likewise, infection of mouse ESCs and human ESCs can be validated by calculating the percentage of cells expressing mCherry using flow cytometry or microscopy.

16. It is recommended to initially test the inducible CRISPRa-sgRNA system in both a pluripotency medium (2i or mTeSR1) and standard differentiation medium. Medium conditions can influence the activation of CRISPRa targets, e.g., by maintaining high levels of pluripotency factors that may compete with induced differentiation transcription factors.
17. Immunofluorescence staining to validate upregulation of target genes is preferred as it provides single-cell resolution via imaging to calculate the percentage of responding cells. If no working antibody is available for your protein of interest, efficiency of the CRISPRa can be tested using RT-qPCR to assess upregulation of the CRISPRa target at the transcript level.
18. Expression of activated cell-fate-determining factors may be observed in the absence of doxycycline due to leakiness of the tetracycline response element in a subset of cells [50]. If persistent background expression is observed of a target gene after several passages, we recommend the following steps.
    (a) Use tetracycline-free serum in the culture medium to minimize basal activity of tet-responsive elements of inducible CRISPR systems, when using serum-based media in experiments.
    (b) Select for lines with minimal or no background expression via clonal expansion.
19. If a low percentage of cells activate the target gene, it is recommended to ensure that cells have both stable integrations of the dCas9-VPR cassette and sgRNA cassette by repeating Hygromycin B and/or Puromycin treatment for at least 3 days. For the sgRNA cassette, mCherry signal can also be measured to calculate the percentage of sgRNA-expressing cells. Next, an additional set of 3–4 gRNAs can be designed and tested. Alternatively, a clonal expansion of cells can be done to isolate responding cells and continue experiments with a high efficiency clone. Response efficiency between cells may vary due to random differences in integration sites in the genome. Various available CRISPRa and CRISPRi systems, using dCas9 fused to other effector domains, can be tested additionally and may be better suited for epigenetic editing of various gene regulatory element types.

Author Contributions G.A.L, B.T, S.A, S.K have developed CRISPR-based epigenome editing protocols. G.A.L, S.K, C.G, B. T, and S.A contributed to writing protocols and proofreading. G.A. L prepared the figures.

Funding Acknowledgments Ali Shariati has received the NIGMS/NIH through a Pathway to Independence Award K99GM126027/ R00GM126027, Maximizing Investigator

Award (R35GM147395, 3R35GM147395-02S1), CORE grant from Office of Research, and a start-up package from the University of California, Santa Cruz (S.A.S). G.A.L. has been supported by a Dutch Research Council (NWO) Rubicon fellowship (019.202EN.029). B.T is supported by the National Institutes of Health (NIH) under award number K12GM139185 and the Institute for the Biology of Stem Cells (IBSC) at UC Santa Cruz. C.G. is supported by the California Institute for Regenerative Medicine Postdoctoral Fellowship award (EDUC4–12759) and the Institute for the Biology of Stem Cells at UC Santa Cruz. The content is solely the responsibility of the authors and does not necessarily represent the official views of the NIH or the IBSC.

References

1. Harrison SE, Sozen B, Christodoulou N, Kyprianou C, Zernicka-Goetz M (2017) Assembly of embryonic and extraembryonic stem cells to mimic embryogenesis in vitro. Science 356. https://doi.org/10.1126/science.aal1810
2. Rivron NC, Frias-Aldeguer J, Vrij EJ, Boisset J-C, Korving J, Vivié J, Truckenmüller RK, van Oudenaarden A, van Blitterswijk CA, Geijsen N (2018) Blastocyst-like structures generated solely from stem cells. Nature 557:106–111. https://doi.org/10.1038/s41586-018-0051-0
3. Sozen B, Amadei G, Cox A, Wang R, Na E, Czukiewska S, Chappell L, Voet T, Michel G, Jing N et al (2018) Self-assembly of embryonic and two extra-embryonic stem cell types into gastrulating embryo-like structures. Nat Cell Biol 20:979–989. https://doi.org/10.1038/s41556-018-0147-7
4. Sozen B, Cox AL, De Jonghe J, Bao M, Hollfelder F, Glover DM, Zernicka-Goetz M (2019) Self-organization of mouse stem cells into an extended potential blastoid. Dev Cell 51:698–712.e8. https://doi.org/10.1016/j.devcel.2019.11.014
5. Amadei G, Lau KYC, De Jonghe J, Gantner CW, Sozen B, Chan C, Zhu M, Kyprianou C, Hollfelder F, Zernicka-Goetz M (2021) Inducible stem-cell-derived embryos capture mouse morphogenetic events in vitro. Dev Cell 56:366–382.e9. https://doi.org/10.1016/j.devcel.2020.12.004
6. Liu X, Tan JP, Schröder J, Aberkane A, Ouyang JF, Mohenska M, Lim SM, Sun YBY, Chen J, Sun G et al (2021) Modelling human blastocysts by reprogramming fibroblasts into iBlastoids. Nature 591:627–632. https://doi.org/10.1038/s41586-021-03372-y
7. Sozen B, Jorgensen V, Weatherbee BAT, Chen S, Zhu M, Zernicka-Goetz M (2021) Reconstructing aspects of human embryogenesis with pluripotent stem cells. Nat Commun 12:5550. https://doi.org/10.1038/s41467-021-25853-4
8. Yu L, Wei Y, Duan J, Schmitz DA, Sakurai M, Wang L, Wang K, Zhao S, Hon GC, Wu J (2021) Blastocyst-like structures generated from human pluripotent stem cells. Nature 591:620–626. https://doi.org/10.1038/s41586-021-03356-y
9. Yanagida A, Spindlow D, Nichols J, Dattani A, Smith A, Guo G (2021) Naive stem cell blastocyst model captures human embryo lineage segregation. Cell Stem Cell 28:1016–1022.e4. https://doi.org/10.1016/j.stem.2021.04.031
10. Tarazi S, Aguilera-Castrejon A, Joubran C, Ghanem N, Ashouokhi S, Roncato F, Wildschutz E, Haddad M, Oldak B, Gomez-Cesar E et al (2022) ¬Post-gastrulation synthetic embryos generated ex utero from mouse naïve ESCs. Cell 185:3290–3306.e25. https://doi.org/10.1016/j.cell.2022.07.028
11. Seong J, Frias-Aldeguer J, Holzmann V, Kagawa H, Sestini G, Heidari Khoei H, Scholte Op Reimer Y, Kip M, Pradhan SJ, Verwegen L et al (2022) Epiblast inducers capture mouse trophectoderm stem cells in vitro and pattern blastoids for implantation in utero. Cell Stem Cell 29:1102–1118.e8. https://doi.org/10.1016/j.stem.2022.06.002
12. Kagawa H, Javali A, Khoei HH, Sommer TM, Sestini G, Novatchkova M, Scholte Op Reimer Y, Castel G, Bruneau A, Maenhoudt

N et al (2022) Human blastoids model blastocyst development and implantation. Nature 601:600–605. https://doi.org/10.1038/s41586-021-04267-8

13. Bao M, Cornwall-Scoones J, Sanchez-Vasquez-E, Cox AL, Chen D-Y, De Jonghe J, Shadkhoo S, Hollfelder F, Thomson M, Glover DM et al (2022) Stem cell-derived synthetic embryos self-assemble by exploiting cadherin codes and cortical tension. Nat Cell Biol 24: 1341–1349. https://doi.org/10.1038/s41556-022-00984-y
14. Amadei G, Handford CE, Qiu C, De Jonghe J, Greenfeld H, Tran M, Martin BK, Chen D-Y, Aguilera-Castrejon A, Hanna JH et al (2022) Embryo model completes gastrulation to neurulation and organogenesis. Nature 610:143–153. https://doi.org/10.1038/s41586-022-05246-3
15. Lau KYC, Rubinstein H, Gantner CW, Hadas R, Amadei G, Stelzer Y, Zernicka-Goetz M (2022) Mouse embryo model derived exclusively from embryonic stem cells undergoes neurulation and heart development. Cell Stem Cell 29:1445–1458.e8. https://doi.org/10.1016/j.stem.2022.08.013
16. Oldak B, Wildschutz E, Bondarenko V, Comar M-Y, Zhao C, Aguilera-Castrejon A, Tarazi S, Viukov S, Pham TXA, Ashouokhi S et al (2023) Complete human day 14 post-implantation embryo models from naïve ES cells. Nature 622:562–573. https://doi.org/10.1038/s41586-023-06604-5
17. Pedroza M, Gassaloglu SI, Dias N, Zhong L, Hou T-CJ, Kretzmer H, Smith ZD, Sozen B (2023) Self-patterning of human stem cells into post-implantation lineages. Nature 622: 574–583. https://doi.org/10.1038/s41586-023-06354-4
18. Weatherbee BAT, Gantner CW, Iwamoto-Stohl LK, Daza RM, Hamazaki N, Shendure J, Zernicka-Goetz M (2023) Pluripotent stem cell-derived model of the post-implantation human embryo. Nature 622: 584–593. https://doi.org/10.1038/s41586-023-06368-y
19. Jana D, Singh P, Sailasree P, Kumar N, Vishnu VV, Kale HT, Lakshmi J, Kumari A, Sowpati DT, Shekar PC (2023) Efficient self-organization of blastoids solely from mouse ESCs is facilitated by transient reactivation of 2C gene network. bioRxiv. https://doi.org/10.1101/2023.04.12.536583
20. Lodewijk GA, Kozuki S, Han C, Topacio BR, Lee S, Nixon L, Zargari A, Knight G, Ashton R, Qi LS, Shariati SA (2025) Self-organization of mouse embryonic stem cells into reproducible pre-gastrulation embryo models via CRISPRa programming. Cell Stem Cell https://doi.org/10.1016/j.stem.2025.02.015
21. Okubo T, Rivron N, Kabata M, Masaki H, Kishimoto K, Semi K, Nakajima-Koyama M, Kunitomi H, Kaswandy B, Sato H et al (2024) Hypoblast from human pluripotent stem cells regulates epiblast development. Nature 626:357–366. https://doi.org/10.1038/s41586-023-06871-2
22. Hislop J, Song Q, Keshavarz FK, Alavi A, Schoenberger R, LeGraw R, Velazquez JJ, Mokhtari T, Taheri MN, Rytel M et al (2024) Modelling post-implantation human development to yolk sac blood emergence. Nature 626:367–376. https://doi.org/10.1038/s41586-023-06914-8
23. Warmflash A, Sorre B, Etoc F, Siggia ED, Brivanlou AH (2014) A method to recapitulate early embryonic spatial patterning in human embryonic stem cells. Nat Methods 11:847–854. https://doi.org/10.1038/nmeth.3016
24. Moris N, Anlas K, van den Brink SC, Alemany A, Schröder J, Ghimire S, Balayo T, van Oudenaarden A, Martinez Arias A (2020) An in vitro model of early anteroposterior organization during human development. Nature 582:410–415. https://doi.org/10.1038/s41586-020-2383-9
25. Sun S, Zheng Y, Kim YS, Zhong Z, Kobayashi N, Xue X, Liu Y, Zhou Z, Xu Y, Zhai J et al (2024) A transgene-free, human peri-gastrulation embryo model with trilaminar embryonic disc-, amnion- and yolk sac-like structures. Dev Biol
26. Shariati SA, Dominguez A, Xie S, Wernig M, Qi LS, Skotheim JM (2019) Reversible disruption of specific transcription factor-DNA interactions using CRISPR/Cas9. Mol Cell 74: 622–633.e4. https://doi.org/10.1016/j.molcel.2019.04.011
27. Qi LS, Larson MH, Gilbert LA, Doudna JA, Weissman JS, Arkin AP, Lim WA (2021) Repurposing CRISPR as an RNA-guided platform for sequence-specific control of gene expression. Cell 184:844. https://doi.org/10.1016/j.cell.2021.01.019
28. Liu Y, Yu C, Daley TP, Wang F, Cao WS, Bhate S, Lin X, Still C 2nd, Liu H, Zhao D et al (2018) CRISPR activation screens systematically identify factors that drive neuronal fate and reprogramming. Cell Stem Cell 23:758–771.e8. https://doi.org/10.1016/j.stem.2018.09.003
29. Nakamura M, Gao Y, Dominguez AA, Qi LS (2021) CRISPR technologies for precise epigenome editing. Nat Cell Biol 23:11–22.

https://doi.org/10.1038/s41556-020-00620-7

30. Gilbert LA, Horlbeck MA, Adamson B, Villalta JE, Chen Y, Whitehead EH, Guimaraes C, Panning B, Ploegh HL, Bassik MC et al (2014) Genome-scale CRISPR-mediated control of gene repression and activation. Cell 159:647–661. https://doi.org/10.1016/j.cell.2014.09.029
31. Pulecio J, Verma N, Mejía-Ramírez E, Huangfu D, Raya A (2017) CRISPR/Cas9-based engineering of the epigenome. Cell Stem Cell 21:431–447. https://doi.org/10.1016/j.stem.2017.09.006
32. Strumpf D, Mao C-A, Yamanaka Y, Ralston A, Chawengsaksophak K, Beck F, Rossant J (2005) Cdx2 is required for correct cell fate specification and differentiation of trophectoderm in the mouse blastocyst. Development 132:2093–2102. https://doi.org/10.1242/dev.01801
33. Niwa H, Toyooka Y, Shimosato D, Strumpf D, Takahashi K, Yagi R, Rossant J (2005) Interaction between Oct3/4 and Cdx2 determines trophectoderm differentiation. Cell 123:917–929. https://doi.org/10.1016/j.cell.2005.08.040
34. Nishiyama A, Xin L, Sharov AA, Thomas M, Mowrer G, Meyers E, Piao Y, Mehta S, Yee S, Nakatake Y et al (2009) Uncovering early response of gene regulatory networks in ESCs by systematic induction of transcription factors. Cell Stem Cell 5:420–433. https://doi.org/10.1016/j.stem.2009.07.012
35. Ralston A, Cox BJ, Nishioka N, Sasaki H, Chea E, Rugg-Gunn P, Guo G, Robson P, Draper JS, Rossant J (2010) Gata3 regulates trophoblast development downstream of Tead4 and in parallel to Cdx2. Development 137:395–403. https://doi.org/10.1242/dev.038828
36. Shimosato D, Shiki M, Niwa H (2007) Extra-embryonic endoderm cells derived from ES cells induced by GATA factors acquire the character of XEN cells. BMC Dev Biol 7:80. https://doi.org/10.1186/1471-213X-7-80
37. Schrode N, Saiz N, Di Talia S, Hadjantonakis A-K (2014) GATA6 levels modulate primitive endoderm cell fate choice and timing in the mouse blastocyst. Dev Cell 29:454–467. https://doi.org/10.1016/j.devcel.2014.04.011
38. Wamaitha SE, del Valle I, Cho LTY, Wei Y, Fogarty NME, Blakeley P, Sherwood RI, Ji H, Niakan KK (2015) Gata6 potently initiates reprograming of pluripotent and differentiated cells to extraembryonic endoderm stem cells. Genes Dev 29:1239–1255. https://doi.org/10.1101/gad.257071.114
39. Chavez A, Scheiman J, Vora S, Pruitt BW, Tuttle M, P R Iyer E, Lin S, Kiani S, Guzman CD, Wiegand DJ et al (2015) Highly efficient Cas9-mediated transcriptional programming. Nat Methods 12:326–328. https://doi.org/10.1038/nmeth.3312
40. Tan X, Worley J, Turunen M, Wong K, Fernández EC, Paull E, Jones S, Wang J, Noh H, Salvatori B et al (2021) Interrogation of genome-wide, experimentally dissected gene regulatory networks reveals mechanisms underlying dynamic cellular state control. https://doi.org/10.1101/2021.06.28.449297
41. Yu C, Liu Y, Ma T, Liu K, Xu S, Zhang Y, Liu H, La Russa M, Xie M, Ding S et al (2015) Small molecules enhance CRISPR genome editing in pluripotent stem cells. Cell Stem Cell 16:142–147. https://doi.org/10.1016/j.stem.2015.01.003
42. Norrander J, Kempe T, Messing J (1983) Construction of improved M13 vectors using oligodeoxynucleotide-directed mutagenesis. Gene 26:101–106. https://doi.org/10.1016/0378-1119(83)90040-9
43. Stewart SA, Dykxhoorn DM, Palliser D, Mizuno H, Yu EY, An DS, Sabatini DM, Chen ISY, Hahn WC, Sharp PA et al (2003) Lentivirus-delivered stable gene silencing by RNAi in primary cells. RNA 9:493–501. https://doi.org/10.1261/rna.2192803
44. Concordet J-P, Haeussler M (2018) CRISPOR: intuitive guide selection for CRISPR/Cas9 genome editing experiments and screens. Nucleic Acids Res 46:W242–W245. https://doi.org/10.1093/nar/gky354
45. Doench JG, Fusi N, Sullender M, Hegde M, Vaimberg EW, Donovan KF, Smith I, Tothova Z, Wilen C, Orchard R et al (2016) Optimized sgRNA design to maximize activity and minimize off-target effects of CRISPR-Cas9. Nat Biotechnol 34:184–191. https://doi.org/10.1038/nbt.3437
46. Sanson KR, Hanna RE, Hegde M, Donovan KF, Strand C, Sullender ME, Vaimberg EW, Goodale A, Root DE, Piccioni F et al (2018) Optimized libraries for CRISPR-Cas9 genetic screens with multiple modalities. Nat Commun 9:5416. https://doi.org/10.1038/s41467-018-07901-8
47. DeWeirdt PC, McGee AV, Zheng F, Nwolah I, Hegde M, Doench JG (2022) Accounting for small variations in the tracrRNA sequence improves sgRNA activity predictions for CRISPR screening. Nat Commun 13:5255.

https://doi.org/10.1038/s41467-022-33024-2

48. Mahy BWJ, Kangro HO (eds) (1996) Virology methods manual. Academic Press. https://doi.org/10.1016/b978-0-12-465330-6.x5000-3
49. Sena-Esteves M, Gao G (2018) Titration of Lentivirus vectors. Cold Spring Harb Protoc 2018:db.prot095695. https://doi.org/10.1101/pdb.prot095695
50. Berens C, Hillen W (2003) Gene regulation by tetracyclines. Constraints of resistance regulation in bacteria shape TetR for application in eukaryotes. Eur J Biochem 270:3109–3121. https://doi.org/10.1046/j.1432-1033.2003.03694.x

Methods in Molecular Biology (2026) 3048: 239–252
DOI 10.1007/7651_2025_648

Published online: 20 May 2025

Direct Reprogramming of Human Fibroblasts into Fully Functional Trophoblast Stem Cells

Meir Azagury and Yosef Buganim

Abstract

Trophoblast stem cells (TSCs), equivalent to first-trimester cytotrophoblasts, serve as valuable models for studying placental diseases and understanding early embryogenesis. Recent studies have demonstrated that human-induced trophoblast stem cells (hiTSCs) can be generated either by overexpressing the pluripotency factors OCT4, SOX2, KLF4, and MYC (OSKM) in fibroblasts or through the transdifferentiation of pluripotent stem cells. In this chapter, we describe a methodology for directly converting fibroblasts into fully functional hiTSCs using the transcription factors GATA3, OCT4, KLF4, and MYC (GOKM). This approach circumvents the need for inducing full pluripotency and avoids the expression of pluripotency factors per se. Moreover, the GOKM method seems to be superior technique, as it yields high number of colonies and a transcriptomic profile closely resembling blastocyst/first trimester-derived TSCs.

Key words Direct conversion, Human trophoblast stem cells, Placenta, Reprogramming, Transcription factors

1 Introduction

Cellular conversion from one cell type to another requires the erasure of the parental cell's epigenetic landscape and the acquisition of the epigenetic signature characteristic of the target cell type [1, 2]. This process involves robust nuclear reprogramming, typically driven by the overexpression of key master regulators, including pioneer transcription factors that can access closed chromatin and facilitate its opening by recruiting chromatin remodeling enzymes [3–5]. Cellular conversion has significant potential for regenerative medicine, cell-based therapies, and disease modeling, offering the ability to generate specialized cells without directly isolating them from patients [6, 7].

Studying placental complications and early embryogenesis remains highly challenging, as trophectoderm and cytotrophoblast progenitor cells, the precursors of the placenta, are difficult to isolate once complications arise. Therefore, developing techniques to recreate these early-stage progenitor cells, such as trophoblast stem cells (TSCs), is of paramount importance [8, 9].

Here, we describe a detailed protocol for generating human-induced trophoblast stem cells (hiTSCs) from fibroblasts. Our methodology relies on the overexpression of GATA3, OCT4, KLF4, and MYC (GOKM) using a lentiviral delivery system to introduce the reprogramming factors [10]. However, it is important to note that similar results can be achieved using non-integrating episomal vectors [10] and possibly mRNA-based approaches.

This protocol covers the entire process, including lentivirus production, viral transduction into fibroblasts, a step-by-step reprogramming procedure, methods for isolating and expanding hiTSC colonies, and strategies for characterizing the resulting cells.

1.1 General Notes

1. The protocol below describes the steps for reprogramming early passages of fibroblasts extracted from foreskin biopsy. Reprogramming fibroblasts from other sources may require additional calibration.
2. Following the same guidelines, fibroblasts can be reprogrammed into human-induced trophoblast stem cells (hiTSCs) using the OSKM reprogramming combination [10, 11].
3. For simplicity, the protocol was designed to prepare a single six-well plate of virus.

2 Materials

2.1 Primary Cultures and Cell Lines

Primary cultures and cell lines were maintained under sterile conditions in a biological safety cabinet. All cells were cultured at 20% O_2 and 6% CO_2.

2.2 Human Fibroblasts

Human fibroblasts can be obtained through commercial sources, such as ATCC (CAT#CRL-2522) and Lonza (CAT#CC-2509), or derived directly from a skin biopsy. Fibroblasts sourced from younger individuals through foreskin biopsy tend to offer higher reprogramming efficiency. Additionally, it is important to use fibroblasts at the earliest possible passage to reduce senescence and the risk of genetic abnormalities that may develop with extended passaging.

2.3 HEK293T Cells

HEK293T cells facilitate the efficient production of lentiviral vectors. These cells can be purchased from various distributors, including ATCC (CAT#CRL-11268).

2.4 Mitotically Inactivated Mouse Embryonic Fibroblasts (Feeder Cells)

Mouse embryonic fibroblasts (MEFs) can be isolated from E13.5 mouse embryos and cultured for a few passages before being mitotically inactivated using radiation or by the addition of mitomycin C. A detailed protocol for the isolation and generation of mouse feeder cells can be found in the Springer Nature protocol online (https://doi.org/10.1007/978-1-0716-1084-8_14).

These inactivated cells can be stored at −80 °C and thawed as required.

2.5 Lentiviral Vectors

The doxycycline (dox)-inducible Tet-on system enables high-efficiency hiTSC generation while allowing precise regulation of integrated transgene expression. Dox-inducible *GATA3* was generated by cloning the open reading frame (ORF) of the transcription factor into the FUW-TetO expression vector [10]. Lentiviral Tet-on vectors for *OCT4*, *KLF4*, and *MYC* are available from Addgene (Plasmids #20726, #20725, and #20723, respectively).

2.6 Reagents and Culture Media

See Tables 1, 2, 3, 4, 5, and 6.

Table 1
General reagents

Component	Final concentration	Provider
Trypsin-EDTA solution	0.25%	Sigma Cat#T4049
Polybrene	8 μg/mL	Santa Cruz Biotechnology Cat#SC-134220
Doxycycline hyclate	2 μg/mL	Sigma Cat#D9891
Polyethylenimine (PEI)	60 μg/mL	Sigma Cat#765090
Dulbecco's phosphate buffered saline without calcium and magnesium (PBS)	NA	Sartorius Cat# 02-023-1A
DAPI	1 μg/mL	Sigma Cat#10236276001

Table 2
Fibroblast culturing medium (FCM)

Component	Final concentration	Provider
DMEM	500 mL	Gibco Cat#11965092
FBS	10%	Gibco Cat#10270106
Penicillin-Streptomycin	1%	Gibco Cat#15140-163
L-Glutamine	2 mM	Gibco Cat# A2916801

Table 3
hTSC medium

Component	Final concentration	Provider
DMEM F-12	500 mL	Gibco Cat#21331020
Bovine serum albumin	1.5 gr	Sigma at#A3803
FBS	1 mL	Gibco Cat#10270106
Penicillin-Streptomycin	1%	Gibco Cat#15140-163
2-Mercaptoethanol	100 μM	Gibco Cat#31350010
Insulin-transferrin-sodium selenite media supplement	1%	Sigma Cat# I1884
L-Ascorbic acid	1.5 μg/mL	Sigma Cat# A8960
Human EGF	50 ng/mL	PeproTech Cat# AF10015
CHIR-99021	2 μM	Focus biomolecules Cat#10-1279
A83-01	0.5 μM	Focus biomolecules CAT#10-1327
SB-431542	1 μM	Focus biomolecules CAT#10-2443
Valproic acid sodium salt	800 μM	Sigma Cat#P4543
Y-27632 (Rock inhibitor)	5 μM	Focus biomolecules Cat#10-2301

3 Methods

Important note: Fibroblasts must express high levels of the doxycycline trans-activator M2rtTA (*see* **Note 1**).

3.1 Three Days Prior to Transfections

1. Split 800,000 HEK293T cells into two 10 cm plates to achieve 80% confluency on the day of transfection (*see* **Note 2**).
2. Replace the medium with fresh fibroblast culturing medium (FCM) 2–3 h prior to transfection.
3. Unfreeze and monitor the growth of your fibroblasts. Ensure they reach 80–90% confluency by Day 2.

Day 1: Transfections

4. In a 15 mL tube, prepare a DNA mix of 4 μg of DNA per transfection, scaled according to the total numbers of transfections planned. *One transfection consists of 4 μg of DNA, mixed with 1.5 × 10^6 HEK293T cells, plated into one well of a six-well plate.* The DNA mix should include, per transfection, 2 μg of

Table 4
qPCR primers

Gene	Application	Primer Sequence (5′ —> 3′)
GAPDH	qPCR analysis of mRNA expression normalization	F: CCTCAACGACCACTTTGTCAG R: TCTTCCTCTTGTGCTCTTGCG
GATA3	qPCR analysis of mRNA expression	F: TCATTAAGCCCAAGCGAAGG R: GTCCCCATTGGCATTCCTC
OCT4	qPCR analysis of mRNA expression	F: GAGAAGGAGAAGCTGGAGCA R: CTTCTGCTTCAGGAGCTTGG
KLF4	qPCR analysis of mRNA expression	F: AGGGAGAAGACACTGCGTCA R: AGTCGCTTCATGTGGGAGAG
MYC	qPCR analysis of mRNA expression	F: AGCGACTCTGAGGAGGAACA R: CTCTGACCTTTTGCCAGGAG
GATA3 5' UTR (endogenous expression)	qPCR analysis of mRNA expression	F: ACGACCCCTCCAAGATAATTTT R: GTCGGGGGTCGTTGAATGAT
TFAP2C	qPCR analysis of mRNA expression	F: GGTTGAATCTTCCGGCCG R: TCTGCCACTGGTTTACTAGGA
KRT7	qPCR analysis of mRNA expression	F: AAGAACCAGCGTGCCAAGT R: TCCAGCTCCTCCTGCTTG
TP63	qPCR analysis of mRNA expression	F: AGAAACGAAGATCCCCAGATGA R: CTGTTGCTGTTGCCTGTACGTT
THY1	qPCR analysis of mRNA expression	F: CCAGAACGTCACAGTGCTCA R: AGGTGTTCTGAGCCAGCAG

Table 5
Antibodies

Antibody	Source	Identifier
Rabbit monoclonal anti VIMENTIN (VIM)	Cell Signaling Technology	#5741
Rabbit polyclonal anti GATA3	Abcam	ab106625
Rabbit polyclonal anti GATA2	Abcam	ab173817
Rabbit polyclonal anti E-Cadherin (CDH1)	Santa Cruz biotechnology	sc-7870
Mouse monoclonal anti cytokeratin 7 (KRT7)	Abcam	ab215855
Mouse monoclonal anti cytokeratin 18 (KRT18)	Santa Cruz biotechnology	sc-51582
Mouse monoclonal anti AP-2γ (TFAP2C)	Santa Cruz biotechnology	sc-12762

Table 6
Other materials

	Source	Identifier
6-well culture plates	Corning	3506
12-well culture plates	Corning	9017
96-well culture plates	Corning	3512
100 mm Culture dish	Corning	430167
50 mL Centrifuge tubes	Corning	430290
15 mL Centrifuge tubes	Corning	430052
Polypropylene cryogenic vial	Corning	8676
Membrane filter, 0.45 μm	Sigma	HVLP08050
RT qPCR 96-well plates	Corning	3752
Fluorescent microscope	Nikon	Eclipse Ti
Binocular stereoscope	Nikon	SMZ1000
Cell counter	DeNovix	CellDrop BF

packaging vectors (psPAX2 and pGDM.2) in a 1:1 ratio, along with a total of 2 μg of DNA for GATA3, OCT4, KLF4, and MYC in a ratio of 3:3:3:1. Gently pipette to ensure thorough mixing (*see* **Note 3**).

5. Per transfection, add 150 μL of serum-free DMEM along with 60 μg/mL of pre-calibrated PEI transfection reagent (*see* **Note 4**). Mix thoroughly by pipetting (*see* **Note 5**).
6. Incubate at room temperature for 15–30 min.
7. Meanwhile, aspirate the medium from the HEK293T plates, rinse carefully with PBS, and add 1 mL of trypsin-EDTA to the plates.
8. Incubate the HEK293T cells for 5 min at 37 °C degrees.
9. Ensure that all cells have detached from the plates, add 5 mL of fresh FCM per plate to neutralize trypsin and pipette the cells thoroughly using a 10 mL stripette before transferring all the content of the plates into a 50 mL tube.
10. Centrifuge the cells for 3 min at 3000 rpm and discard the supernatant.
11. Add 10 mL of fresh FCM and thoroughly pipette the cells with a 10 mL stripette.
12. Take 10 μL of the sample and determine the cell count using a cell counter.
13. Dilute the cells to three million cells per ml.

14. Add 500 uL of HEK293T cells, multiply by the number of transfections performed, gently mix by pipetting, and divide the whole volume equally to the wells of a six-well plate.
15. On top of the viruses, add 1 mL of fresh FCM to each well.

Day 2:

16. 24 h later, replace the medium of the transfections with 1 mL of FCM per well.
17. Split 120 k human fibroblasts per well into six-well plates in FCM. *Ensure to allocate extra wells of infected and non-infected fibroblasts for verifying transgene expression* (*see* **Note 6**).

Day 3: Infection

18. Collect the medium filled with viruses from the HEK293T wells into a 15 mL tubes and replace it with 1 mL of fresh FCM per well.
19. Add polybrene into the tube at a concentration of 8 μg/mL.
20. Filter the supernatant through a 0.45 μM filter in order to avoid contaminating your fibroblasts with any floating HEK293T cells (*see* **Note 7**).
21. Remove the current medium of the fibroblasts and replace it with a mix of filtrated viruses and fresh FCM at a ratio of 1:3. *Based on the efficiency of your transfection, 0.5 mL to 1 mL virus should be added per well.*

Day 4: Transgene Expression

22. Collect once again the virus from the HEK293T cells and store it in −80 °C freezer for future infections (*see* **Note 8**).
23. Replace the medium in the fibroblast plates with a fresh FCM supplemented with 2 μg/mL doxycycline.
24. Change medium every other day.

Day 7:

Following the addition of doxycycline, significant cell death along with morphological changes should be observed in the wells compared to the non-infected control cells.

25. On the fifth day after doxycycline addition, trypsinize and collect cells from three separate wells of both infected and non-infected cultures.
26. Proceed with RNA extraction and qRT-PCR to evaluate transgene expression [10] (*see* **Note 9**) (Fig. 1).
27. Change medium every other day to the rest of the reprogramming wells.

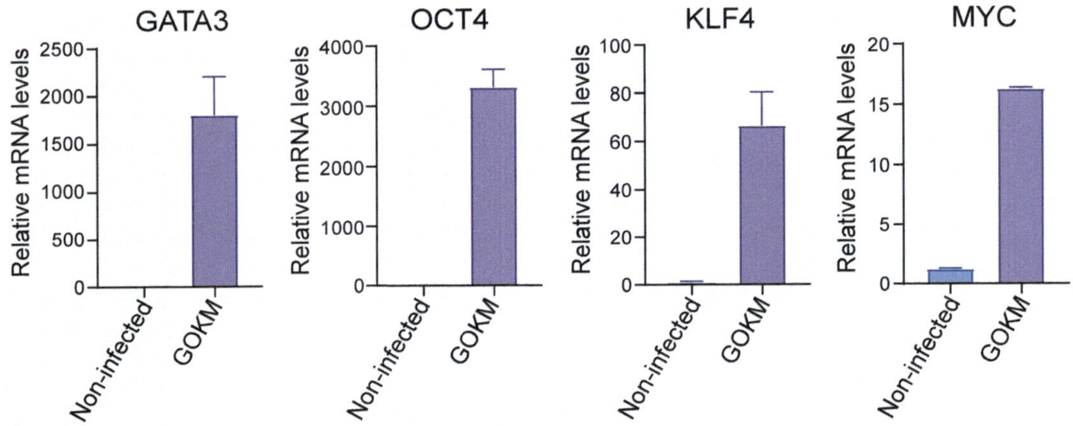

Fig. 1 Transgene expression in GOKM transduced fibroblast. qPCR analysis of mRNA levels of the indicated transgenes in GOKM-transduced fibroblasts following 5 days of dox. Results are normalized to the mRNA levels of the housekeeping control gene *GAPDH*. For each sample, three replicates were used

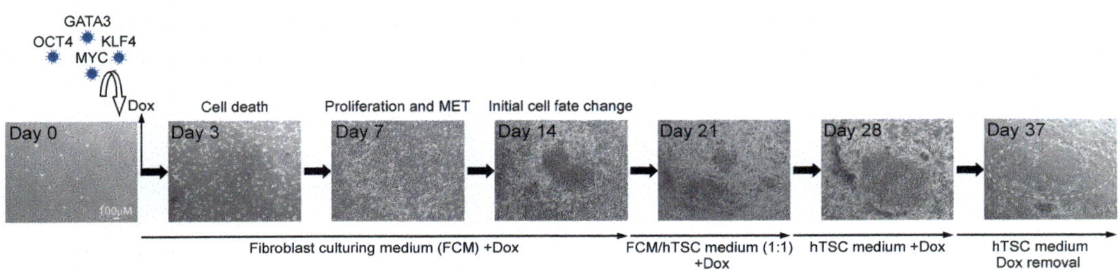

Fig. 2 Morphological characteristics of GOKM reprogramming. Representative bright-field images illustrating the entire reprogramming process, from initial infection to the formation of final hiTSC colonies, highlighting key morphological changes such as cell death, proliferation, epithelial foci formation, and the emergence of mature hiTSC colonies within the indicated reprogramming days

3.2 Continued Reprogramming and Monitoring (Days 7–28)

Reprogramming to human iTSCs is a lengthy process marked by numerous morphological changes such as cell death, proliferation, mesenchymal to epithelial transition (EMT), and cell fate change (Fig. 2). However, clear signs of hiTSC colony generation typically do not emerge until after doxycycline removal, approximately 1 month into the reprogramming process.

1. 14 days after the addition of doxycycline, replace the medium with hTSC medium and FCM in a ratio of 1:1 supplemented with 2 μg/mL doxycycline.
2. 7 days later, replace the medium with hTSC medium supplemented with 2 μg/mL doxycycline.
3. On day 28 of the experiment, 4 weeks post-transgene activation, replace the medium with hTSC without doxycycline. *Following the removal of doxycycline, transgene overexpression will cease.*

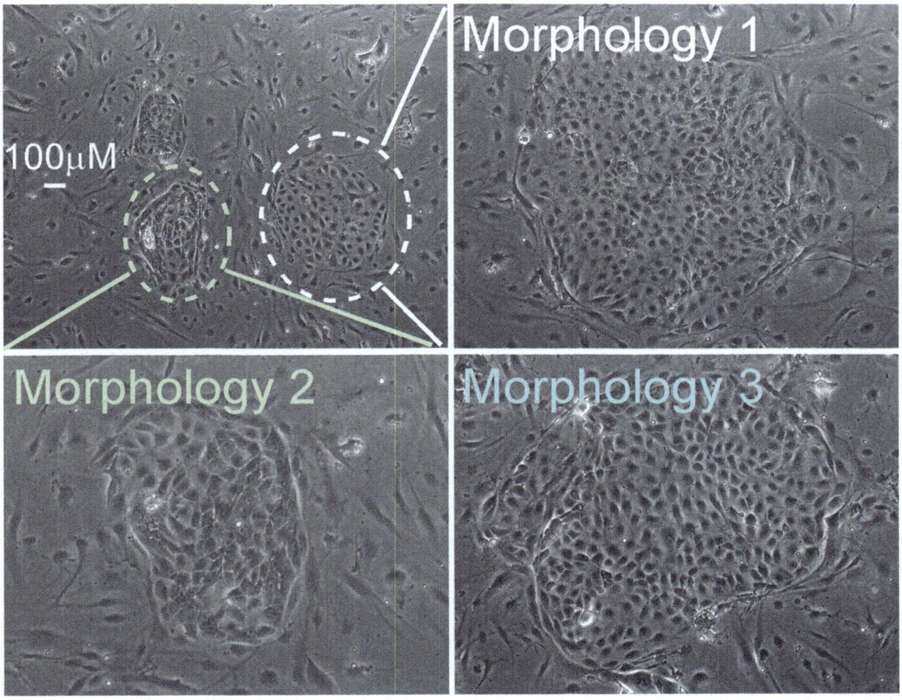

Fig. 3 Distinct morphology of human iTSC in early passages. Representative bright-field images showing the diverse epithelial morphologies of hiTSC colonies at early passage stages, particularly in terms of cell density and colony compaction

3.3 Appearance of iTSC Colonies (Days 30–40)

hiTSC colonies are typically expected to emerge in wells 5 to 12 days after doxycycline removal. Although early signs of potential hiTSC colonies formation are difficult to discern (*see* **Note 10**), fully formed hiTSC colonies display a distinct epithelial morphology, allowing for quick identification once they emerge (Fig. 2).

Once distinct hiTSC colonies are formed, isolate individual colonies for further characterization.

Important note: In the early stages of hiTSC colony formation, various morphologies can be observed within the wells (Fig. 3). However, after several passages, only a single hiTSC morphology persists (Fig. 3 *(morphology 3)* and 4a). It remains to be determined whether these early colonies have yet to reach their final morphology or if they represent distinct hiTSC subtypes that do not survive continuous passaging.

3.4 Isolation of hiTSC Colonies

1. Prior to the isolation, thaw mitotically inactivated feeders (*see* **Note 11**) and plate them in 12-well plates with FCM.
2. On the day of the isolation, replace the medium with 0.5 mL hTSC medium.
3. Prepare a 96-well plate and add 50 μL of Triple E to several wells.

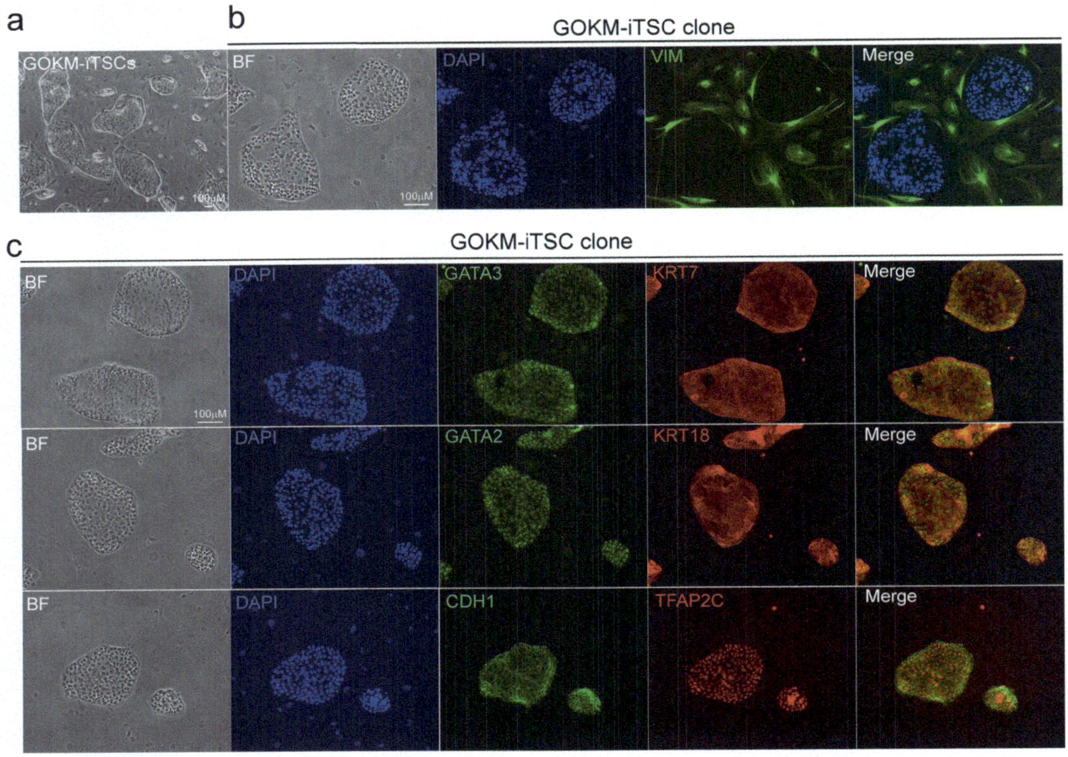

Fig. 4 Protein-level characterization of newly generated hiTSCs. (**a**) Bright-field image of a stable GOKM-derived hiTSC colony. (**b**) Immunofluorescence staining for the fibroblast-specific marker *VIM* in hiTSC clones. (**c**) Immunofluorescence staining for hTSC-specific markers (*TFAP2C, GATA2, GATA3*) and epithelial markers (*CDH1, KRT18, KRT7*) in newly generated hiTSC clones.

4. Identify well-formed epithelial hiTSC colonies. Carefully isolate each colony by gently detaching it from the plate using a 10 μL pipette tip, then aspirate it, and deposit it into one of the wells of the 96-well plate.
5. After isolating several colonies, incubate the plate at 37 °C for 5 min.
6. Add 100 μL of hTSC medium to each well, pipette thoroughly, then transfer the contents of each well into the 12-well plate containing feeders.
7. Place the plates in the incubator. hiTSC colonies should appear 3 to 7 days later.
8. Once homogeneous hiTSC colonies appear, proceed with further validation.
9. Replace the medium every other day.
10. Stable colonies can be frozen in a freezing medium composed of FBS and DMSO (9:1 ratio) and can be stocked at −80 °C for future uses (*see* **Note 12**).

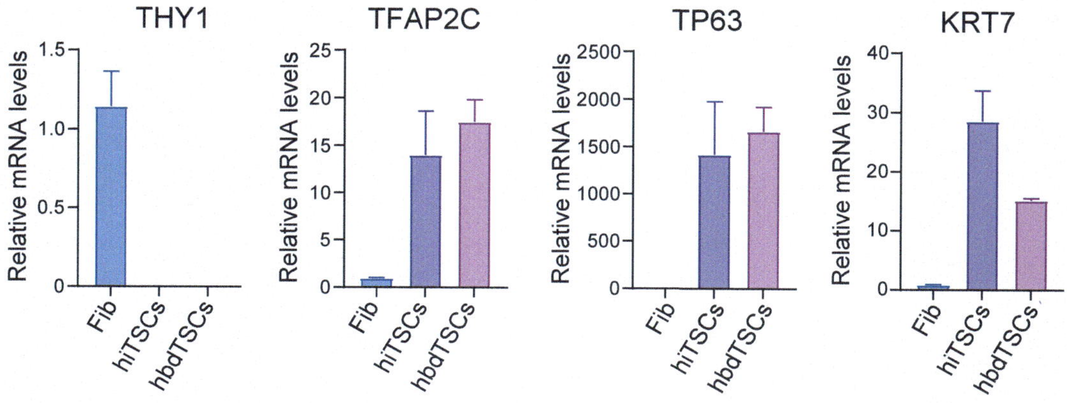

Fig. 5 Characterization of newly generated hiTSCs at the mRNA level. qPCR analysis of mRNA levels for hTSC-specific markers *KRT7*, *TP63*, and *TFAP2C* and mesenchymal-specific markers *THY1* in hiTSC clone compared to hbdTSC clone and primary fibroblast (Fib) culture. Results are normalized to the mRNA levels of the housekeeping control gene *GAPDH*. For each sample, two replicates were used

3.5 Validation of hiTSC

hiTSC colonies should undergo thorough validation to confirm their identity and functionality. We recommend validating hiTSCs through a combination of techniques. Protein expression analysis (Fig. 4b, c) and complementary RNA expression analysis (Fig. 5) should first be conducted to compare newly generated hiTSCs with blastocyst-derived trophoblast stem cells (hbdTSCs). These analyses will highlight the molecular similarities between hiTSCs and hbdTSCs. Advanced transcriptomic analyses can further validate the unique identity of hiTSCs (*see* **Note 13**).

To validate hiTSC functionality, in vitro differentiation should be performed to demonstrate their capacity to develop into various trophoblast cell types, such as multinucleated syncytiotrophoblasts (STs) and extravillous trophoblasts (EVTs) [12]. Additionally, in vivo validation can involve injecting hiTSC colonies subcutaneously into NOD-SCID mice to partially recapitulate trophoblast stem cell differentiation and confirm their ability to form trophoblastic lesions [10, 12].

3.6 Troubleshooting

Severe cell death following factor transduction:

- Your HEK293 cells produce too much of viruses, infect your cells with lower amounts of viruses, or seed the fibroblasts at a higher confluency.

The conversion process is not initiated:

- Make sure your cells are on their log phase of proliferation and that they express the M2rtTA transactivator.

- Make sure your viruses are effective. As shown above, infection with GOKM should lead to some cell death in the first days of the reprogramming.

hiTSC colonies do not emerge in the plate:

- Examine the levels of the reprogramming factors. They should be highly expressed.
- Make sure that your hTSC medium supports the growth of blastocyst-derived hTSC.

4 Notes

1. Fibroblasts must be transduced with a lentiviral vector encoding the doxycycline trans-activator *M2rtTA*, along with a selection marker, to enable the expression of dox-inducible factors. The transfection and infection process can be carried out as outlined in **steps 1** to **3**, substituting GOKM with 2 μg of the FUW-M2rtTA vector (Addgene #20342).
2. HEK293T cells can be continuously maintained in culture. A 10 cm plate of HEK293T cells at 80–90% confluency typically contains around ten million cells.
3. Ensure plasmids are pure and of high quality before use. The 260/280 ratio should be ≥1.8, and the 260/230 ratio should be ≥2.0. It is recommended to run 200 ng of plasmid DNA on an electrophoresis gel prior to transfection to confirm integrity.
4. The PEI transfection reagent should be pre-calibrated to determine the optimal concentration for transfection. Calibration can be performed by transfecting HEK293T cells with a constitutively expressed GFP or RFP vector and assessing transfection efficiency via flow cytometry. An efficiency of 95–100% is expected 48 h post-transfection.
5. While this protocol utilizes PEI, any other suitable transfection reagent should be effective.
6. To confirm successful transgene overexpression, analyze infected cells 48–72 h after doxycycline addition and compare transgene expression to non-infected controls. Transgene expression typically peaks at 48 h post-induction before declining during the reprogramming process.
7. Avoid using 0.2 μm filters to purify viruses from HEK293T cells, as they increase the risk of viral aggregation and clogging.
8. Virus can be collected twice from each transfected well at 24-h intervals. The collected virus can be stored at −80 °C for up to 1 year.

9. In this protocol, primers were designed to detect both endogenous and exogenous transcripts. To specifically highlight the expression of transgenes whose endogenous counterparts are also expressed in fibroblasts, such as *KLF4* and *MYC*, primers can be designed to target only the transgenes. This can be achieved by positioning the reverse primer on the FUW-TetO element WPRE, while placing the forward primer in the last exon of the transgene of interest. This strategy ensures specific amplification of the transgene while avoiding endogenous gene expression interference.
10. To the best of our knowledge, no clear correlation has been established between morphological changes during reprogramming and the emergence of final colonies. Final hiTSC colonies appear to form randomly across the plate following doxycycline removal.
11. While variations may occur due to the quality of the preparation, one 15 cm plate of frozen feeders can typically be thawed to seed six wells of a six-well plate.
12. Freeze at least one full well of a six-well plate per cryogenic vial to ensure sufficient cell recovery after thawing.
13. Recent studies have identified a transcriptomic overlap between trophoblast stem cells and amnion-like cells [13, 14]. Therefore, it is strongly recommended to distinguish hiTSCs from amnion-like cells using a well-defined panel of specific gene expression markers such as *ISL1*, *GCM1*, and *HAVCR1* [15].

Acknowledgments

Y.B. is supported by research grants from EMBO Young Investigator Programme (YIP), Israel Science Foundation (ISF, 161/23), Leo Foundation (LF-OC-24-001624), Israel Cancer Research Funds (ICRF, 1276940), and by a generous gift from Ms. Nadia Guth Biasini. Prof. Buganim holds a patent regarding the production of human TSCs with GOKM factors.

References

1. Maherali N, Sridharan R, Xie W, Utikal J, Eminli S, Arnold K, Stadtfeld M, Yachechko R, Tchieu J, Jaenisch R, Plath K, Hochedlinger K (2007) Directly reprogrammed fibroblasts show global epigenetic remodeling and widespread tissue contribution. Cell Stem Cell 1:55–70
2. Mikkelsen TS, Hanna J, Zhang X, Ku M, Wernig M, Schorderet P, Bernstein BE, Jaenisch R, Lander ES, Meissner A (2008) Dissecting direct reprogramming through integrative genomic analysis. Nature 454:49–55
3. Buganim Y, Faddah DA, Jaenisch R (2013) Mechanisms and models of somatic cell reprogramming. Nat Rev Genet 14:427–439
4. Lasry R, Maoz N, Cheng AW, Yom Tov N, Kulenkampff E, Azagury M, Yang H, Ople C, Markoulaki S, Faddah DA, Makedonski K,

Orzech D, Sabag O, Jaenisch R, Buganim Y (2023) Complex haploinsufficiency in pluripotent cells yields somatic cells with DNA methylation abnormalities and pluripotency induction defects. Stem Cell Rep 18:2174–2189

5. Apostolou E, Hochedlinger K (2013) Chromatin dynamics during cellular reprogramming. Nature 502:462–471
6. Cherry AB, Daley GQ (2013) Reprogrammed cells for disease modeling and regenerative medicine. Annu Rev Med 64:277–290
7. Kuang J, Huang T, Pei D (2022) The art of reprogramming for regenerative medicine. Front Cell Dev Biol 10:927555
8. Azagury M, Buganim Y (2024) Unlocking trophectoderm mysteries: in vivo and in vitro perspectives on human and mouse trophoblast fate induction. Dev Cell 59:941–960
9. Naama M, Buganim Y (2023) Human trophoblast stem cell-state acquisition from pluripotent stem cells and somatic cells. Curr Opin Genet Dev 81:102084
10. Naama M, Rahamim M, Zayat V, Sebban S, Radwan A, Orzech D, Lasry R, Ifrah A, Jaber M, Sabag O, Yassen H, Khatib A, Epsztejn-Litman S, Novoselsky-Persky M, Makedonski K, Deri N, Goldman-Wohl D, Cedar H, Yagel S, Eiges R, Buganim Y (2023) Pluripotency-independent induction of human trophoblast stem cells from fibroblasts. Nat Commun 14:3359
11. Liu X, Ouyang JF, Rossello FJ, Tan JP, Davidson KC, Valdes DS, Schroder J, Sun YBY, Chen J, Knaupp AS, Sun G, Chy HS, Huang Z, Pflueger J, Firas J, Tano V, Buckberry S, Paynter JM, Larcombe MR, Poppe D, Choo XY, O'Brien CM, Pastor WA, Chen D, Leichter AL, Naeem H, Tripathi P, Das PP, Grubman A, Powell DR, Laslett AL, David L, Nilsson SK, Clark AT, Lister R, Nefzger CM, Martelotto LG, Rackham OJL, Polo JM (2020) Reprogramming roadmap reveals route to human induced trophoblast stem cells. Nature 586:101–107
12. Okae H, Toh H, Sato T, Hiura H, Takahashi S, Shirane K, Kabayama Y, Suyama M, Sasaki H, Arima T (2018) Derivation of human trophoblast stem cells. Cell Stem Cell 22(50–63):e56
13. Shao Y, Taniguchi K, Gurdziel K, Townshend RF, Xue X, Yong KMA, Sang J, Spence JR, Gumucio DL, Fu J (2017) Self-organized amniogenesis by human pluripotent stem cells in a biomimetic implantation-like niche. Nat Mater 16:419–425
14. Shao Y, Taniguchi K, Townshend RF, Miki T, Gumucio DL, Fu J (2017) A pluripotent stem cell-based model for post-implantation human amniotic sac development. Nat Commun 8: 208
15. Zheng Y, Yan RZ, Sun S, Kobayashi M, Xiang L, Yang R, Goedel A, Kang Y, Xue X, Esfahani SN, Liu Y, Resto Irizarry AM, Wu W, Li Y, Ji W, Niu Y, Chien KR, Li T, Shioda T, Fu J (2022) Single-cell analysis of embryoids reveals lineage diversification roadmaps of early human development. Cell Stem Cell 29: 1402–1419 e1408

Methods in Molecular Biology (2026) 3048: 253–263
DOI 10.1007/7651_2025_646

Published online: 22 May 2025

3D In Vitro Culture of Early Mouse Embryos and Trophoblast Tissue Explants

Devila Prit, Shashank Jaitly, Niraimathi Govindasamy, Adrian Ranga, and Ivan Bedzhov

Abstract

The cellular dynamics during peri-implantation embryogenesis and the concurrent interactions at the embryo-maternal interface are inherently difficult to study due to intrauterine development in mammals. To model certain aspects of these processes in vitro, we have generated a biomimetic environment resembling the mechanical properties of the murine uterine stroma. Here we describe a step-by-step methodology for 3D culture of mouse embryos and ectoplacental cone explants in synthetic hydrogels that allow ex utero trophoblast invasion.

Key words 3D culture, Blastocyst, Ectoplacental cone, Egg cylinder, Hydrogel, Implantation, Invasion, Mouse embryo, Trophoblast, Trophoblast giant cells

1 Introduction

After fertilization, during the first days of pre-implantation embryogenesis, the embryo develops within its glycoprotein envelope, the zona pellucida. Only after hatching, the blastocyst makes its first direct contact with the uterine tissue, initiating implantation. This process establishes an embryo-maternal interface that supports the development of the conceptus in the early stages of pregnancy.

In mice, implantation is initiated at embryonic day four and a half (E4.5). At this stage, the blastocyst consists of three cell lineages—the pluripotent epiblast (Epi), which gives rise to all the tissues of the fetus, and two supportive tissues—the primitive endoderm (PrE) and the trophectoderm (TE). The TE forms two compartments—the polar TE, which is in direct contact with the epiblast, and the mural TE, which surrounds the blastocyst cavity [1].

The mural TE mediates the initial contact with the uterine tissues and gives rise to terminally differentiated polyploid cells known as trophoblast giant cells (TGCs) or primary TGCs. These cells undergo multiple rounds of DNA endoreduplication, upregulate vascular markers, and secrete matrix metalloproteinases (MMPs) that promote their invasive behavior [2, 3]. As the embryo attaches to the uterine wall, the TGCs eliminate the luminal epithelial cells by entosis and invade the underlying stroma [4]. At the same time, the stromal cells rapidly proliferate and transform into decidual cells, forming a specialized compartment (decidua) that completely engulfs the embryo to support its further development.

During the peri-implantation stages, the blastocyst transforms into an early post-implantation conceptus (egg cylinder). In contrast to the terminal differentiation of the mural TE, the polar TE expands after implantation to form the extraembryonic ectoderm (ExE) [5]. The ExE consists of multipotent precursor cells (trophoblast stem cells) that give rise to the ectoplacental cone (EPC), a source of a second wave of TGCs (secondary TGCs) that invade the surrounding decidua [6].

While lineage specification in the early embryo has been extensively characterized, the intrauterine development in mammals makes it inherently challenging to study cellular dynamics at the embryo-maternal interface. To model certain aspects of this process in vitro, we have established a synthetic environment based on methacrylated dextran (DexMA) or polyethylene glycol (PEG) that resembles the mechanical properties of the decidualized stroma, namely degradability, stiffness, and adhesion. Briefly, we incorporated MMP-cleavable peptide sequences to generate degradable hydrogels and integrated RGD peptide sequences to allow integrin adhesion and trophoblast invasion. We also measured the stiffness of the decidua using atomic force microscopy and adjusted the hydrogel compositions to achieve decidua-like stiffness [2].

This system is a reductionist approximation of physiological conditions and can serve as a platform for further modifications, such as tuning of mechanical properties, 3D shaping to create tissue architectures, and incorporation of mammalian embryos or embryo-like structures. The synthetic environment enables the co-culture of embryos and uterine cells, providing a system for modeling embryo-maternal interactions. For studies requiring microfluidics embedded in the substrate, DexMA hydrogels are advantageous as they do not swell or contract, thus ensuring the integrity of the microfluidic channels [7]. We have previously described a detailed protocol for co-culturing embryos and endothelial cells in a DexMA-based microfluidic device [8]. Here we describe a step-by-step methodology for 3D culture of embryos and EPC explants in PEG-based hydrogels that allow ex utero trophoblast invasion.

2 Materials

2.1 Dissection Tools

1. Stereomicroscope equipped with a transillumination base and a light source.
2. Bunsen burner.
3. Glass Pasteur pipettes.
4. Rubber tubing.
5. Iris scissors.
6. Fine scissors.
7. Fine forceps.
8. 1 mL syringes
9. 27 gauge needles
10. Petri dishes, 35 × 10 mm.
11. Four-well plates.
12. Ibidi μ-slide 15 well 3D plates.
13. Mouth pipette assembly. Attach a 200 μL filter tip (mouth-piece) to one end of the tubing and a 1000 μL tip without filter to the other end. After pulling the glass pipette over the flame of the Bunsen burner, attach the Pasteur pipette to the 1000 μL tip. The diameter of the open end of the glass pipette should be such that the embryo can move freely, taking into account the stage of development.

2.2 Media Stock Solutions

1. Advanced DMEM/F-12 (Life Technologies, Cat. No. 12634-010).
2. Fetal calf serum (FCS), embryonic stem cell grade. Serum is heat-inactivated by incubation at 56 °C for 30 min.
3. Knock-out Serum Replacement—KSR (Life Technologies, Cat. No. 10828010).
4. L-glutamine stock: 200 mM in ddH_2O.
5. Penicillin-Streptomycin stock: 5000 units/mL penicillin and 5000 μg/mL streptomycin.
6. ITS-X stock: Insulin 1 g/L, transferrin 0.55 g/L, sodium selenite 0.00067 g/L, ethanolamine 0.20 g/L.
7. β-estradiol stock: 10 μM solution in DMSO. Store at −20 °C for up to 2 months.
8. Progesterone stock: 1 mg/mL solution in DMSO. Store at −20 °C for up to 2 months.
9. N-Acetyl-L-cysteine stock: 50 mM solution in ddH_2O. Store at −20 °C for up to 2 months.

2.3 Media

1. M2 medium: Warm (37 °C) before use.
2. IVC1 medium: supplement Advanced DMEM/F12 with 20% (vol/vol) FCS, 2 mM L-glutamine, penicillin (25 units/mL), streptomycin (25 μg/mL), 1× ITS-X (10 mg/L insulin, 5.5 mg/L transferrin, 0.0067 mg/L sodium selenite, and 2 mg/L ethanolamine), 8 nM β-estradiol, 200 ng/mL progesterone, and 25 μM NAC. The medium can be kept at −20 °C for 3 months or at 4 °C for up to 2 weeks. Warm the medium (37 °C) and equilibrate in a humidified atmosphere of 5% CO_2 for at least half an hour before use.
3. IVC2 medium: supplement Advanced DMEM/F12 with 30% (vol/vol) KSR, 2 mM L-glutamine, penicillin (25 units/mL), streptomycin (25 μg/mL), 1× ITS-X (10 mg/L insulin, 5.5 mg/L transferrin, 0.0067 mg/L sodium selenite, and 2 mg/L ethanolamine), 8 nM β-estradiol, 200 ng/mL progesterone, and 25 μM N-acetyl-L-cysteine. The medium can be kept at −20 °C for 3 months or at 4 °C for up to 2 weeks. Warm the medium (37 °C) and equilibrate in a humidified atmosphere of 5% CO_2 for at least half an hour before use.

2.4 PEG Hydrogel Stock Solutions

1. PEG precursor preparation: Functionalize 8-arm PEG-vinylsulfone, 40 kDa (PEG-VS) (NOF Corporation) with FXIIIa-peptide substrates via Michael-type addition to yield a glutamine-PEG precursor (Q-PEG) and a lysine-PEG precursor (K-PEG) as described in detail in Gjorevski et al. [9]. The amino acid sequence of the glutamine-containing peptide is NQEQVSPLERCG-NH2 and that of the lysine-containing peptide with the MMP-sensitive sequence is AcFKGGGPQ-GIWGQ -ERCG-NH2.
2. TG-RGD: Reconstitute TG-RGD peptide powder to 500 μM concentration stock solution in MilliQ water. Store aliquots of 100 μL or 500 μL at −80 °C for up to 3 years. Aliquots can be freeze-thawed. Peptide solution is stable over time and has been tested to maintain bioactivity over at least 10 freeze-thaw cycles.
3. 10× Buffer: Prepare a solution containing 500 mM Tris and 500 mM $CaCl_2$. Adjust the pH of the buffer to 7.6. Prepare fresh buffer every 3 months. Store at room temperature. Do not freeze. Long-term storage can alter the pH. It is recommended to make a fresh buffer every 3 months.
4. Fibrogammin: Reconstitute 1250 U of Fibrogammin in 6.25 mL of water provided by the manufacturer. Prepare 1 mL aliquots and store them at −80 °C for up to 12 months.
5. Thrombin: Reconstitute thrombin to 20 U/mL final concentration in a buffer containing 10 mM Tris, 150 mM NaCl, and

25 mM $CaCl_2$, pH 7.4. Prepare 100 μL aliquots and store them at −80 °C for up to 12 months.

6. Factor XIII (FXIII) activation: Prepare 200 U/mL of activated FXIII by mixing 1 mL of the Fibrogammin solution with 100 μL of the thrombin solution. Incubate the solution at 37 °C for 30 min. Prepare 10 μL aliquots and store them at −80 °C for up to 12 months. Keep on ice after thawing. Do not refreeze and reuse the activated FXIII. Enzyme activity is reduced after the freeze-thaw cycle, thereby altering the mechanical properties of the gel. Thawed FXIII aliquot should be kept on ice during gel preparation and can be used in this form for 2 h.

2.5 PEG Hydrogel Preparation

Mix the ingredients at room temperature in the following order. Add 10% (vol/vol) of the 10× Buffer and 10% (vol/vol) of the RGD peptide to 60% (vol/vol) of water in a 1.5 mL Eppendorf tube. After that, add 10% (vol/vol) of the PEG solution to a final concentration of 1% (vol/vol). Mix the solution by gentle pipetting (pipetting full volume approximately 10 times). Thaw the frozen FXIII stock solution and keep it on ice. Initiate the crosslinking reaction by adding 5% (vol/vol) of the activated FXIII (200 U/mL to a final concentration of 10 U/mL) or, alternatively, dilute the FXIII stock solution (200 U/mL) 1:2 in water and add 10% (vol/vol). Mix the solution vigorously for approximately 1 min by pipetting (pipetting full volume approximately 10 times). The gel should solidify in approximately 5 min after adding the FXIII. An example recipe for the preparation of 100 μL PEG hydrogel is indicated in Table 1 (*see* **Note 1**).

Table 1
Recipe preparation of PEG hydrogel

Reagent	Final concentration	Volume
Water		60 μL
Buffer	1×	10 μL (10× stock)
PEG	1% (wt/vol)	10 μL (10% wt/vol stock)
RGD	50 μM	10 μL (500 μM stock)
FXIII	10 U/mL	10 μL (100 U/mL stock)
Total		100 μL

3 Methods

Embryo isolation, EPC microdissection, and hydrogel embedding are performed at room temperature. Subsequent embryo and EPC culture must be performed at 37 °C and 5% CO_2 in air in a humidified incubator.

3.1 Isolation of E3.5 Blastocysts and Removal of Zona Pellucida

1. Humanely euthanize pregnant female mice at 3.5 days post-coitum (d.p.c.), corresponding to the early blastocyst stage. Cut through the skin and peritoneum and push away the abdominal organs to expose the reproductive tract. Make an incision between the cervix and the uterus. Dissect and collect the uterus in a pre-warmed M2 medium.
2. Fill a syringe with a pre-warmed M2 medium and attach the needle. Insert the needle into the uterine horn and flush the lumen at least 3 times with M2 medium. Also, flush the connection between the uterus and the oviduct to ensure that all embryos are released. Collect the blastocysts and transfer them to a 4-well plate filled with pre-warmed M2 medium using a mouth pipette. Keep the embryos at 37 °C, protected from light.
3. Removal of zona pellucida. Invert the lid of a 35 mm dish and place 30 μL drops of pre-warmed Tyrode's solution and 30 μL drops of pre-warmed M2 medium. Transfer small groups of 5–10 embryos from the 4-well plate to a drop of M2 medium. Remove the remaining medium from the pipette and refill the pipette with Tyrode's solution. Using the same pipette, transfer the embryos to the drop of Tyrode's solution. Discard the remaining liquid and refill the pipette with fresh Tyrode's solution. Transfer the embryos to the second drop of Tyrode's solution (*see* **Note 2**). Repeat the procedure and observe the rapid disappearance of the zona pellucida (within 30 s) in the third drop of Tyrode's solution. Refill the pipette with M2 medium and immediately transfer the zona pellucida-free blastocysts to a drop of M2 medium for washing and then transfer the embryos to a 4-well plate filled with pre-warmed M2 medium (*see* **Note 3**).

3.2 Isolation of E6.5 Embryos, Removal of the Reicherts Membrane, and Dissection of the EPC

1. Humanely euthanize pregnant female mice at 6.5 days post-coitum (d.p.c.). Collect the uterus in a pre-warmed M2 medium.
2. At E6.5, the embryos are "hidden" in the decidua. Individual deciduae are visible as bulges of the uterus. Cut between the bulges and remove the outer layers of the uterus (perimetrium and myometrium) to expose the decidua using fine scissors (*see* **Note 4**).

3. Use one pair of forceps to hold the decidua in place and another pair to open the decidua. If the embryo is not visible, cut the decidua completely in half. Carefully scoop out the embryo. A more detailed protocol for the isolation of early post-implantation embryos has been described previously [10].
4. Removal of the Reichert's membrane. If the whole embryo is used for 3D culture, the Reichert's membrane must be removed as it hinders embryo growth. A pair of very sharp, fine forceps is required for this procedure. Holding the forceps open, insert one tip into the space between the visceral endoderm and Reichert's membrane. Using the tip as a scalpel, cut and peel away the membrane to expose the egg cylinder (*see* **Note 5**).
5. Dissection of the EPC. Holding the forceps open, use one of the tips as a scalpel to cut between the ExE and the EPC. The EPC is not surrounded by the Reichert's membrane, but its removal may help to locate the ExE—EPC junction. Transfer the embryos and the EPCs to a 4-well plate filled with pre-warmed M2 medium.

3.3 In Vitro Culture of Mouse Blastocysts

1. Transfer the E3.5 blastocysts from the M2 medium into a 4-well plate containing ICV1 medium (*see* **Note 6**) and culture the embryos for 16–20 h in a humidified incubator at 37 °C, 5% CO_2 (*see* **Note 7**).
2. Prepare PEG hydrogel solution as described in the Materials. Pipette 10 μL of the solution into the inner well of the Ibidi μ-slide 15-well 3D plate (*see* Fig. 1a, b) (*see* **Note 8**).

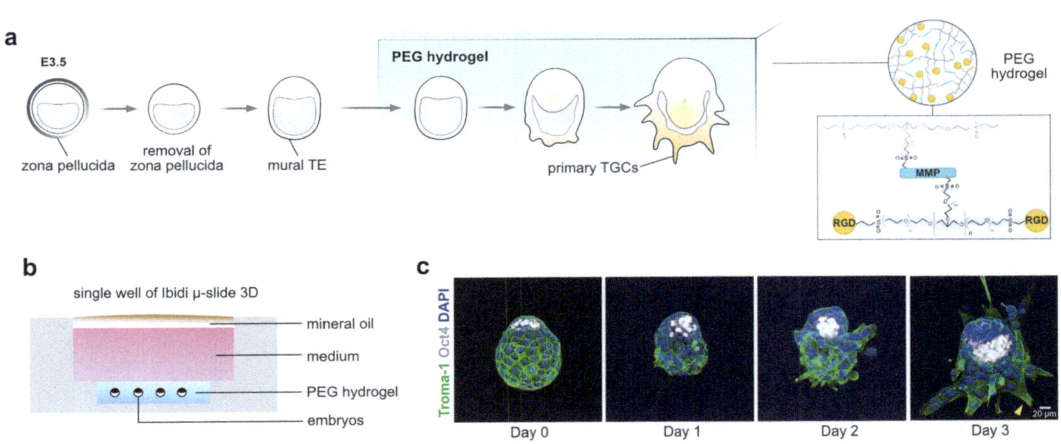

Fig. 1 In vitro culture of mouse blastocysts. (**a**) Schematic representation of zona pellucida removal, 3D culture of mouse blastocysts in PEG hydrogel and primary TGCs invasion; (**b**) schematic representation of single well of Ibidi μ-slide 15-well 3D plate containing embryos embedded in hydrogel, IVC medium, and mineral oil; (**c**) 3D culture of mouse embryos, the embryos are stained for Oct4, Troma-1, and DAPI, arrowhead indicates invasive TGCs, scale bar 20 μm

3. Transfer 5–10 blastocysts into the PEG hydrogel solution. Place the embryos at an equal distance from each other. Maintain the embryos approximately in the middle of the hydrogel depth by sequential mouth pipetting while the PEG solution gradually increases its viscosity (*see* **Note 9**). Within approximately 5 min, the position of the embryos should remain fixed as the gel polymerizes.
4. After positioning the blastocysts, place the μ-slide plate in a humidified incubator at 37 °C, 5% CO_2 for 15 min allowing the hydrogel to polymerize completely.
5. Add 40 μL of a 1:1 mixture of IVC1 and IVC2 media to the upper chamber of the well, then add 10 μL of mineral oil to prevent evaporation of the medium (*see* Fig. 1b) (*see* **Note 10**). Culture the embryos in a humidified incubator at 37 °C, 5% CO_2. If serum-free conditions are required by the experiment, a sequential IVC1, IVC2 media culture approach can be followed as previously described [11].
6. By the end of day 1, the mural region of the TE should thicken and, on day 2, the TGCs should start invading the hydrogel (*see* Fig. 1c).

3.4 In Vitro Culture of Egg Cylinder Stage Embryo and EPC Explants

1. Prepare PEG hydrogel solution as described in the Materials. Pipette 10 μL of the solution into the inner well of the Ibidi μ-slide 15-well 3D plate.
2. Transfer 1–2 Reicherts membrane-free E6.5 embryos or 1–4 EPC explants into the PEG hydrogel solution, at an equal distance from each other (*see* Fig. 2a, b).
3. After positioning the embryo or the EPCs explants, place the μ-slide plate in a humidified incubator at 37 °C, 5% CO_2 for 15 min allowing the hydrogel to polymerize completely.
4. Add 40 μL of a 1:1 mixture of IVC1 and IVC2 media and then add 10 μL of mineral oil to prevent evaporation of the medium. Alternatively, if a high-quality serum is available, only ICV1 can be used for the in vitro culture. Culture the embryos or the EPC explants in a humidified incubator at 37 °C, 5% CO_2.
5. Between day 1 and day 2, the secondary TGCs of the embryos' EPC or the EPC explants should start invading the hydrogel (*see* Fig. 2c–e).

4 Notes

1. Gelation is initiated approximately 5 min after the addition of FXIII, and gelation is complete after approximately 25 min.

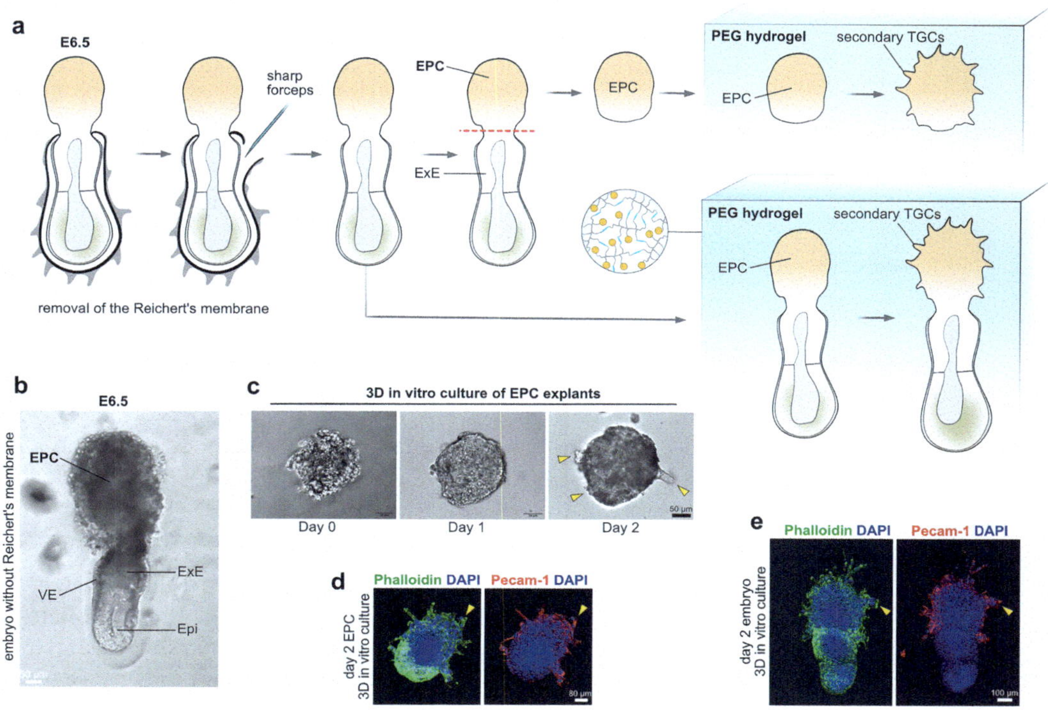

Fig. 2 In vitro culture of egg cylinder stage embryo and EPC explants. (**a**) Schematic representation of Reichert's membrane removal, EPC dissection, 3D culture of EPC explants or egg cylinder stage embryos in PEG hydrogel and secondary TGCs invasion; (**b**) E6.5 embryo without Reichert's membrane, scale bar 50 μm; (**c**) 3D culture of EPC explants, arrowheads indicate trophoblast invasion, scale bar 50 μm; (**d**) day 2 EPC explants, stained for phalloidin and the vascular marker Pecam-1, arrowhead indicates secondary TGCs invasion, scale bar 80 μm; (**e**) day 2 embryos, stained for phalloidin and Pecam-1, arrowhead indicates secondary TGCs invasion, scale bar 100 μm

2. The removal of the zona pellucida can be significantly slowed by traces of M2 medium in the Tyrode's solution. For this reason, the pipette should be refilled with fresh Tyrode's solution when embryos are sequentially transferred in the Tyrode's solution drops.
3. The zona pellucida must be completely removed. Observe the embryos carefully and if a thin layer of zona pellucida remains, repeat the treatment with Tyrode's solution. As an alternative to removing the zona pellucida, the E3.5 blastocysts can be cultured in IVC1 medium to allow the embryos to hatch naturally. Only the zona pellucida-free embryos can then be used for 3D culture, as this glycoprotein envelope can block embryo growth and trophoblast invasion.
4. The decidua is fragile, try not to tear it apart while removing the myometrium.

5. The removal of the Reichert's membrane is a critical step. As the space between the visceral endoderm and the Reichert's membrane is narrow, the chances of damaging the embryo are high.
6. The serum component of the ICV1 medium is required to stimulate embryo growth and trophoblast differentiation. The quality of the serum is therefore critical and each batch of sera must be individually tested.
7. Do not culture the embryos for more than 24 h in the 4-well plate, as the blastocysts may start to attach to the surface of the plate. Alternatively, zona pellucida-free E3.5 embryos can be directly embedded in PEG hydrogel in the presence of an IVC1 medium.
8. The optical plastic at the bottom of the μ-slide plate is thin and prone to cracking due to temperature differences between the slide and the microscope stage. To avoid cracking, place the slide in a 10 cm tissue culture plate.
9. The blastocysts should be fully embedded in the hydrogel. If the blastocysts reach the bottom of the plate, they will flatten out and form an outgrowth that spreads over the plastic surface. To avoid this, the embryos can be placed in the hydrogel at the third minute after pipetting, allowed to sink, and repositioned at the fourth minute.
10. Equilibrate the temperature and pH of the medium at 37 °C, 5% CO_2 for 30 min before pipetting onto the well.

Acknowledgments

This work was supported by the European Research Council (ERC) Consolidator grant (MORPHEUS, 101043753) and the German Research Foundation (DFG) grant BE 5800/4-1 to I.B. Further, D.P. and S.J. are supported by the International Max Planck Research School—Molecular Biology and Medicine, Münster, Germany.

Funded/Co-funded by the European Union (ERC, MORPHEUS, 101043753). Views and opinions expressed are, however, those of the author(s) only and do not necessarily reflect those of the European Union or the European Research Council. Neither the European Union nor the granting authority can be held responsible for them.

References

1. Kim YS, Bedzhov I (2022) Mechanisms of formation and functions of the early embryonic cavities. Semin Cell Dev Biol 131:110–116. https://doi.org/10.1016/j.semcdb.2022.04.020
2. Govindasamy N et al (2021) 3D biomimetic platform reveals the first interactions of the embryo and the maternal blood vessels. Dev Cell 56(23):3276–3287. e3278. https://doi.org/10.1016/j.devcel.2021.10.014
3. Zhu JY, Pang ZJ, Yu YH (2012) Regulation of trophoblast invasion: the role of matrix metalloproteinases. Rev Obstet Gynecol 5(3–4): e137–e143
4. Li Y, Sun X, Dey SK (2015) Entosis allows timely elimination of the luminal epithelial barrier for embryo implantation. Cell Rep 11(3): 358–365. https://doi.org/10.1016/j.celrep.2015.03.035
5. Ozguldez HO et al (2023) Polarity inversion reorganizes the stem cell compartment of the trophoblast lineage. Cell Rep 42(4):112313. https://doi.org/10.1016/j.celrep.2023.112313
6. Thowfeequ S, Hanna CW, Srinivas S (2024) Origin, fate and function of extraembryonic tissues during mammalian development. Nat Rev Mol Cell Biol. https://doi.org/10.1038/s41580-024-00809-w
7. Trappmann B et al (2017) Matrix degradability controls multicellularity of 3D cell migration. Nat Commun 8(1):371. https://doi.org/10.1038/s41467-017-00418-6
8. Govindasamy N et al (2023) 3D biomimetic environment enabling ex utero trophoblast invasion and co-culture of embryos and somatic cells. STAR Protoc 4(3):102456. https://doi.org/10.1016/j.xpro.2023.102456
9. Gjorevski N, Lutolf MP (2017) Synthesis and characterization of well-defined hydrogel matrices and their application to intestinal stem cell and organoid culture. Nat Protoc 12(11):2263–2274. https://doi.org/10.1038/nprot.2017.095
10. Govindasamy N, Bedzhov I (2019) Isolation and culture of periimplantation and early postimplantation mouse embryos. Methods Mol Biol 2006:373–382. https://doi.org/10.1007/978-1-4939-9566-0_25
11. Bedzhov I, Leung CY, Bialecka M, Zernicka-Goetz M (2014) In vitro culture of mouse blastocysts beyond the implantation stages. Nat Protoc 9(12):2732–2739. https://doi.org/10.1038/nprot.2014.186

Methods in Molecular Biology (2026) 3048: 265–275
DOI 10.1007/7651_2025_651

Published online: 25 July 2025

Generation of Porcine Blastoids from Embryonic Stem Cells

Jinzhu Xiang, Hanning Wang, and Duanqing Pei

Abstract

Stem cell-based blastocyst-like structure models (blastoids) that mimic preimplantation blastocysts can be used to study embryogenesis and key early embryonic developmental events. Large animals may benefit from blastoid models for purposes such as improving and accelerating breeding. We developed a three-dimensional (3D) two-step differentiation strategy to generate porcine blastoids from embryonic stem cells (ESCs). The ESC-derived blastoids exhibit similar morphology, cell lineage composition, and single-cell transcriptome characteristics to porcine blastocysts. These porcine blastoids can survive and expand for more than two weeks in vitro under two different culture conditions. Here, we describe a step-by-step protocol for the generation of porcine blastoids from ESCs.

Key words A 3D two-step differentiation strategy, Blastoids, ESCs, Porcine

1 Introduction

Stem cell-based embryo-like structures, such as blastoids, gastruloids, embryoids, and assembloids, provide useful and convenient in vitro models for exploring biological events of embryonic development. Among them, blastoids that mimic preimplantation blastocysts are valuable models for studying embryogenesis and early embryonic developmental events. Human [1–8] and mouse [9–13] blastoids have been generated using different types of stem cells and induction strategies. Similarly, blastoids from nonhuman primates, such as cynomolgus monkeys [14], and livestock, such as bovine [15], were generated by using naïve monkey ESCs and aggregating bovine extended pluripotent stem cells (EPSCs) and trophoblast stem cells (TSCs), respectively.

The pig is considered a preferable and valuable model for investigating human disease and a very suitable surrogate for humanized organs as it shares similarities with humans in physiology, immunology, and anatomical size. Recently, we derived porcine ESCs from blastocysts [16, 17] and developed a 3D two-step method to generate porcine blastoids from ESCs [17]. In this

Jinzhu Xiang and Hanning Wang contributed equally with all other contributors.

chapter, we describe a detailed protocol for the derivation of porcine ESCs and ESC-derived blastoids.

2 Materials

2.1 Reagents and Equipment

Reagents	*Source*	*Identifier*
TrypLE™ Select	Thermo Fisher Scientific	# 12563029
Neurobasal medium	Thermo Fisher Scientific	# 21103049
DMEN/F12	Thermo Fisher Scientific	# 10565018
DMEM	HyClone	# SH30022.01B
N2 Supplement	Thermo Fisher Scientific	# 17502048
B27 Supplement	Thermo Fisher Scientific	# 17504044
FBS	Thermo Fisher Scientific	# 10099141C
KnockOut™ Serum Replacement (KOSR)	Thermo Fisher Scientific	# 10828028
Nonessential amino acids (NEAA)	Thermo Fisher Scientific	# 11140050
Penicillin and streptomycin	Thermo Fisher Scientific	# 15140122
GlutaMAX	Thermo Fisher Scientific	# 35050061
2-Mercaptoethanol	Sigma-Aldrich	# M3148
Recombinant Human IL-6	Peprotech	#AF-200-06
Recombinant Human sIL-6 Receptor α	Peprotech	#200-06RC
Recombinant Human IGF1	MCE	# HY-P7018
Recombinant Human LIF	Peprotech	# 300-05
Human activin A	Peprotech	# 120-14-1000
bFGF	Peprotech	# 100-18B
2-Phospho-L-ascorbic acid	Sigma-Aldrich	# 49752
TSA	Solarbio	# IT1250
BMP4	R&D systems	# 314-BP
DZNep	Selleck	# S7120

(continued)

XAV939	Selleck	# S1180
IWR1	Selleck	# S7086
SB431542	Selleck	# S1067
CHIR99021	Tocris	# 4423
Y-27632	TargetMol	# T1725
Pronase	Sigma-Aldrich	# P8811
Mitomycin C	Selleck	# S8146
Equipment	*Source*	*Identifier*
Ultra-Low Attachment Multiple Well	Corning	# CLS3471-24EA
8 well μ-slides	ibidi	# 80826
4-well plate	Thermo Fisher Scientific	# 176740
96-well plate	Corning	# 3599
12-well plate	Corning	# 3513

2.2 Antibodies

Antibodies	Source	Identifier
GATA-4 (G-4)	Santa Cruz	# sc-25310
Gata-3 Monoclonal Antibody	Thermo Fisher Scientific	# 14-9966-82
Human/Mouse/Rat SOX2 Antibody	R&D systems	# MAB2018
Alexa Fluor 594 anti-mouse	Cell Signaling Technology	# 8890S
Alexa Fluor 488 anti-rabbit	Cell Signaling Technology	# 4412S

2.3 Mouse Embryonic Fibroblast (MEF) Medium

The MEF medium was composed of Dulbecco's Modified Eagle Medium (DMEM), 10% FBS, and penicillin/streptomycin (1×).

2.4 Porcine ESC Maintenance Medium (4FIXY Medium)

Component	Final concentration	Amount (50 mL)
Neurobasal medium	n/a	22.5 mL
DMEM/F12	n/a	22.5 mL
N2 (100×)	0.5%	250 μL
B27 (50×)	1%	500 μL

(continued)

Component	Final concentration	Amount (50 mL)
NEAA (100×)	1×	500 μL
GlutaMAX (100×)	1×	500 μL
Penicillin/streptomycin (100×)	1×	500 μL
KOSR	5%	2.5 mL
FBS	0.15%	75 μL
2-mercaptoethanol (1 M)	100 μM	5 μL
Human IL-6 (20 μg/mL)	20 ng/mL	50 μL
Human sIL-6 Receptor α (20 μg/mL)	20 ng/mL	50 μL
Activin A (20 μg/mL)	20 ng/mL	50 μL
Human IGF1 (100 μg/mL)	50 ng/mL	25 μL
XAV939 (10 mM)	2.5 μM	12.5 μL
IWR1 (25 mM)	2.5 μM	5 μL
2-Phospho-L-ascorbic acid trisodium salt (50 mg/mL)	50 μg/mL	50 μL
Y-27632 (10 mM)	5 μM	25 μL

2.5 Porcine ESC Derivation Medium

The porcine ESC derivation medium was the 4FIXY medium plus 5% FBS.

2.6 Stage I Culture Medium (4FXY Medium)

The stage I culture medium (4FXY medium) was the 4FIXY medium minus IWR1.

2.7 Stage II Culture Medium (iBlastoid Medium)

Component	Final concentration	Amount (50 mL)
Neurobasal medium	n/a	22.5 mL
DMEM/F12	n/a	22.5 mL
N2 (100×)	0.5%	250 μL
B27 (50×)	1%	500 μL
NEAA (100×)	1×	500 μL
GlutaMAX (100×)	1×	500 μL
Penicillin/streptomycin (100×)	1×	500 μL
KOSR	5%	2.5 mL
FBS	0.075%	37.5 μL

(continued)

Component	Final concentration	Amount (50 mL)
2-mercaptoethanol (1 M)	100 μM	5 μL
2-Phospho-L-ascorbic acid trisodium salt (50 mg/mL)	50 μg/mL	50 μL
Human IGF1 (100 μg/mL)	50 ng/mL	25 μL
Human IL-6 (20 μg/mL)	10 ng/mL	25 μL
Human sIL-6 Receptor α (20 μg/mL)	10 ng/mL	25 μL
Activin A (20 μg/mL)	10 ng/mL	25 μL
XAV939 (10 mM)	2.5 μM	12.5 μL
BMP4 (50 μg/mL)	5 ng/mL	5 μL
bFGF (100 μg/mL)	10 ng/mL	5 μL
CHIR99021 (6 mM)	1.5 μM	12.5 μL
Human LIF (100 μg/mL)	5 ng/mL	2.5 μL
TSA (100 μM)	2.5 nM	1.25 μL
DzNep (50 μM)	2.5 nM	2.5 μL
SB431542 (50 mM)	1 μM	1 μL
Y-27632 (10 mM)	7.5 μM	37.5 μL

3 Methods

3.1 Generation of Porcine ESCs from Embryos

3.1.1 Preparation of MEF Feeder Cells

1. Prewarm MEF medium at 37 °C.
2. Thaw the MEF feeder cells in a 37 °C water bath and transfer cells in a 15 mL conical tube containing 5 mL of MEF medium.
3. Centrifuge the cells at 250 × g for 5 min and remove the supernatant, and resuspend the pellet in the MEF medium.
4. Transfer cells into a 4-well plate at a density of about 150,000 cells per well.
5. Incubate the plate at 37 °C in a 5% CO_2 incubator overnight.

3.1.2 Derivation of Porcine ESCs

1. Take out the MEF feeder 4-well plate from 37 °C incubator, aspirate the MEF medium, and wash each well with 500 μL 1 × DPBS.
2. Aspirate the 1 × DPBS and add the 500 μL of porcine ESC derivation medium (*see* **Note 1**).
3. Digest E6 parthenogenetic blastocysts with 0.5% pronase for about 3 min (*see* **Note 2**).

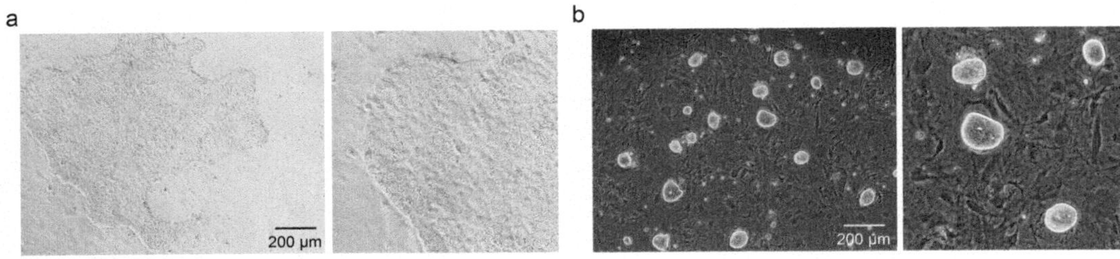

Fig. 1 Derivation of porcine ESCs. (**a**) Representative morphology of outgrowths. (**b**) Representative morphology of porcine ESCs

4. Remove the zona pellucida using a mouth pipette.
5. Seed the blastocysts without zona pellucida onto the MEF feeder 4-well plate containing the porcine ESC derivation medium.
6. Incubate the plate in a 38.5 °C incubator with 5% O_2 and 5% CO_2 for over 3 days until the blastocysts grow with adherence (*see* **Note 3**).
7. Change the medium every other two days.
8. After about 10 days, the outgrowths appear and grow (Fig. 1a).
9. Digest the outgrowths with TrypLE™. Select at 38.5 °C for 5 min (*see* **Note 4**).
10. Cut the outgrowths into small clumps using a mouth pipette.
11. Transfer the cell clumps to a new MEF feeder-coated 4-well plate containing the 4FIXY medium.
12. Incubate the plate in a 38.5 °C hypoxia incubator with 5% O_2 and 5% CO_2 overnight (*see* **Note 5**).
13. Change the 4FIXY medium every day.
14. After 3–4 days, porcine ESCs can be harvested.

3.2 Culturing of Porcine ESCs

3.2.1 Preparation of MEF Feeder Cells

1. Prewarm MEF medium at 37 °C.
2. Thaw the MEF feeder cells in a 37 °C water bath and transfer cells in a 15 mL conical tube containing 5 mL of MEF medium.
3. Centrifuge the cells at 250 × g for 5 min and aspirate the supernatant, and resuspend the pellet in the MEF medium.
4. Transfer cells into a 12-well plate at a density of about 300,000 cells per well (*see* **Note 6**).
5. Incubate the plate at 37 °C in a 5% CO_2 incubator overnight.

3.2.2 Culturing of Porcine ESCs

1. Prewarm 4FIXY medium at 37 °C or 38.5 °C.
2. Seed 30,000–50,000 ESCs onto the MEF feeder 12-well plate containing the 4FIXY medium (*see* **Notes 1** and 7).

3. Incubate the plate in a 38.5 °C hypoxia incubator with 5% O_2 and 5% CO_2 overnight.
4. Change the 4FIXY medium every day (Fig. 1b).
5. After 3–4 days, porcine ESCs can be harvested and passaged.

3.2.3 Passaging of Porcine ESCs

1. Dissociate porcine ESCs with TrypLE™. Select into single cells at 38.5 °C for 5 min (*see* **Note 4**).
2. Add 1 × DPBS to dilute the digested solution, and use a P1000 pipette to dissociate into single cells.
3. Centrifuge the cells at 250 × g for 5 min (*see* **Note 8**).
4. Remove the supernatant and resuspend the pellet using a prewarmed 4FIXY medium.
5. Count the cell number and seed into new 12-plates.

3.3 Generation of Porcine Blastoids from ESCs

3.3.1 Stage I

1. Prewarm 4FXY medium at 37 °C or 38.5 °C.
2. Dissociate porcine ESCs with TrypLE™. Select into single cells at 38.5 °C for 5 min (*see* **Notes 4** and **9**).
3. Add 1 × DPBS to dilute the digested solution, and use a P1000 pipette to dissociate colonies into single cells.
4. Centrifuge the cells at 250 × g for 5 min.
5. Remove the supernatant and resuspend the pellet using a prewarmed 4FXY medium.
6. Count the cell number, and seed 40,000 cells per well into ultra-low attachment multiple well plates in 4FXY medium (*see* **Note 10**).
7. Incubate at 38.5 °C in a 5% CO_2 incubator (*see* **Note 11**).
8. After 1–2 days, cell clusters can be observed (Fig. 2a).

3.3.2 Stage II

1. Prewarm the iBlastoid medium at 37 °C or 38.5 °C.
2. Gently collect cell clusters into a 15-mL conical tube and centrifuge the cells at 250 × g for 5 min.
3. Remove the supernatant and resuspend the pellet using prewarmed iBlastoid medium (*see* **Note 12**).
4. Incubate at 38.5 °C in a 5% CO_2 incubator (*see* **Note 11**).
5. After 3–5 days, porcine blastoids can be observed (Fig. 2b).

3.4 Immunofluorescent Staining of Porcine Blastoids

1. Use a mouth pipette to collect blastoids and transfer them into 96-well plates for immunofluorescent staining.
2. Fix blastoids with 4% paraformaldehyde for about 30 min at room temperature and wash three times with PBS.

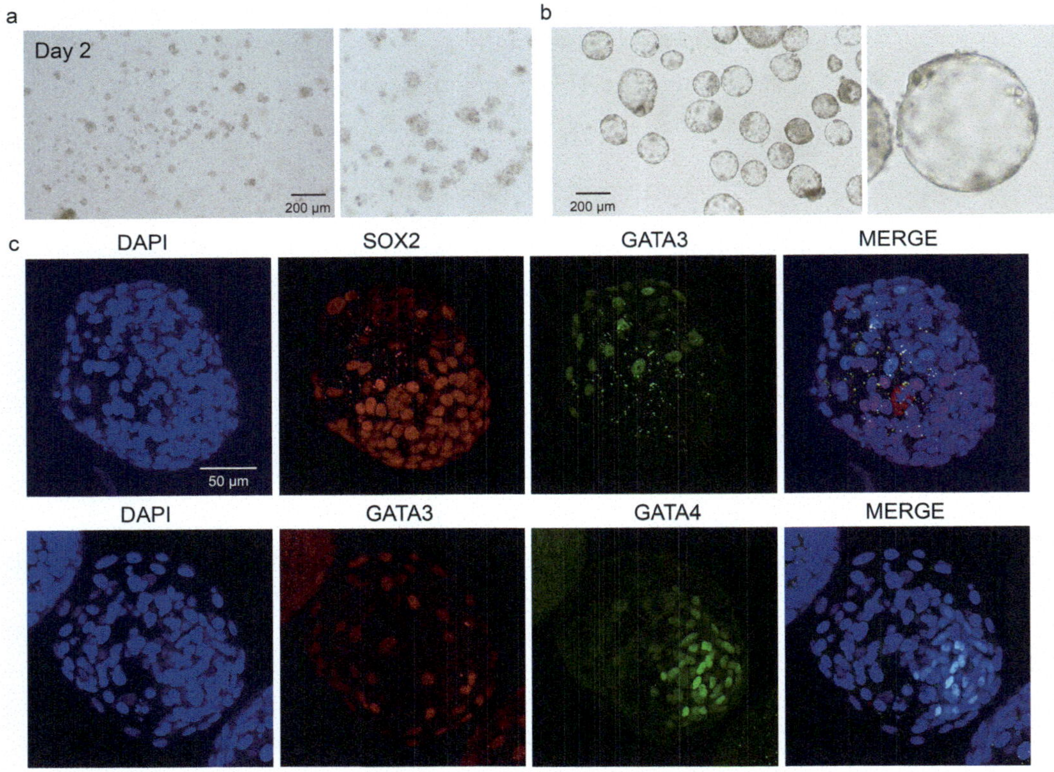

Fig. 2 Generation of porcine blastoids. (**a**) Representative images of cell clusters at indicated time points. (**b**) Representative images of porcine blastoids generated from ESCs on day 6 using a 3D two-step method. (**c**) Representative immunofluorescent staining for EPI marker SOX2, TE marker GATA3, and HYPO marker GATA4 in ESC-derived blastoids

3. Permeabilize blastoids with 0.2% Triton X-100 in PBS for 30 min and wash three times with PBS.
4. Block with 3% BSA in PBS for 40 min at room temperature.
5. Incubate blastoids with primary antibody diluted in primary antibody dilution buffer overnight at 4 °C and washed three times with PBS (*see* **Note 13**).
6. Incubate blastoids with fluorescent-dye conjugated secondary antibodies diluted in secondary antibody dilution buffer for 1 h at room temperature and washed three times with PBS (*see* **Note 12**).
7. Counterstain blastoids with DAPI at room temperature for 3 min.
8. Transfer blastoids on 8 well μ-slides and image using confocal microscopes (Fig. 2c).

4 Notes

1. The 5% FBS in the porcine ESC derivation medium is important for blastocyst adherence. The porcine ESC derivation medium minus IWR1 can also be applied for deriving ESCs. For the maintenance of ESCs, 4FIXY is required.
2. Longer digestion time may destroy the blastocysts.
3. After 3 days, the blastocysts without adherence can be transferred into a new MEF feeder-coated 4-well plate containing the porcine ESC derivation medium. They may grow with adherence.
4. TrypLE™ Express can also be used for digesting cells.
5. The hypoxia incubator is necessary for porcine ESC derivation and maintenance.
6. The density of MEF feeder cells is important for porcine ESC maintenance. If the feeder cells are sparse, porcine ESCs may differentiate. Ideally, the MEF feeder cells should be plated for 24–48 h before use.
7. The initial porcine ESC density is crucial, as overcrowded cultures will differentiate and sparse cultures will crash.
8. If the cells are not completely pelleted after centrifugation, they can be centrifuged again.
9. The optimal cell state is critical; otherwise, the induction efficiency will be significantly reduced.
10. The initial porcine ESC density for the generation of blastoids is crucial, as overcrowded or sparse cultures will reduce efficiency.
11. The hypoxia incubators are not appropriate for blastoid generation, which is different from the derivation and maintenance of ESCs.
12. Gently resuspend the pellet so as not to destroy the cell clusters.
13. The appropriate antibodies for the pig are critical; otherwise, lineage cells cannot be clearly detected.

Acknowledgments

This work was supported by the National Natural Science Foundation of China (31830060, 92068201).

References

1. Yanagida A, Spindlow D, Nichols J, Dattani A, Smith A, Guo G (2021) Naive stem cell blastocyst model captures human embryo lineage segregation. Cell Stem Cell 28(6):1016–1022 e1014. https://doi.org/10.1016/j.stem.2021.04.031
2. Yu L, Wei Y, Duan J, Schmitz DA, Sakurai M, Wang L, Wang K, Zhao S, Hon GC, Wu J (2021) Blastocyst-like structures generated from human pluripotent stem cells. Nature 591(7851):620–626. https://doi.org/10.1038/s41586-021-03356-y
3. Kagawa H, Javali A, Khoei HH, Sommer TM, Sestini G, Novatchkova M, Scholte Op Reimer Y, Castel G, Bruneau A, Maenhoudt N, Lammers J, Loubersac S, Freour T, Vankelecom H, David L, Rivron N (2022) Human blastoids model blastocyst development and implantation. Nature 601(7894):600–605. https://doi.org/10.1038/s41586-021-04267-8
4. Fan Y, Min Z, Alsolami S, Ma Z, Zhang E, Chen W, Zhong K, Pei W, Kang X, Zhang P, Wang Y, Zhang Y, Zhan L, Zhu H, An C, Li R, Qiao J, Tan T, Li M, Yu Y (2021) Generation of human blastocyst-like structures from pluripotent stem cells. Cell Discov 7(1):81. https://doi.org/10.1038/s41421-021-00316-8
5. Sozen B, Jorgensen V, Weatherbee BAT, Chen S, Zhu M, Zernicka-Goetz M (2021) Reconstructing aspects of human embryogenesis with pluripotent stem cells. Nat Commun 12(1):5550. https://doi.org/10.1038/s41467-021-25853-4
6. Liu X, Tan JP, Schroder J, Aberkane A, Ouyang JF, Mohenska M, Lim SM, Sun YBY, Chen J, Sun G, Zhou Y, Poppe D, Lister R, Clark AT, Rackham OJL, Zenker J, Polo JM (2021) Modelling human blastocysts by reprogramming fibroblasts into iBlastoids. Nature 591(7851):627–632. https://doi.org/10.1038/s41586-021-03372-y
7. Tu Z, Bi Y, Zhu X, Liu W, Hu J, Wu L, Mao T, Zhou J, Wang H, Wang H, Gao S, Wang Y (2023) Modeling human pregastrulation development by 3D culture of blastoids generated from primed-to-naive transitioning intermediates. Protein Cell 14(5):337–349. https://doi.org/10.1093/procel/pwac041
8. Yu LQ, Logsdon D, Pinzon-Arteaga CA, Duan JL, Ezashi T, Wei YL, Orsi AER, Oura S, Liu LZ, Wang L, Liu K, Ding XY, Zhan LF, Zhang JF, Nahar A, Stobbe C, Katz-Jaffe M, Schoolcraft WB, Tan T, Hon GC, Yuan Y, Wu J (2023) Large-scale production of human blastoids amenable to modeling blastocyst development and maternal-fetal cross talk. Cell Stem Cell 30(9):1246-+. https://doi.org/10.1016/j.stem.2023.08.002
9. Rivron NC, Frias-Aldeguer J, Vrij EJ, Boisset JC, Korving J, Vivie J, Truckenmuller RK, van Oudenaarden A, van Blitterswijk CA, Geijsen N (2018) Blastocyst-like structures generated solely from stem cells. Nature 557(7703):106–111. https://doi.org/10.1038/s41586-018-0051-0
10. Li R, Zhong C, Yu Y, Liu H, Sakurai M, Yu L, Min Z, Shi L, Wei Y, Takahashi Y, Liao HK, Qiao J, Deng H, Nunez-Delicado E, Rodriguez Esteban C, Wu J, Izpisua Belmonte JC (2019) Generation of blastocyst-like structures from mouse embryonic and adult cell cultures. Cell 179(3):687–702 e618. https://doi.org/10.1016/j.cell.2019.09.029
11. Sozen B, Cox AL, De Jonghe J, Bao M, Hollfelder F, Glover DM, Zernicka-Goetz M (2019) Self-organization of mouse stem cells into an extended potential blastoid. Dev Cell 51(6):698–712. https://doi.org/10.1016/j.devcel.2019.11.014
12. Liu K, Xu X, Bai D, Li Y, Zhang Y, Jia Y, Guo M, Han X, Liu Y, Sheng Y, Kou X, Zhao Y, Yin J, Liu S, Chen J, Wang H, Wang Y, Liu W, Gao S (2023) Bilineage embryo-like structure from EPS cells can produce live mice with tetraploid trophectoderm. Protein Cell 14(4):262–278. https://doi.org/10.1093/procel/pwac029
13. Kime C, Kiyonari H, Ohtsuka S, Kohbayashi E, Asahi M, Yamanaka S, Takahashi M, Tomoda K (2019) Induced 2C expression and implantation-competent blastocyst-like cysts from primed pluripotent stem cells. Stem Cell Rep 13(3):485–498. https://doi.org/10.1016/j.stemcr.2019.07.011
14. Li J, Zhu Q, Cao J, Liu Y, Lu Y, Sun Y, Li Q, Huang Y, Shang S, Bian X, Li C, Zhang L, Wang Y, Nie Y, Fu J, Li W, Mazid MA, Jiang Y, Jia W, Wang X, Sun Y, Esteban MA, Sun Q, Zhou F, Liu Z (2023) Cynomolgus monkey embryo model captures gastrulation and early pregnancy. Cell Stem Cell 30(4):362–377 e367. https://doi.org/10.1016/j.stem.2023.03.009
15. Pinzon-Arteaga CA, Wang Y, Wei Y, Ribeiro Orsi AE, Li L, Scatolin G, Liu L, Sakurai M, Ye J, Hao M, Yu L, Li B, Jiang Z, Wu J (2023) Bovine blastocyst-like structures derived from stem cell cultures. Cell Stem Cell 30(5):611–616 e617. https://doi.org/10.1016/j.stem.2023.04.003

16. Wang HN, Zhong L, Wang ZF, Xiang JZ, Pei DQ (2025) Wnt inhibition safeguards porcine embryonic stem cells from the acquisition of extraembryonic endoderm cell fates. Adv Sci 12. https://doi.org/10.1002/advs.202416802

17. Xiang JZ, Wang HN, Shi BB, Li JJ, Liu D, Wang KP, Wang ZF, Min QK, Zhao CC, Pei DQ (2024) Pig blastocyst-like structure models from embryonic stem cells. Cell Discov 10(1):72. https://doi.org/10.1038/s41421-024-00693-w

Methods in Molecular Biology (2026) 3048: 277–289
DOI 10.1007/7651_2025_654

Published online: 22 July 2025

A Deep Learning Approach to Assessing Cell Identity in Stem Cell–Based Embryo Models

Nazmus Salehin, Martin Proks, and Joshua M. Brickman

Abstract

Since the generation of embryoid bodies from embryonic stem cells (ESCs), three-dimensional differentiation has been used to mimic developmental processes. To what extent do these in vitro cell types reflect the cells generated by the embryo? We used deep learning (DL) to develop an integrated model of early human development leveraging existing single-cell RNA-seq (scRNA-seq) and using scvi-tools to both integrate and classify cell types. This tool can interrogate in vitro cell types and assign them both identity and provide an entropy score for the reliability of this classification. In this protocol we explain how to use state-of-the-art tools and our associated, publicly available DL models for early embryonic development to explore phenotypes and cell types derived in vitro. Our tools represent an important new resource to interrogate stem cell-based embryo models and the fidelity with which they recapitulate development.

Key words scRNA-seq, Variational Autoencoders, Integration, Interpretability, Classification

1 Introduction

With the advent of an increasing number of different stem cell-based embryo models, it has become essential to find new ways to evaluate the extent to which they recapitulate development. The existence of stem cell cultures that represent the three principal cell types of the preimplantation human embryo has led to a variety of three-dimensional differentiation approaches that mimic aspects of primate development. These cell culture models include naïve embryonic stem cells (ESCs) representing early epiblast [1–3], naïve extra-embryonic endoderm (nEnd) representing primitive endoderm [4], and trophoblast stem cells (TSCs) representing trophectoderm [5–7]. These cell types can be differentiated alone or in combination to give rise to structures that resemble elements of embryos [8–12]. To what extent do these three-dimensional

Nazmus Salehin and Martin Proks contributed equally with all other contributors.

Supplementary Information The online version contains supplementary material available at https://doi.org/10.1007/7651_2025_654.

differentiations truly recapitulate development? The tool described in this chapter facilitates addressing this question.

Single-cell RNA-seq of early human development has created a huge data resource [8, 13–17] which facilitates attempts to benchmark these models. It provides an abundance of data that both informs on heterogeneity in cell culture and enables the direct comparison of in vivo and in vitro cell types. However, utilizing entire datasets for these comparisons has been complex, initially involving simplistic approaches to align datasets or via normalization strategies that risk losing important facets of information with in a dataset [18, 19].

To reduce the complexity of these analyses and assign cell identities, it is common to select specific sets of genes that are thought to be key markers of a cell type. While this approach has been effective, it risks missing key unknown differences between in vivo and in vitro cell types. One can also simplify populations by defining clusters and comparing their overall gene expression and though this has been very effective at comparing different in vitro protocols, it does not make individual cell type assignments but rather provides a similarity score [20].

To circumvent these problems and avoid the distortion of data introduced when integrating datasets from multiple different technologies, we employ deep learning integration techniques [21]. We have used these to build a model for early mammalian development which can be used to query the identity of cell types in embryo models (Fig. 1a). Human development begins with fertilization of an egg forming a zygote which undergoes rounds of cleavage divisions until zygotic transcription is activated around the 8-cell stage [22]. From the 16-cell stage, the outer cells become polarized and give rise to the trophoblast (future placenta) and the inner cells, the ICM [23, 24]. This specification event is followed by blastulation, where the embryo generates a fluid-filled cavity which pushes the ICM to one side. The next cell specification event occurs quite rapidly in primates and, therefore, has been somewhat controversial [15, 17, 25]. It involves the differentiation of ICM cells into the epiblast (future embryo) and primitive endoderm (primarily future yolk sac) in a process that is FGF/ERK dependent [4, 26, 27].

Our model was built by collating all single scRNA-seq datasets available to date and utilizing scvi-tools [28] to create a probabilistic model for early development (Fig. 1b). All processed reference datasets and models are made available on both Zenodo and HuggingFace. Based on our initial mouse model, we used scANVI [29] to train a classifier which predicts cell labels and can be used to query unknown cell types and identify them. Finally, we developed scANVIExplainer [30–32], which estimates and ranks the contributions of different features in cell assignment and provide a set of putative markers as Shapley values [31, 32]. Our current model is available online at https://huggingface.co/spaces/brickmanlab/

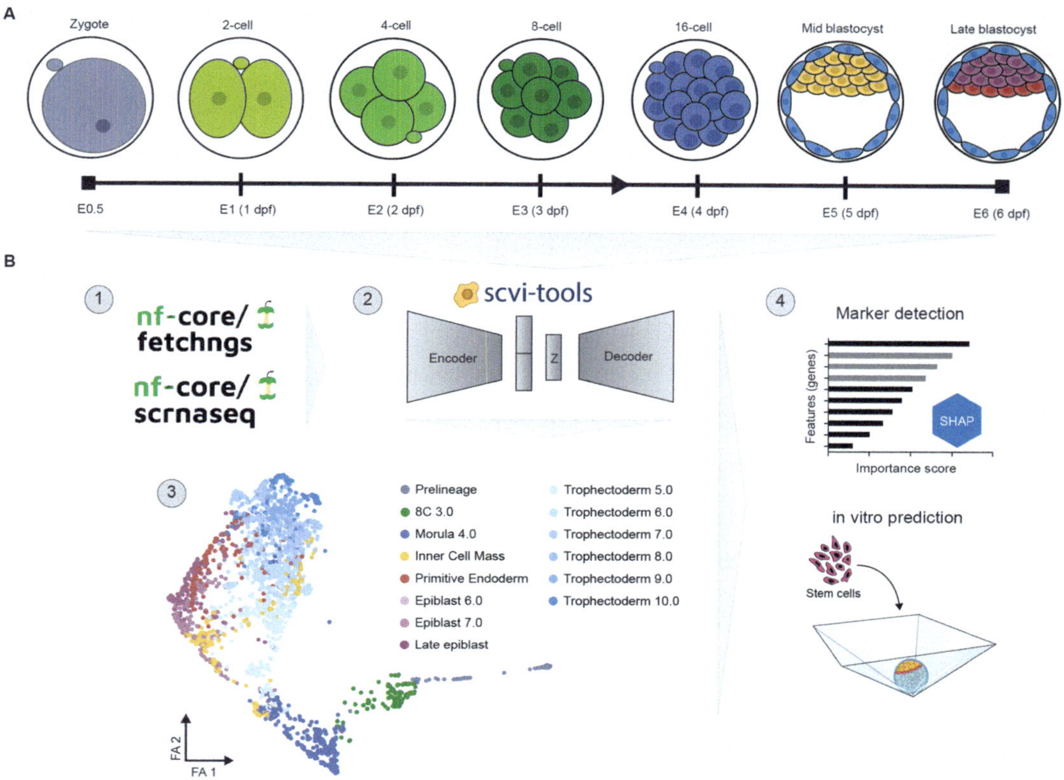

Fig. 1 Schematic illustration of our deep learning model for early human development. (**a**) Overview of preimplantation human development. (**b**) Workflow for model training and application: Downloading and preprocessing data (step 1). Integration and training of the model (step 2). Downstream analysis and visualization using a forced directed graph (step 3). Derivation of unique marker sets and predicting cell identity in a stem cell-based embryo models (step 4). Figure adapted from [30, 35]

hf-preimplantation-portal and we have used it to assess the identity of 8C-like cells [33] and one existing stem cell–based embryo model [9].

This protocol explains how to download and apply our model to quantify the extent to which a stem cell-based embryo model recapitulates development. We have extended the human model to the gastrulation stage, and these updates will be available online in the near future.

2 Materials

High performance computing (HPC) machine with graphics card(s) for optimal performance (*see* **Note 1**).

3 Methods

3.1 Software

1. For preprocessing and alignment of FASTQ reads: nextflow and nf-core (optional).
2. For downstream analysis.
   (a) Interface: Jupyter Notebook or Visual Studio Code with Jupyter extension.
   (b) Processing of scRNA-seq dataset: scanpy.
   (c) General data manipulation: numpy, pandas.
   (d) Plotting and visualization: matplotlib, seaborn.
   (e) Integration and cell type prediction: scvi-tools (scVI, scANVI).
   (f) Model interpretation (scANVIExplainer)
3. We recommend creating a Conda environment and installing the required dependencies with the Python package manager pip. This recipe can be defined in YAML format as shown below (environment.yml).

```
environment.yml

name: integration
channels:
  - anaconda
  - conda-forge
  - bioconda
  - defaults
dependencies:
  - python=3.10.11
  - pip
  - pip:
    - ipykernel     # required by Jupyter
    - ipywidgets    # additional widgets for Jupyter
    - scanpy[leiden]
    - scvi-tools
    - huggingface-hub
```

Executing conda env create -f environment.yml will create a conda environment called integration. Enabling GPU requires manual installation of an additional package, JAX. Please follow the instructions from https://docs.jax.dev/en/latest/installation.html#installation. You can test your setup by running the following commands.

```
$ source activate integration
$ python -m "import jax,scanpy,scvi; print(jax.devices())"
```

The command above should not return any import error message and should display available GPUs in the format: [cuda (id = 0), cuda(id = 1)] (*see* **Note 2**).

3.2 Preparing Genome References Manually

1. Download the human GRCh38 reference genome and version 110 of Ensembl gene annotations (*see* **Note 3**).

```
$ mkdir -p references/GRCh38_110
$ wget -P references/GRCh38_110 https://ftp.ensembl.org/pub/release-
110/fasta/homo_sapiens/dna/Homo_sapiens.GRCh38.dna_sm.primary_assembly.fa.gz
$ wget -P references/GRCh38_110 https://ftp.ensembl.org/pub/release-
110/gtf/homo_sapiens/Homo_sapiens.GRCh38.110.gtf.gz
$ gunzip references/GRCh38_110/*
```

2. Generate a STAR v2.7.10a index, required for alignment of raw reads.

```
$ STAR \
  --runMode genomeGenerate \
  --runThreadN 30 \
  --genomeDir references/GRCh38_110/star_2.7.10a \
  --genomeFastaFiles
references/GRCh38_110/Homo_sapiens.GRCh38.dna_sm.primary_assembly.fa \
  --sjdbGTFfile references/GRCh38_110/Homo_sapiens.GRCh38.110.gtf \
  --sjdbOverhang 99
```

3.3 Alignment of Raw Data

1. Align the raw sequencing reads (FASTQ files) using the Brickman lab fork [34] of the single-cell RNA sequencing nf-core pipeline (brickmanlab/scrnaseq) (*see* **Note 4**).
2. Create a samplesheet file (see Electronic Supplementary Material: samplesheet.csv) where you define all your samples and corresponding raw FASTQ files. For processing of libraries that are not created using SMART-seq or 10× Chromium technologies, you will have to create a custom samplesheet file with an additional configuration file for STAR aligner. Below we provide examples for single-end (SMART-seq) and paired-end (10× Chromium) sequencing experiments.

3.3.1 SMART-Seq Example

```
samplesheet.singleEnd.csv (SMART-seq)

sample,fastq_1,fastq_2
blastoid_1,/scratch/fastq/READ_1.fastq.gz,
```

(*see* **Note 5**)

```
smartseq.config (SMART-seq)

params {
    fasta = "references/GRCh38_110/Homo_sapiens.GRCh38.dna_sm.primary_assembly.fa"
    gtf = "references/GRCh38_110/Homo_sapiens.GRCh38.110.gtf"
    star_index = "references/GRCh38_110/star_2.7.10a"
    aligner = "star"
    protocol = "smartseq"
}
process {
    withName: STAR_ALIGN {
        ext.args = "--readFilesCommand zcat --soloUMIdedup NoDedup --soloStrand
Unstranded"
    }
}
```

3.3.2 10× Example

```
samplesheet.pairend.csv (10X)

sample,fastq_1,fastq_2
pbmc8k,pbmc8k_S1_L007_R1_001.fastq.gz,pbmc8k_S1_L007_R2_001.fastq.gz
```

```
10X.config (10X)

params {
    fasta = "references/GRCh38_110/Homo_sapiens.GRCh38.dna_sm.primary_assembly.fa"
    gtf = "references/GRCh38_110/Homo_sapiens.GRCh38.110.gtf"
    star_index = "references/GRCh38_110/star_2.7.10a"
    aligner = "star"
    protocol = "10XV2" // can be also "10XV1" or "10XV3"
}
process {
    withName: STAR_ALIGN {
        ext.args = "--readFilesCommand zcat"
    }
}
```

3. Execute the nextflow pipeline to produce a cell by gene count matrix in h5ad format for downstream processing (*see* **Notes 6** and **7**).

```
$ nextflow run brickmanlab/scrnaseq
    -r feature/smartseq \
    -profile <profile> \
    -c smartseq.config \
    --input samplesheet.singleEnd.csv \
    --output aligned_GRCh38v110/
```

or

```
$ nextflow run brickmanlab/scrnaseq
    -r feature/smartseq \
    -profile <profile> \
    -c 10X.config \
    --input samplesheet.paired.csv \
    --output aligned_GRCh38v110/
```

3.4 Integration and Cell Type Prediction

For demonstration, we will be using a SMART-seq2 dataset of human blastoids generated in the Rivron lab [9].

1. Download raw FASTQ files (see Electronic Supplementary Material: ids_Kagawa.csv).

```
$ nextflow run nf-core/fetchngs -profile <profile> --input ids_Kagawa.csv --outdir
Kagawa_2022
```

2. Align reads.

```
$ nextflow run brickmanlab/scrnaseq \
    -r feature/smartseq \
    -c smartseq.config \
    -profile <profile> \
    --input samplesheet_Kagawa.csv \
    --outdir Kagawa_2022
```

3. Load libraries with helper functions in a JupyterLab Python environment.

```
%matplotlib inline

import numpy as np
import pandas as pd
import seaborn as sns

import scvi
import scanpy as sc
import matplotlib.pyplot as plt
from scvi.hub import HubModel


def normalize_smartseq(
  adata: anndata.AnnData,
  gene_len_file: str,
  column: Literal['mean', 'median'] = 'mean')
-> anndata.AnnData:

    print("SMART-SEQ: Normalization")
    gene_len = pd.read_table(gene_len_file).set_index('gene')
    common_genes = adata.var_names.intersection(gene_len.index)
    print(f"SMART-SEQ: Common genes {common_genes.shape[0]}")
    lengths = gene_len.loc[common_genes, column].values
    normalized = adata[:, common_genes].copy()
    if issparse(normalized.X):
        normalized.X = normalized.X.A
    normalized.X = normalized.X / lengths * np.median(lengths)
    normalized.X = np.rint(normalized.X)

    return normalized
```

4. Load in your data from step 1.2 (*see* **Note 8**).

```
adata = sc.read_h5ad('Kagawa_2022/star/mtx_conversions/combined_matrix.h5ad')
adata
```

5. Normalize SMART-seq raw reads to mean gene length. To extract gene lengths, we apply gtftools to the Ensembl annotation file.

```
$ wget -P references/GRCh38_110
https://raw.githubusercontent.com/RacconC/gtftools/refs/heads/master/gtftools/gtftools.p
y
$ python3.10 references/GRCh38_110 ./gtftools.py -l
references/GRCh38_110/Homo_sapiens.GRCh38.110.gene_length.tsv \
references/GRCh38_110/Homo_sapiens.GRCh38.110.gtf
```

6. Use the helper function provided in 4.2 for normalization.

```
normalize_smartseq(adata,
'references/GRCh38_110/Homo_sapiens.GRCh38.110.gene_length.tsv')
```

7. Calculate quality control metrics and visualize the mitochondrial read percentage per cell (Fig. 2) (*see* **Note 9**).

```
adata.var['mt'] = adata.var.gene_symbol.str.startswith('MT-')
sc.pp.calculate_qc_metrics(adata, qc_vars=['mt'], percent_top=None, log1p=False,
inplace=True)

sns.violinplot(y=adata.obs['pct_counts_mt'], orient='v')
```

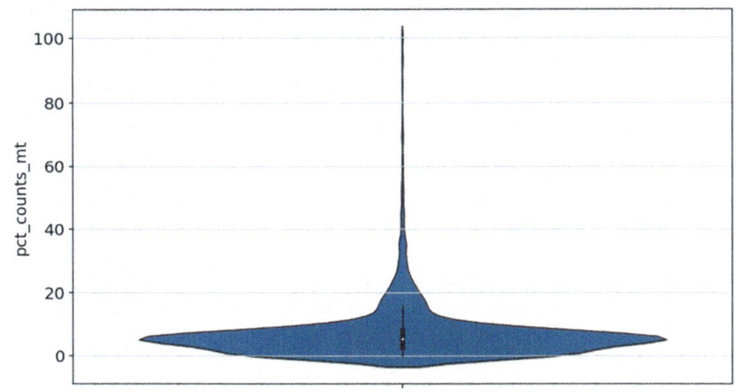

Fig. 2 Distribution of mitochondrial percentages. Violin plot depicting the percentage of mitochondrial content on y-axis for each cell

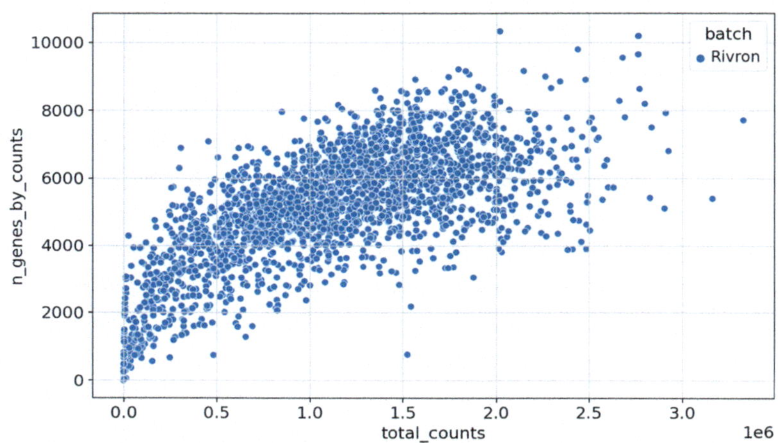

Fig. 3 Quality control using total reads and the number of genes detected. Scatter plot, where each dot represents an individual cell, determined by the total number of transcripts (x-axis) vs the number of expressed genes (y-axis)

8. Filter out cells with a high mitochondrial read percentage. We chose 12.5% as a general cutoff in this dataset—the value will change according to your data.

```
adata = adata[adata.obs.pct_counts_mt < 12.5].copy()
```

9. Visualize the distribution of read depth to number genes quantified to determine cutoffs to remove both under- and over-sequenced cells (Fig. 3).

```
sns.scatterplot(x='total_counts', y='n_genes_by_counts', data=adata.obs, hue='batch')
```

For this example, we decided to remove cells with fewer than 250,000 or more than 2,500,000 counts. In addition, we remove cells in which fewer than 2,000 genes were quantified.

```
sc.pp.filter_cells(adata, min_counts=2.5e5)
sc.pp.filter_cells(adata, max_counts=2.5e6)
sc.pp.filter_cells(adata, min_genes=2_000)
```

10. Before depth normalizing the data, the current counts are stored in a `counts` layer of the data structure.

```
adata.layers["counts"] = adata.X.copy()
```

11. The counts are then depth normalized to 10,000 reads and then log transformed (Log(reads+1)) (*see* **Note 10**).

```
sc.pp.normalize_total(adata, target_sum=10_000)
sc.pp.log1p(adata)
adata.raw = adata
```

12. Before integrating the data to the reference, mitochondrial and ribosomal reads are removed (*see* **Note 11**).

```
# remove mitochondrial genes
adata = adata [:, adata.var[~adata.var.gene_symbol.str.startswith('MT-')].index].copy()

# remove ribosomal genes
adata = adata[:, adata.var[~adata.var.gene_symbol.str.startswith(('RPS',
'RPL'))].index].copy()
```

13. Load pre-trained scANVI model for cell-type classification from Hugging Face (*see* **Note 12**).

```
hmo = HubModel.pull_from_huggingface_hub(
    repo_name="brickmanlab/human-scanvi",
    revision="v1.0.1",
)
lvae = hmo.model
```

14. Run the preparation step of scANVI which checks how many genes in the in vitro dataset overlap with the reference model (*see* **Note 13**).

```
query = adata.copy()
scvi.model.SCVI.prepare_query_anndata(query, lvae)
```

15. Perform query integration with the reference scANVI model for cell type classification.

```
lvae_q = scvi.model.SCANVI.load_query_data(query, lvae)
lvae_q.train(
    max_epochs=100,
    plan_kwargs=dict(weight_decay=0.0),
    check_val_every_n_epoch=10,
    early_stopping=True
)
```

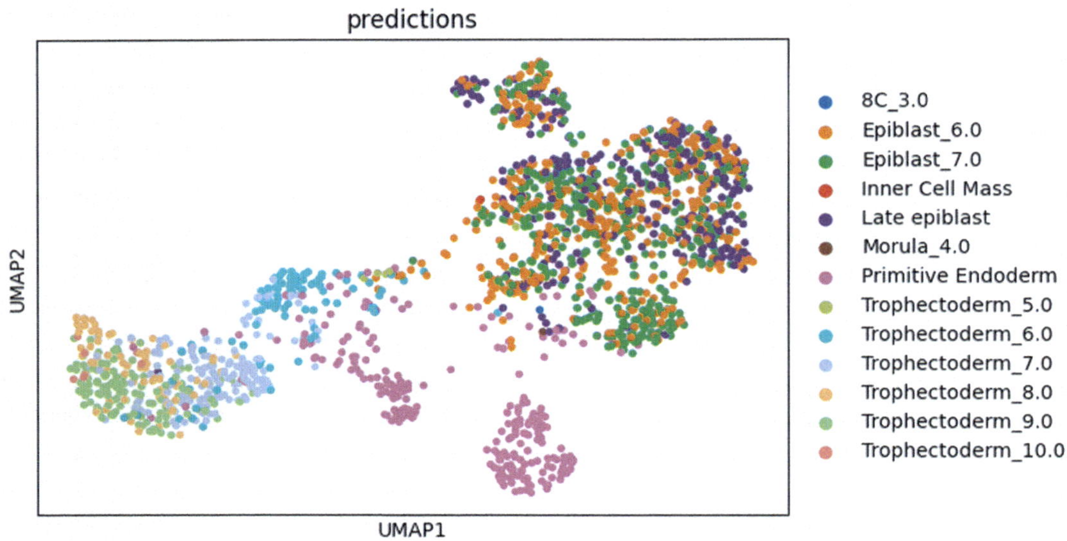

Fig. 4 Dimensional reduction visualization. Representation of the scRNA-seq experiment using UMAP with colors depicting cell types as predicted by the scANVI classifier

16. Store the following outputs from the model—the scANVI reduced dimensional space with your dataset, cell type, and entropy (or uncertainty) score of the predictions.

```
query.obsm["X_scANVI"] = lvae_q.get_latent_representation()
query.obs["predictions"] = lvae_q.predict()
query.obs['entropy'] = 1 - lvae_q.predict(soft=True).max(axis=1)
```

17. The query data can now be visualized using standard methods.

```
sc.pp.highly_variable_genes(
    query,
    flavor="seurat_v3",
    n_top_genes=5_000,
    layer="counts",
    batch_key="batch",
    subset=True,
)

sc.pp.neighbors(query, use_rep="X_scANVI")
sc.tl.umap(query)
```

18. The predicted cell identity and entropy can now be visualized on an UMAP (Figs. 4 and 5).

```
sc.pl.umap(query, color=['predictions',  'entropy'], ncols=1)
```

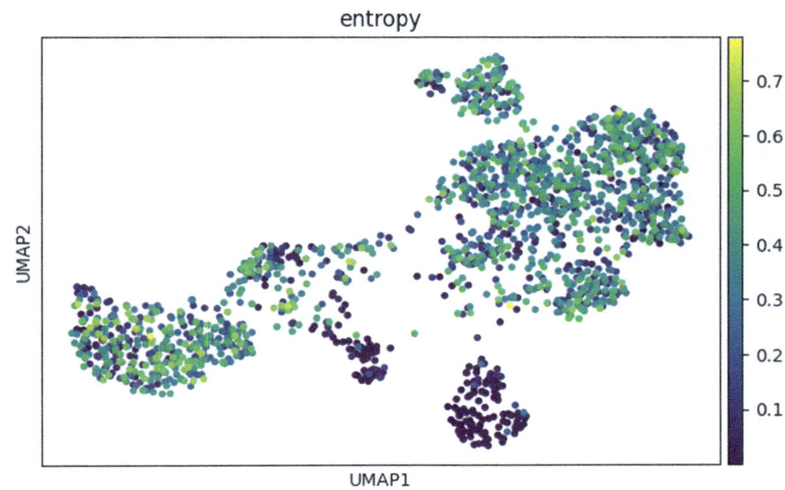

Fig. 5 Certainty score. UMAP visualization of scRNA-seq experiment where the color of each cell depicts the entropy score (range 0 to 1) or confidence. Cells with low entropy represent high confidence, while high entropy represents low certainty

4 Notes

1. Integration and cell type prediction can be executed without GPUs but might not scale well for large datasets.
2. In case Jupyter Lab is installed in a separate conda environment, you may additionally run conda install nb_conda_kernels, which will scan and detect all available conda environments that contain ipykernel.
3. In this example, we download and decompress the references to the references/GRCh38_110 folder. Please make sure to generate the reference in your scratch folder or drive in order not to overwhelm your network drives. The indexes need to be the same as used by the reference, in this case GRCh38 with Ensembl version 110, built using version 2.7.10a of STAR.
4. The Brickman lab fork of the pipeline introduces modifications for processing of data generated with full-length protocols such as SMART-seq.
5. Make sure the line ends with comma, otherwise the pipeline will terminate with an error message.
6. Some research institutions have university profiles for nextflow and nf-core. You can find these at https://nf-co.re/configs/.
7. In case you want to track your pipeline execution online in real time, you can use the Sequera Platform following steps at https://training.nextflow.io/latest/basic_training/seqera_platform/. Remember to export TOWER_WORKSPACE_ID in

terminal and including -with-tower flag in the nextflow execution above.

8. In this step, you can include additional metadata information which can better describe the individual cells.
9. High levels of mitochondrial genes are associated with dying cells. You can also include additional metrics such as ERCC spike-ins if used, or ribosomal content ("RPS", "RPL").
10. By not setting target_sum, the cells will be normalized by median sum.
11. We remove mitochondrial and ribosomal genes as these genes should not be used by classifier to determine a cell type.
12. We constantly improve the models and therefore it is important to keep track of which revision is used when predicting a cell type. For more info, see the documentation at https://huggingface.co/brickmanlab/human-scanvi.
13. The human model expects Ensembl gene ids as names. The helper function prints the percentage of common genes, which will give you an indication if your data is incorrectly formatted.

References

1. Gafni O, Weinberger L, Mansour AA et al (2013) Derivation of novel human ground state naive pluripotent stem cells. Nature 504:282–286. https://doi.org/10.1038/nature12745
2. Takashima Y, Guo G, Loos R et al (2015) Resetting transcription factor control circuitry toward ground-state pluripotency in human. Cell 162:452–453. https://doi.org/10.1016/j.cell.2015.06.052
3. Bredenkamp N, Stirparo GG, Nichols J et al (2019) The cell-surface marker sushi containing domain 2 facilitates establishment of human naive pluripotent stem cells. Stem Cell Reports 12:1212–1222. https://doi.org/10.1016/j.stemcr.2019.03.014
4. Linneberg-Agerholm M, Wong YF, Romero Herrera JA et al (2019) Naïve human pluripotent stem cells respond to Wnt, nodal and LIF signalling to produce expandable naïve extra-embryonic endoderm. Development 146:dev180620. https://doi.org/10.1242/dev.180620
5. Dong C, Beltcheva M, Gontarz P et al (2020) Derivation of trophoblast stem cells from naïve human pluripotent stem cells. elife 9:10.7554/elife.52504
6. Cinkornpumin JK, Kwon SY, Guo Y et al (2020) Naive human embryonic stem cells can give rise to cells with a trophoblast-like transcriptome and methylome. Stem Cell Reports 15:198–213. https://doi.org/10.1016/j.stemcr.2020.06.003
7. Okae H, Toh H, Sato T et al (2018) Derivation of human trophoblast stem cells. Cell Stem Cell 22:50–63.e6. https://doi.org/10.1016/j.stem.2017.11.004
8. Yanagida A, Spindlow D, Nichols J et al (2021) Naive stem cell blastocyst model captures human embryo lineage segregation. Cell Stem Cell 28:1016–1022.e4. https://doi.org/10.1016/j.stem.2021.04.031
9. Kagawa H, Javali A, Khoei HH et al (2022) Human blastoids model blastocyst development and implantation. Nature 601:600–605. https://doi.org/10.1038/s41586-021-04267-8
10. Liu X, Tan JP, Schröder J et al (2021) Modelling human blastocysts by reprogramming fibroblasts into iBlastoids. Nature 591:627–632. https://doi.org/10.1038/s41586-021-03372-y
11. Yu L, Wei Y, Duan J et al (2021) Blastocyst-like structures generated from human pluripotent stem cells. Nature 591:620–626. https://doi.org/10.1038/s41586-021-03356-y
12. Fan Y, Min Z, Alsolami S et al (2021) Generation of human blastocyst-like structures from pluripotent stem cells. Cell Discov 7:1–14. https://doi.org/10.1038/s41421-021-00316-8

13. Xue Z, Huang K, Cai C et al (2013) Genetic programs in human and mouse early embryos revealed by single-cell RNA sequencing. Nature 500:593–597. https://doi.org/10.1038/nature12364
14. Yan L, Yang M, Guo H et al (2013) Single-cell RNA-Seq profiling of human preimplantation embryos and embryonic stem cells. Nat Struct Mol Biol 20:1131–1139. https://doi.org/10.1038/nsmb.2660
15. Meistermann D, Bruneau A, Loubersac S et al (2021) Integrated pseudotime analysis of human pre-implantation embryo single-cell transcriptomes reveals the dynamics of lineage specification. Cell Stem Cell 28:1625–1640.e6. https://doi.org/10.1016/j.stem.2021.04.027
16. Xiang L, Yin Y, Zheng Y et al (2020) A developmental landscape of 3D-cultured human pre-gastrulation embryos. Nature 577:537–542. https://doi.org/10.1038/s41586-019-1875-y
17. Petropoulos S, Edsgärd D, Reinius B et al (2016) Single-cell RNA-Seq reveals lineage and X chromosome dynamics in human preimplantation embryos. Cell 165:1012–1026. https://doi.org/10.1016/j.cell.2016.03.023
18. Luecken MD, Büttner M, Chaichoompu K et al (2022) Benchmarking atlas-level data integration in single-cell genomics. Nat Methods 19:41–50. https://doi.org/10.1038/s41592-021-01336-8
19. Brendel M, Su C, Bai Z et al (2022) Application of deep learning on single-cell RNA sequencing data analysis: a review. Genomics Proteomics Bioinformatics 20:814–835. https://doi.org/10.1016/j.gpb.2022.11.011
20. Rothová MM, Nielsen AV, Proks M et al (2022) Identification of the central intermediate in the extra-embryonic to embryonic endoderm transition through single-cell transcriptomics. Nat Cell Biol 24:833–844. https://doi.org/10.1038/s41556-022-00923-x
21. Lopez R, Gayoso A, Yosef N (2020) Enhancing scientific discoveries in molecular biology with deep generative models. Mol Syst Biol 16:e9198. https://doi.org/10.15252/msb.20199198
22. Yuan S, Zhan J, Zhang J et al (2023) Human zygotic genome activation is initiated from paternal genome. Cell Discovery 9:13
23. Nishioka N, Inoue K, Adachi K et al (2009) The hippo signaling pathway components lats and yap pattern tead4 activity to distinguish mouse trophectoderm from inner cell mass. Dev Cell 16:398–410. https://doi.org/10.1016/j.devcel.2009.02.003
24. Gerri C, McCarthy A, Alanis-Lobato G et al (2020) Initiation of a conserved trophectoderm program in human, cow and mouse embryos. Nature 587:443–447. https://doi.org/10.1038/s41586-020-2759-x
25. Radley A, Corujo-Simon E, Nichols J et al (2023) Entropy sorting of single-cell RNA sequencing data reveals the inner cell mass in the human pre-implantation embryo. Stem Cell Reports 18:47–63. https://doi.org/10.1016/j.stemcr.2022.09.007
26. Dattani A, Simon EC, Radley A et al (2023) Human naive stem cell models reveal the role of FGF signalling in hypoblast specification in the human embryo:2023.11.30.569161
27. Simon CS, McCarthy A, Woods L et al (2024) Suppression of ERK signalling promotes pluripotent epiblast in the human blastocyst:2024.02.01.578414
28. Gayoso A, Lopez R, Xing G et al (2022) A python library for probabilistic analysis of single-cell omics data. Nat Biotechnol 40:163–166. https://doi.org/10.1038/s41587-021-01206-w
29. Xu C (2021) Probabilistic harmonization and annotation of single-cell transcriptomics data with deep generative models. Mol Syst Biol 17:e9620. https://doi.org/10.15252/msb.20209620
30. Proks M, Salehin N, Brickman JM (2025) Deep learning-based models for preimplantation mouse and human embryos based on single-cell RNA sequencing. Nat Methods 22:207–216. https://doi.org/10.1038/s41592-024-02511-3
31. Lundberg SM, Lee S-I (2017) A unified approach to interpreting model predictions. In: Advances in neural information processing systems. Curran Associates, Inc.
32. Lundberg SM, Erion G, Chen H et al (2020) From local explanations to global understanding with explainable AI for trees. Nat Mach Intell 2:56–67. https://doi.org/10.1038/s42256-019-0138-9
33. Mazid MA, Ward C, Luo Z et al (2022) Rolling back human pluripotent stem cells to an eight-cell embryo-like stage. Nature 605:315–324. https://doi.org/10.1038/s41586-022-04625-0
34. Peltzer A, de Almeida FM, Sturm G, et al (2023) Nf-core/scrnaseq: nf-core/scrnaseq 2.4.1
35. Linneberg-Agerholm M, Sell AC, Redó-Riveiro A et al (2024) The primitive endoderm supports lineage plasticity to enable regulative development. Cell 187:4010–4029.e16. https://doi.org/10.1016/j.cell.2024.05.051

Methods in Molecular Biology (2026) 3048: 291–303
DOI 10.1007/7651_2025_666

Published online: 20 September 2025

An Optimized Protocol for the Extended Culture of Gastruloids from Mouse Embryonic Stem Cells

Marloes Blotenburg, Shruthi Buddharaju, Beril Esin, and Peter Zeller

Abstract

Gastruloids are three-dimensional aggregates, derived from mouse embryonic stem cells, that recapitulate key events of early embryogenesis. Protocols to generate gastruloids are highly sensitive to aggregation conditions, often resulting in variability. Here, we present an optimized method for generating gastruloids and extending their culture up to 168 h post-aggregation by embedding them in 10% Matrigel at 96 h. This protocol enables reproducible generation of gastruloids with derivatives of all three germ layers, providing an extended experimental window to study post-gastrulation developmental processes in vitro.

Key words Early embryonic development, Extended culture, Gastruloids, Matrigel, Mouse embryonic stem cells

1 Introduction

Gastruloids are three-dimensional aggregates derived from mouse embryonic stem cells (mESCs) that mimic key events of early post-implantation development, including germ layer formation, symmetry breaking, and axial elongation [1–3]. As an in vitro model system, they offer a tractable and reproducible platform to study mammalian embryogenesis under controlled conditions [4].

The formation and developmental progression of gastruloids are highly sensitive to culture conditions, including mESC pre-culture [see chapter "Optimization of gastruloid pre-culture conditions for reproducible germ layer composition"], aggregation parameters, and the timing of signaling molecule exposure [5]. Careful optimization and standardization of these steps are critical to minimize variability across and within experiments.

Here, we describe an extended protocol for gastruloid generation and culture up to 168 h after aggregation (AA). At 96 h AA, gastruloids are embedded in 10% Matrigel to facilitate differentiation, following previously established methods [6, 7]. We further modify this protocol to allow extended culture up to 168 h AA

Authors Marloes Blotenburg and Shruthi Buddharaju have equally contributed to this chapter.

(7 days AA). This approach yields gastruloids containing cell types from all three germ layers and reproducible morphologies [8].

2 Materials

2.1 Reagents

2.1.1 Pre-culture

1. Gelatin (0.1%, e.g. diluted from Sigma G1393).
2. ESLIF medium (500 mL):
    1. 450 mL G-MEM (Gibco 21710025).
    2. 50 mL Fetal bovine serum (Embryonic stem cell FBS, Gibco 16141079).
    3. 5 mL Sodium pyruvate (100 mM, Gibco 11360070).
    4. 5 mL non-essential amino acids (100×, Gibco 11140035).
    5. 5 mL GlutaMAX (100×, Gibco 35050061).
    6. 5 mL PenStrep (100×, Gibco 15140122).
    7. 1 mL β-mercaptoethanol (50 mM in PBS; Sigma PHR9274).
    8. 50 μL Leukemia inhibitory factor (mLIF) (stock: ten million units/mL; final concentration in medium: 1000 units/mL, ESGRO ESG1106).
3. 2i medium (500 mL):
    1. 250 mL DMEM/F12 (Gibco 31331028).
    2. 250 mL Neurobasal medium (Gibco 21103049).
    3. 2.5 mL N-2 supplement (Gibco, 17502048).
    4. 5 mL B-27 supplement (50×, serum free, Gibco, 17504044).
    5. 5 mL GlutaMAX (100×, Gibco 35050061).
    6. 5.5 mL PenStrep (100×, Gibco 15140122).
    7. 1 mL β-mercaptoethanol (50 mM in PBS; Sigma PHR9274).

       The following supplements should be added to the medium prior to use. The volumes listed below are for 500 mL of medium (*see* **Note 1**).
    8. 50 μL Leukemia inhibitory factor (mLIF) (stock: ten million units/mL; final concentration in medium: 1000 units/mL, ESGRO ESG1106).
    9. 150 μL Chiron (CHIR99021, 3 μM final concentration, from 10 mM stock in DMSO, Tocris, 4423).
    10. 50 μL PD0325901 (1 μM final concentration, from 10 mM stock in DMSO, Sigma, PZ0162).
4. PBS- (without Ca^{2+} and Mg^{2+}, Gibco 14190144).
5. TrypLE (Gibco 12605010).

2.1.2 Gastruloid Generation

1. ESLIF medium.
2. TrypLE (Gibco 12605010).
3. N2B27 medium (Takara, Y40002).
4. PBS+ (with Ca^{2+} and Mg^{2+}, Gibco 14040133).
5. Chiron (10 mM in DMSO, CHIR99021, Tocris 4423).
6. Matrigel (Corning, 356231).
7. Cell recovery solution (Corning, 354253).

2.2 FACS Staining

1. DAPI (1 mg/ml, Thermo Fisher 62248).
2. Trypsin–EDTA, no phenol red (Gibco 15400054).

2.3 Immunofluorescence [9]

1. DAPI (Thermo Fisher Scientific, D1306).
2. Triton X-100 (Sigma-Aldrich, T8787).
3. 4% Paraformaldehyde (PFA) in PBS (Alfa Aesar, J61899-AK).
4. Permount mounting medium (Fisher Scientific, SP15-500).
5. PBS+ (with Ca2+ and Mg2+, Gibco 14040133).
6. PBS- (without Ca2+ and Mg2+, Gibco 14190144).
7. Fetal bovine serum (Embryonic stem cell FBS, Gibco 16141079).
8. Fluoromount-G mounting Medium (Southern Biotech, 0100-01).
9. PBS-FT (PBS+ with 10% FBS and 0.2% Triton X-100).

2.4 Plastics

1. Non-adherent U-bottomed 96-well plates (Greiner Bio-one; 650185).
2. Sterile dispension tray (e.g., Heathrow Scientific, HS120641-100EA).
3. Low retention pipette tips (e.g., Greiner bio-one; 775363).
4. 1.5 mL DNA LoBind tubes (Eppendorf 022431021).
5. 5 mL DNA LoBind tubes (Eppendorf 0030108310).
6. Standard tissue culture plastics (6-well plates, 96-well plates, culture flasks, etc.).
7. FACS tubes with 35 μm cell strainer (for e.g., Falcon 08-771-23).

2.5 Equipment

1. Standard tissue culture equipment (cell culture hood, centrifuge, inverted microscope for cell counting, etc.).
2. Humidified incubator (37 °C and 5% CO_2).
3. Hemocytometer (cell counting chamber, e.g., Bürker-Türk).
4. P1000 and P200 multichannel pipettes.

3 Methods

3.1 Culture Conditions After Thawing mESCs (Optimized for 129S1/SvImJ/C57BL/6; See Notes 2 and 3)

1. Maintain the mESCs in 0.1% gelatine-coated tissue culture dishes in ESLIF medium for 2–3 passages before starting the pre-culture of mESCs.
2. Refresh the media every day and split the cells every alternate day.
3. For the pre-culture conditions in this protocol, PBS- was used for washes, TrypLE was used for 3–5 min for dissociation of cells, and cells were centrifuged at 300 g for 3 min.
4. Ensure that the cells are 80% confluent at the time of the final passage.

3.2 Culture Conditions Before Aggregation (Pre-culture, See Note 4)

1. Day -4: When the cells are at 80% confluency, split at a 1:5 to 1:8 ratio and resuspend in 2i medium.
2. Day -3: Refresh with 2 mL 2i medium.
3. Day -2: Dissociate cells with TrypLE, quench with ESLIF medium, and resuspend in 5 mL of fresh ESLIF medium. Plate 1.5 mL, 1.2 mL, 1 mL, 0.8 mL, and 0.5 mL of the cell suspension in each well of a 6-well plate. Top up to 2 mL with ESLIF medium and swirl to ensure even coverage of the wells.
4. Day -1: Refresh with 2 mL ESLIF medium.
5. Day 0: Check all positions in all the wells thoroughly and choose the well that is consistently 80% confluent through the entire area of the well for the start of gastruloid aggregation.

3.3 Generation of Gastruloids from mESCs (Fig. 1)

Preparation prior to gastruloid generation (*see* **Notes 5–7**):

1. Thaw N2B27 medium at least 2 days before the start of aggregate generation at 4 °C (*see* **Note 5**).
2. Add the thawed medium to a T175 flask and place in the humidified incubator 1 day before staring the protocol.

   Day 0: Aggregate preparation (*see* **Notes 8–13**):

3. Pre-warm PBS+, ESLIF medium, and TrypLE to 37 °C in the waterbath.
4. Choose the well with ideal confluency of mESCs from the previously prepared 6-well plate. Aspirate the medium and add 1 mL of PBS+. Swirl the PBS around the entire well gently and aspirate. Repeat one time.
5. Add 1 mL of pre-warmed TrypLE to the well and place the plate in the incubator for 3 min or until the cells detach from the plate.

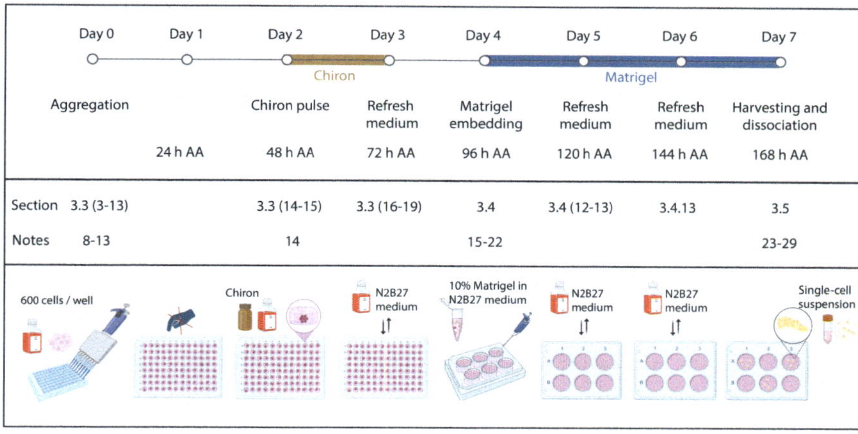

Fig. 1 Overview of gastruloid culture and processing. Created with Biorender

6. Make a single-cell suspension by pipetting up and down with a P1000 pipette (*see* **Note 8**).
7. Add 4 mL of ESLIF medium to neutralize the TrypLE and transfer the cell suspension to a 50 mL falcon tube.
8. Centrifuge at 300 g for 3 min. Aspirate medium and add 5 mL of PBS+. Swirl the tube to thoroughly wash cells. Repeat once. Ensure that all PBS+ is completely removed prior to **step 9** (*see* **Note 9**).
9. Resuspend cells in 1 mL of N2B27 using a P1000 pipette to make a homogeneous single-cell suspension. Add 4 mL of fresh N2B27 medium (total of 5 mL for one well of a 6-well plate) to facilitate counting (*see* **Note 10**).
10. Cell counting and calculation for aggregates of 600 cells (*see* **Note 11**):
    1. The aggregates are made in a U-bottomed non-adherent 96-well plate, in which one gastruloid will be generated in each single well. For 129S1/SvImJ/C57BL/6 mESCs, we prepared aggregates of 600 cells (*see* **Note 12**).
    2. Each well in a 96-well plate should contain 40 μL of cell suspension containing 600 cells. Therefore, we require a final concentration of 600/40 = 15 cells/μL. For an entire 96 well plate, we prepare 5 mL of fresh N2B27 medium containing 15 cells/μL ×5000 μL = 75.000 cells.
    3. For cell counting, add 10 μL of the cell suspension to both sides of the hemocytometer, count the cells, and calculate accordingly.
11. To calculate the volume of diluted cells that needs to be added to 5 mL of medium, to get a concentration of 15 cells/μL, we use the following formula. Volume to be added (μL) = (total

number of cells required (in this case, 75.000)/number of cells counted per μL)).

12. Add the required volume of cell suspension to 5 mL of fresh N2B27 medium in a new falcon tube. Swirl the tube to make a homogeneous cell suspension.
13. Transfer the suspension to a sterile dispension tray and distribute 40 μL of the suspension to each well of the plate using a P200 multi-channel P200 pipette and low retention pipette tips. Make sure that the drop settles at the bottom and place the plate in the incubator for 48 h (*see* **Note 13**).

    Day 2: Chiron pulse (48 h after aggregation)
14. Prepare medium containing a final concentration of 3 μM Chiron by adding 4.5 μL 10 mM Chiron to 15 mL pre-warmed N2B27 medium in a falcon tube per 96-well plate.
15. Transfer the medium to a sterile dispension tray and add 150 μL of the medium to each well of the plate using a P200 multi-channel pipette. Hold the pipette at an angle toward the side of the walls and pipette the medium out with force without spilling. This is done to detach the aggregate off the bottom. Place the plate in the incubator for 24 h (*see* **Note 14**).

    Day 3: Refresh medium (72 h after aggregation)
16. Carefully remove 150 μL of medium from each well using a P200 multi-channel pipette, ensuring that a small volume (~10–20 μL) remains at the bottom of each well to avoid disturbing the gastruloid. Note: The gastruloids tend to be loosely adhered to the bottom of the plate at this stage.
17. Transfer 15 mL pre-warmed N2B27 medium to a sterile tray.
18. Add 150 μL N2B27 medium to each well using the P200 multi-channel pipette. Pipette with some force to detach the gastruloids from the bottom of the well.
19. Place the plate in the incubator for 24 h.

3.4 Embedding of Gastruloids in Matrigel (96 Hours After Aggregation, See Fig. 2)

Before embedding (*see* **Notes 15–16**):

1. Thaw Matrigel on ice. This could take 1 h or longer.
2. For each 96-well plate of gastruloids, four wells of a 6-well plate are prepared for Matrigel embedding (24 gastruloids, or ¼ of the 96-well plate, per well of the 6-well plate). Typically, the 2 leftmost and 2 rightmost wells of the 6-well plate are used. This simplifies evenly distributing the gastruloids across the well later (*see* **Note 15**). Transfer N2B27 medium from the incubator to a 50 mL falcon tube and place on ice to cool down. Make 4 mL of 10% Matrigel by adding 400 μL Matrigel to 3600 μL of ice-cold N2B27 medium and keep on ice

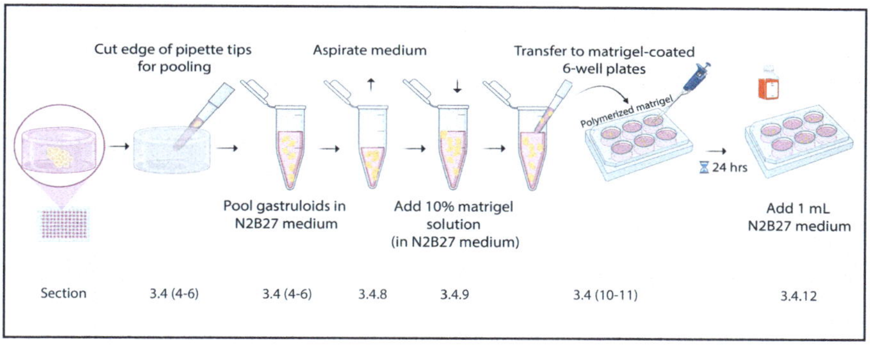

Fig. 2 Pooling and embedding of gastruloids in Matrigel on day 4 (96 h after aggregation). Created with Biorender

(prepare more solution based on the required number of gastruloids to be embedded) (*see* **Note 15**).

3. Coat each well of the 6-well plate with 900 μL of 10% Matrigel solution and leave the Matrigel to polymerize at room temperature for 1–1.5 h. Keep the remaining 10% Matrigel solution on ice for later steps.

   Pooling the gastruloids (*see* **Notes 17–20**):

4. Prepare four 5 mL low-bind Eppendorf tubes per 96-well plate. This means that ¼ of a 96-well plate will be pooled into one 5 mL Eppendorf tube.
5. Aspirate 150 μL of medium from the full 96-well plate either with a P1000 or a multichannel pipette without disturbing the gastruloids.
6. Pool 24 gastruloids into each Eppendorf tube using a P200 pipette. Place the pipette tip on top of the gastruloid and pipette up with some force to dislodge and pipette up the gastruloid (*see* **Note 20**).

   Embedding and plating of gastruloids (*see* **Notes 21** and **22**)

7. Prepare 500 μL of ice-cold 10% Matrigel solution for each well of the 6-well plate. The 10% Matrigel solution that was saved in step 3.4.3 can also be used in this step.
8. Aspirate as much old medium as possible from the Eppendorf tube containing gastruloids without disturbing them.
9. Add 500 μL of 10% Matrigel solution to each Eppendorf tube with gastruloids and mix carefully by pipetting up and down using a P1000 pipette (*see* **Note 21**).
10. Distribute each Eppendorf tube (containing 24 gastruloids) across one 6-well and swirl the plate to distribute the gastruloids evenly across each well (*see* **Note 22**).

11. Place the 6-well plate carefully into the incubator and do not move for at least 3–4 h to allow gastruloids to settle and Matrigel to polymerize.
12. After 24 h (120 h after aggregation), add 1 mL of warm N2B27 medium on top of the Matrigel/gastruloid mix.
13. Refresh the medium every 24 h by carefully aspirating 1 mL of old N2B27 medium without disturbing the Matrigel and adding 1 mL of fresh N2B27 medium. Repeat until 168 h after aggregation.

3.5 Harvesting and Dissociation of Gastruloids from Matrigel

1. Prepare an ice bucket and place a flat metal plate on top to cool down (e.g., inverted 96-well cool block). It is important that the surface is flat to transfer the cold efficiently. Cool down aliquots of PBS- and Corning cell recovery solution on ice.
2. Aspirate the medium from each well with Matrigel and gastruloids and add 1 mL of ice-cold PBS- to each well. Swirl the plate around to wash the gastruloids and aspirate the solution. Repeat 2 more times.
3. Add 1 mL of ice-cold Corning cell recovery solution to each well. Place the 6-well plate on the cold plate and incubate for 15 min on ice. This step activates the Corning solution to dissolve the Matrigel.
4. Pipette the Corning cell recovery solution against the gastruloids to remove them from the Matrigel. When gastruloids start to dissociate from the Matrigel, pipette more carefully to prevent them from completely disassembling into pieces and entering the pipette tip.
5. After the gastruloids are dissociated from the Matrigel, follow the respective preparation protocols for the specific downstream experiment to be performed (Subheadings 3.6 and 3.7).

3.6 Single Gastruloid Processing for FACS (See Notes 23–26)

1. Transfer single gastruloids to a flat-bottomed 96-well plate.
2. Add 50 μL of trypsin–EDTA without phenol red. Incubate at room temperature for 3–5 min (*see* **Note 24**).
3. Make a single-cell suspension by pipetting the solution up and down using a P200 pipette, check progress under the microscope.
4. Add 100 μL of the FACS buffer (PBS- with 10% FBS).
5. Prepare the FACS filters by pre-wetting them with 50 μL of FACS buffer. Pass 150 μL of cell suspension in FACS buffer through the filter. Wash the well with an additional 100 μL of FACS buffer and pass through the filter as well.
6. Add 700 μL of FACS buffer to the FACS tube to obtain a final volume of 1 mL per gastruloid. (Part of) This volume can also

be used to additionally wash the wells. Pass this volume through the filter again to thoroughly retrieve all the cells (*see* **Note 25**).

7. Add DAPI (1:1000) and keep the cell suspension on ice.

3.7 Gastruloid Processing for Immunofluorescence Staining (See Notes 27–29) [9]

1. Transfer the dissociated gastruloids to a 6-well plate using a P1000 or P200 pipette with the tip cut off.
2. Aspirate the medium that was carried over with the gastruloids.
3. Add 2 mL of 4% paraformaldehyde (PFA) to each well and incubate for 2 h at room temperature (*see* **Note 27**).
4. Aspirate the PFA and wash the gastruloids with PBS for 10 min. Repeat the wash two more times.
5. Transfer gastruloids into a well with 2 mL PBS-FT and block for 1 h at room temperature on a shaker (*see* **Note 28**).
6. Prepare 500 μL of primary antibody solution in PBS-FT for each condition in separate Eppendorf tubes. Include DAPI at 1:500 dilution.
7. Transfer gastruloids to a 48-well plate and add the corresponding primary antibody solutions to each well. Cover the plate with aluminum foil and incubate for 24 h at 4 °C.
8. Wash the gastruloids 3 times with PBS-FT for 20 min each at room temperature.
9. Prepare 500 μL of secondary antibody solution in PBS-FT for each condition.
10. Transfer the gastruloids to a new 48-well plate and add the secondary antibody solutions to the corresponding wells. Cover the plate and incubate for 24 h at 4 °C.
11. Wash the gastruloids 3 times with PBS-FT for 20 min each.
12. Add a 30 μL drop of mounting medium to the center of a glass microscope slide.
13. Transfer the stained gastruloids to the mounting medium drop.
14. Gently place a coverslip over the sample to avoid air bubbles.
15. Remove excess liquid around the coverslip using a Kimwipe and seal all edges with Permount or an appropriate hardening resin.
16. Store the slides at 4 °C in the dark until imaging (*see* **Note 29**). Representative morphologies at different developmental timepoints are shown in Fig. 3.

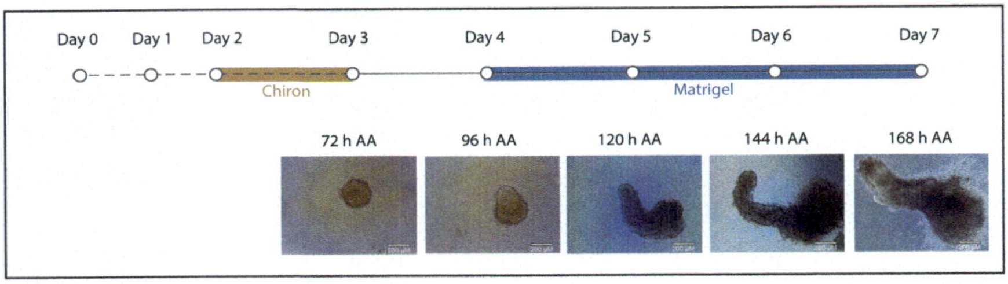

Fig. 3 Timeline of gastruloid development and representative morphologies at each time point

4 Notes

1. Small molecules such as CHIR99021 (Chiron) are unstable in solution and lose activity over time. Prepare fresh working solutions before use, based on daily volume requirements, and avoid freeze/thaw cycles of aliquots. Similarly, media supplemented with β-mercaptoethanol should be used within 6 weeks, as its activity diminishes over time. While the protocol lists volumes for 500 mL preparations, smaller batches can be prepared as needed to minimize waste and maintain reagent stability.
2. It is recommended to use cells that have a lower passage number. Gastruloids made from cells with higher passage number (above 30) may show a higher degree of variability [10].
3. Cultures should be routinely tested and kept free of mycoplasma.
4. Multiple factors in the pre-culture conditions can be adjusted and optimized based on the cell line used [see chapter "Optimization of gastruloid pre-culture conditions for reproducible germ layer composition"].
5. Medium can be aliquoted into 50 mL falcon tubes and stored at −20 °C. Before use, thaw aliquots gradually at 4 °C to avoid protein aggregates that appear as white strings. Thaw in an upright position to prevent leakage and minimize the risk of contamination.
6. N2B27 medium can be passed through a 0.22 μm membrane to ensure sterility. N2B27 does not contain antibiotics and is therefore sensitive to contamination. N2B27 medium is sensitive to pH changes. Placing the medium in the incubator before use allows the pH to adjust to optimal conditions. Make sure that the color of the medium is always red, this indicates its ideal pH. If the color changes to orange, the pH is off which inhibits gastruloid formation, and medium needs

to be recalibrated by placing the aliquot back into the incubator.

7. It is recommended to keep the N2B27 medium in the incubator in between steps of the protocol to prevent changes in its pH.
8. Make sure that all cells are dissociated into a homogeneous single-cell suspension after trypsinization. If required, rock the plate to aid cell detachment.
9. Make sure to remove all the left-over PBS before aggregation with a P200 pipette if required. PBS+ carryover could inhibit the formation of aggregates.
10. Resuspend the pellet thoroughly into a single-cell suspension and swirl the tube before counting to get an accurate result. This also ensures reproducible aggregates of the same size.
11. Cells were counted manually using a Bürker-Türk hemacytometer. Alternatively, any other validated method or automated cell counter may be used. Cell counting can be performed twice for more accurate results.
12. The number of cells needed per aggregate depends on the cell line used and is experimentally determined by performing a titration between 300 and 600 cells.
13. The diluted cell suspension is gently swirled and poured into a reservoir before transferring to a 96-well plate to minimize cell loss due to adhesion to plastic surfaces during handling. This is done to avoid excessive pipetting after cell counting and helps maintain cell number accuracy during seeding.
14. Gastruloids tend to adhere to the bottom. To dislodge the aggregates without medium spillage, it is recommended to add 75 μL medium first with more force, followed by a gentler addition of the remaining 75 μL medium. Dislodging gastruloids from the bottom of the well is crucial for symmetry breaking and retaining a 3D structure.
15. One 96-well plate of gastruloids can be typically distributed across 3–4 wells in a 6-well plate. It is recommended to not use more than 4 wells in a 6-well plate to limit processing time and to allow for an even distribution of gastruloids in Matrigel. If even distribution of gastruloids in all wells proves to be challenging, it is possible to use 2 wells of 2 6-well plates each.
16. Always keep the Matrigel stock and 10% solution on ice to avoid polymerization. A sterile and empty pipette tip box can be used to take ice into the cell culture hood. Do not prepare more than required and use carefully.
17. Always keep Eppendorf tubes containing gastruloids on ice.

18. Do not aspirate the N2B27 medium from the tube until right before adding Matrigel.
19. Work quickly to minimize the time gastruloids spend on ice, as prolonged exposure may induce a stress response.
20. If the gastruloids are bigger in size, cut off the edge of the pipette tip for gastruloid transfer.
21. When handling gastruloids beyond this point, make sure to only use low-retention (low-bind) pipette tips, as gastruloids are highly adhesive and can easily stick to standard tips. Alternatively, pre-coat standard pipette tips by pipetting up and down in 100% fetal bovine serum (FBS) prior to gastruloid manipulation to minimize sample loss.
22. Make sure that the gastruloids are not located too close to each other as this can allow them to grow into each other.
23. If processing a pool of gastruloids is the aim, it is recommended to pool directly in a protein low-bind 15 mL falcon tube, wash the sample carefully with PBS- and dissociate with TrypLE. Wash with FACS buffer after centrifugation to remove residual TrypLE and resuspend in FACS buffer.
24. Trypsin is used for dissociation here as it is deactivated efficiently by the FBS in the FACS buffer. In contrast, TrypLE remains active even on ice, which can lead to reduced cell viability. Cells are diluted in FACS buffer and transferred directly into FACS tubes, preventing additional cell loss through a separate centrifugation step. Since the solution is directly transferred to FACS tubes, it is necessary to use trypsin without phenol red to prevent autofluorescence during sorting.
25. Resuspending the sample in a total volume of 1 mL improves sorting efficiency. A less concentrated sample results in fewer cells processed per second, reducing the fraction of cells being lost between sorting events. However, this also extends the sorting time.
26. During the sorting, when the FACS tube with cell suspension appears empty, add more FACS buffer to the tube and vortex well. This releases additional cells from the tube and allows sorting of up to one additional 384-well plate per gastruloid.
27. The fixed gastruloids can be stored in PBS at 4 °C, protected from light for several months.
28. Permeabilization is critical for staining. Because the gastruloids can be quite dense, they might not get stained on the inside if permeabilization is not efficient.
29. Immuno-stained slides can be kept at 4 °C in the dark for months before imaging with very little loss of signal. However,

it is recommended to image slides as soon as possible after staining.

30. It is recommended to place the gastruloids in a humidity chamber during longer incubation steps to prevent drying out.

References

1. Van Den Brink SC et al (2014) Symmetry breaking, germ layer specification and axial organisation in aggregates of mouse embryonic stem cells. Development 141:4231–4242
2. Beccari L et al (2018) Multi-axial self-organization properties of mouse embryonic stem cells into gastruloids. Nature 562:272–276
3. Turner DA, Rue P, Mackenzie JP, Davies E, Martinez Arias A (2014) Brachyury cooperates with Wnt/β-catenin signalling to elicit primitive-streak-like behaviour in differentiating mouse embryonic stem cells. BMC Biol 12: 8–11
4. Moris N et al (2020) An in vitro model of early anteroposterior organization during human development. Nature 582:410–415
5. Blotenburg M et al (2024) Stem cell culture conditions affect in vitro differentiation potential and mouse gastruloid formation. bioRxiv:2024.05.09.593319
6. van den Brink SC et al (2020) Single-cell and spatial transcriptomics reveal somitogenesis in gastruloids. Nature 582:405–409
7. Veenvliet JV et al (2020) Mouse embryonic stem cells self-organize into trunk-like structures with neural tube and somites. Science (80) 370:eaba4937
8. Zeller P et al (2024) T-ChIC: multi-omic detection of histone modifications and full-length transcriptomes in the same single cell. bioRxiv 2024.05.09.593364
9. S. Vianello, M. Girgin, G. Rossi, M. Lutolf (2020) Protocol to immunostain Gastruloids (LSCB, EPFL) v1, 10.17504/protocols.io.7tzhnp6
10. Avni L et al (2023) Gastruloid optimization. Emerg Top Life Sci 7:409–415

Methods in Molecular Biology (2026) 3048: 305–319
DOI 10.1007/7651_2025_667

Published online: 11 October 2025

Optimization of Gastruloid Pre-culture Conditions for Reproducible Germ Layer Composition

Marloes Blotenburg, Beril Esin, Shruthi Buddharaju, and Peter Zeller

Abstract

Gastruloids are an in vitro model that mimics key aspects of gastrulation and can be used to study implantation-stage embryonic development. Even though current protocols have contributed significantly to our understanding of developmental biology, further improvements in terms of consistency and reproducibility are still required. In this chapter, we present a detailed pre-culture protocol optimized for the generation of gastruloids from 129S1/SvImJ/ C57BL/6 mouse embryonic stem cells (mESCs). Additionally, we provide information on the steps of the protocol open to optimization and present a workflow for selecting optimal conditions and evaluating the gastruloid formation outcome when using different cell lines. Overall, this workflow facilitates a step-by-step approach toward generating gastruloids from any cell line in a highly reproducible manner by optimizing the mESC pre-culture conditions.

Key words 2i medium, ESLIF medium, Gastrulation, Gastruloids, Germ layer formation, Pre-culture

1 Introduction

Gastrulation is a fundamental process in embryonic development, in which the body plan of the embryo is formed. During gastrulation, all three germ layers (ectoderm, mesoderm, and endoderm) are established, and precursors of all major cell types are formed [1]. In recent years, this process has been recapitulated utilizing three-dimensional in vitro models, such as gastruloids, to study embryonic development [2]. Gastruloids are a type of embryonic organoids formed by the aggregation of approximately 300–600 mouse embryonic stem cells (mESCs) [2–5]. Under defined culture conditions, aggregated mESCs undergo controlled proliferation, symmetry breaking, and initiate expression of markers associated with the three germ layers. This developmental progression is triggered through the addition of the Wnt agonist Chiron (CHIR99021), introduced for 24 h between 48 and 72 h post aggregation [6, 7]. By 120 h, the resulting gastruloids are

Authors Marloes Blotenburg and Beril Esin have equally contributed to this chapter.

comprised of cell types from all germ layers, organized along the three body axes and exhibiting collinear *Hox* gene expression [2–4, 8].

Protocols for gastruloid formation have significantly advanced developmental biology research; however, current models can still be improved. One limitation of gastruloids is the variability of cell type composition between different cell lines, individual experiments, and individual gastruloids, which compromises reproducibility. This variability in gastruloid formation has been specifically demonstrated in organoid-based perturbation studies [9]. Variability in gastruloids can be affected by variations in culture conditions to which the cells are subjected prior to the gastruloid protocol, defined as pre-culture conditions [10]. In pre-culture, ESLIF medium (serum-containing medium supplemented with leukemia inhibitory factor (LIF)) is conventionally used to maintain ESCs in a naïve pluripotency state [11, 12]. During the pre-culture period, incorporating 2i medium, which contains two inhibitors against glycogen synthase kinase 3-beta (GSK3b) and MAPK/ERK kinase (MEK), can de-differentiate ESCs further toward ground-state pluripotency [13–15]. The short-term incorporation of 2i medium can benefit the gastruloid formation process in a cell line-dependent manner. In addition to the culture medium itself [10], additional parameters can be adjusted, such as seeding density and cell confluency, as well as cell numbers used per gastruloid [3, 8, 16]. By optimizing these key factors for the used mESC cell line during the pre-culture period, a higher consistency in gastruloid morphology can be achieved, and the efficiency of symmetry breaking, reflected in the extent of axial elongation, can be enhanced [8].

Here, we present a detailed protocol containing optimized conditions for the generation of gastruloids from 129S1/SvImJ/C57BL/6 mESCs and summarize recommendations for other frequently used ESC cell lines. Additionally, we provide a general workflow highlighting which factors can be modulated to achieve optimal conditions for gastruloid formation. We further explain how bright field microscopy, immunofluorescence, and single-cell RNA sequencing can be used to determine efficient axial elongation quantified by the aspect ratio (AR) [17], germ layer abundance, and cell type complexity [10].

2 Materials

2.1 Pre-culture

1. ESLIF Medium (for 500 mL):

   450 mL G-MEM (Gibco, 21710025).

   50 mL Fetal bovine serum (Embryonic stem cell FBS, Gibco, 16141079).

   5 mL Sodium pyruvate (100 mM, Gibco, 11360070).

   5 mL non-essential amino acids (100×, Gibco, 11140035).

5 mL GlutaMAX (100×, Gibco, 35050061).

5 mL Penicillin-streptomycin (100×, Gibco, 15140122).

1 mL β-mercaptoethanol (50 mM in PBS, Sigma PHR9274; store at 4 °C for a maximum of 1 month).

50 μL Leukemia inhibitory factor (mLIF) (stock: ten million units/mL; final concentration in medium: 1000 units/mL, ESGRO ESG1106).

2. 2i medium (For 500 mL):

   250 mL DMEM/F12 (Gibco, 31331028).

   250 mL Neurobasal medium (Gibco, 21103049).

   2.5 mL N-2 supplement (Gibco, 17502048).

   5 mL B-27 supplement (50×, Gibco, 17504044).

   5 mL GlutaMAX (100×, Gibco, 35050061).

   5.5 mL Penicillin-streptomycin (100x, Gibco, 15070063).

   1 mL β-mercaptoethanol (50 mM50 mM in PBS; Sigma PHR9274).

   The following supplements should be added to the medium prior to use. The volumes listed below are for 500 mL of medium.

   50 μL mouse leukemia inhibitory factor (mLIF, stock: ten million units/mL; final concentration in medium: 1000 units/mL, ESGRO ESG1106 mLIF).

   150 μL Chiron (CHIR99021, 3 μM final concentration, from 10 mM stock in DMSO, Tocris, 4423).

   50 μL PD0325901 (1 μM final concentration, from 10 mM stock in DMSO, Sigma, PZ0162).

3. PBS- (without Ca^{2+} and Mg^{2+}, Gibco 14190144).
4. TrypLE (with phenol red, Gibco 12605010).
5. Gelatin (0.1% in H_2O from 2% Sigma Aldrich, G1393).

2.2 Gastruloid Formation

1. ESLIF medium.
2. TrypLE (Gibco, 12605010).
3. N2B27 medium (Takara, Y40002).
4. PBS+ (with Ca^{2+} and Mg^{2+}, Gibco, 14040133).
5. Chiron (10 mM in DMSO, CHIR99021, Tocris, 4423).

   Optional

6. Matrigel (Corning, 356231).
7. Cell recovery solution (for gastruloids embedded in Matrigel, Corning, 354253).

2.3 Brightfield Microscopy

1. Leica THUNDER Imager Live Cell & 3D Assay with a 10× dry lens.
2. Python and MorgANA Software.
3. Image J Fiji (v.2.1.0/1.53c).

2.4 Plasticware

1. Sterile dispension tray (e.g., Heathrow Scientific, HS120641-100EA).
2. DNA low retention tips (e.g., Greiner bio-one, 738265).
3. 1.5 mL DNA LoBind tubes (Eppendorf, 30108051).
4. 5 mL DNA LoBind tubes (Eppendorf, 30108310).
5. Standard tissue culture plastics (6-well plates, culture flasks, Falcon tubes, etc.).

2.5 Equipment

1. Standard tissue culture equipment (cell culture hood, centrifuge, inverted microscope for cell counting, etc.).
2. Humidified incubator (5% CO2, 37 °C).
3. Hemocytometer (cell counting chamber, e.g., *Bürker-Türk, Marienfeld*).
4. P1000 and P200 multichannel pipettes.

3 Methods

3.1 Pre-culture

The following pre-culture methodology is specific for 129S1/ SvImJ/ C57BL/6 (B6) mouse embryonic stem cells (mESCs) to serve as an example (Fig. 1). Please refer to Subheading 3.3 for alternative culture conditions recommended for other cell lines.

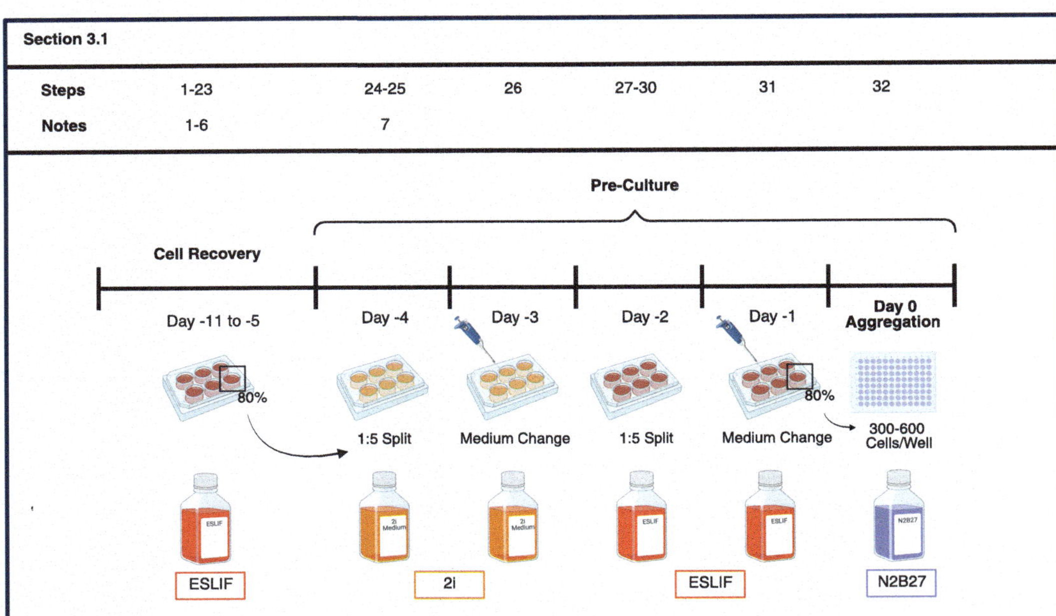

Fig. 1 Schematic representation of pre-culturing 129S1/SvImJ/C57BL/6 mESCs for gastruloid formation. (Created with Biorender.com)

- *Cell Recovery: (7 days)*

  Day -11

1. Prepare a gelatin-coated 6-well plate by adding 0.1% gelatin to cover the base of each required well. Incubate at RT for 5 to 30 min.
2. Take one vial of 129S1/SvImJ/ C57BL/6 mESCs (*see* **Notes 1–2**) and thaw the cryovial in a 37 °C water bath.
3. Transfer cells into a 15 mL Falcon immediately once thawed. Wash the cryovial with 1 mL *ESLIF medium* to collect leftover cells and transfer to the 15 mL Falcon tube.
4. Centrifuge the cells for 3 min at 300 g. Remove the supernatant. Resuspend the pellet in 1 mL ESLIF with a P1000. Add 3 mL *ESLIF medium* (*see* **Note 3**) to the 15 mL Falcon to a total volume of 4 mL.
5. Aspirate the gelatin from the 6-well plate before seeding the cells.
6. Transfer the cell suspension to 2 wells of the gelatin-coated 6-well plate (2 mL per well).
7. Swirl the plates in crosses or 8-motions to evenly distribute the cells across the wells. Culture the cells in the incubator (5% CO_2, 37 °C).

   Day -10

8. 24 h after thawing, remove the medium by suctioning using a sterilized glass pipette. Replace with 2 mL fresh *ESLIF medium* per well.

   Day -9

9. 48 h after thawing, cells should have reached 80% confluency and should be ready for splitting.
10. Prepare gelatin-coated 6-well plate (*see* **step 1**).
11. Remove medium by suctioning.
12. Wash cells once with 1 mL PBS- and remove by suctioning.
13. Add 0.5 mL of TrypLE to each well (*see* **Note 4**). Incubate the cells for a maximum of 5 min in the incubator (5% CO2, 37 °C), or until cells detach.
14. Pipette up and down with a P1000 micropipette to break cell clumps into a single-cell suspension. Add 2 mL *ESLIF medium* (4 times the TrypLE volume) and transfer the cells to a 15 mL Falcon tube.
15. Centrifuge cells for 3 min at 300 g. Remove the supernatant.

16. Add 1 mL *ESLIF medium* to the cell pellet and resuspend to a homogeneous single-cell suspension using a P1000 micropipette.
17. Add *ESLIF medium* to a total volume of 5 mL and mix the cell suspension.
18. Aspirate the gelatin from the 6-well plate prepared in **step 10** before seeding.
19. Transfer the cells to the gelatin-coated 6-well plate at a 1:5 splitting ratio (*see* **Note 5**). This is done by adding 1 mL fresh *ESLIF medium* per well, followed by 1 mL of the cell suspension.
20. Swirl the plate to evenly distribute the cells and culture in the incubator (5% CO2, 37 °C).

Day -8 to -5

21. 24 h later, change the medium with 2 mL fresh *ESLIF medium* per well.
22. Split the cells at a 1:5 splitting ratio (*see* **Note 5**) every other day until the end of their recovery time. Make sure to refresh the medium every day in between splits.
23. At the end of the recovery period, assess cell confluency 1 day before the planned start of pre-culture. If cells are too confluent, adjust by splitting cells at a 1:2 ratio, as explained in **steps 10–20**, to achieve 80% confluency at the start of the pre-culture the following day (*see* **Note 6**).

- *Pre-culture: (4 days)*

Day -4:

24. Prepare a 6-well plate coated with gelatin, as explained in **step 1**.
25. The cells should be 80% confluent at this step. Split cells at a 1: 5 ratio into *2i medium* following **steps 10–20**. Be careful to use the correct medium; resuspending and plating should be done in *2i* (corresponding to **steps 14–17**). We are moving to *2i medium* for the first 48 h (*see* **Note 7**); however, there are different cell line-specific options for medium choice. Please see Subheading 3.3 for alternative conditions.

Day -3:

26. Change the medium with 2 mL fresh *2i medium* for each well.

Day -2:

27. Prepare 5 wells of a 6-well plate coated with gelatin, as explained in **step 1.**

28. Harvest cells following **steps 10–18**. Be careful to use the correct medium; resuspending and plating should be done in *ESLIF medium* (corresponding to **steps 14–17**).
29. To obtain optimal confluency on the day of aggregation, replate a dilution series of cells in *ESLIF medium* in the gelatinized 6-well plate. Add an appropriate volume of medium to adjust the total volume to 2 mL per well first. Add 1.5 mL of the 5 mL (1:3) final mixture to one well, then 1.2 mL (1:4), 1 mL (1:5), 0.8 mL (1:6), and 0.5 mL (1:10) to wells containing 0.5, 0.8, 1, 1.2, and 1.5 mL of ESLIF medium, respectively.
30. Swirl the plates to distribute cells and culture in the incubator (5% CO_2, 37 °C).

Day -1:

31. Change the medium with 2 mL fresh *ESLIF medium* per well.

Day 0:

32. Choose the well with an optimal confluency of 80% for the aggregation step. Consider scanning the whole well to ensure cells are equally distributed throughout the well.

3.2 Gastruloid Formation

Please refer to Chapter *An optimized protocol for the extended culture of gastruloids from mouse embryonic stem cells* for more detailed methodology.

Day 0: Preparing Aggregates

1. Dissociate and pellet cells from the well with optimal confluency using TrypLE (*see* Subheading 3.1, **steps 11–13**).
2. Wash the cells twice with PBS +.
3. Centrifuge the cells for 3 min at 300 g. Remove the supernatant.
4. Resuspend the final cell pellet with N2B27 medium and count the cells. For each aggregate, 300–600 cells in 40 μL of medium are needed (7.5–15 cells per μL). Prepare 5 mL of diluted cells per 96-well plate. For detailed information on cell numbers per aggregate, please refer to Subheading 3.3.2.
5. Distribute 40 μL of the final cell suspension into a round-bottomed, non-adherent 96-well plate.
6. Place the plate in the incubator (5% CO2, 37 °C) for 48 h.

Day 1 (24 h After Aggregation)

7. Keep the cells under standard culture conditions in the incubator. No procedures are required on this day.

Day 2 (48 h After Aggregation): Chiron Pulse

8. Prepare 15 mL N2B27 medium with 4.5 μL of Chiron (3 μM final concentration from 10 mM stock in DMSO) per 96-well plate.
9. Add 150 μL of Chiron-supplemented N2B27 medium to each well, ensure that the aggregates are dislodged.
10. Place the plate in the incubator for 24 h.

Day 3 (72 h After Aggregation):

11. Refresh medium with 150 μL of N2B27 medium per well.
12. Place the plate in the incubator for 24 h.

Day 4 (96 h After Aggregation):

13. Refresh medium with 150 μL of N2B27 medium per well.
14. *Optional*: Embed gastruloids in Matrigel. Please refer to Chapter *An optimized protocol for the extended culture of gastruloids from mouse embryonic stem cells* for details.

Day 5 (120 h After Aggregation):

15. At this stage, the gastruloid formation is complete, and gastruloid development can be evaluated.
16. Depending on the downstream read-out, gastruloids can be dissociated for FACS sorting and sequencing or fixed for staining and imaging analysis. For detailed protocols, please refer to Chapter *An optimized protocol for the extended culture of gastruloids from mouse embryonic stem cells* and the gastruloid immunostaining protocol by Vianello et al. [18].

3.3 Optimization of Gastruloid Formation Through Pre-culture Conditions

To achieve the optimal ESC state after pre-culturing for gastruloid formation, it is important to consider several variables. In this section, we describe the possible variables to optimize, such as medium conditions, cell number, and cell confluency, and describe approaches to determine optimal conditions (Fig. 1).

3.3.1 Pre-culture Medium Conditions of mESCs

- *2i Pulse Length*

In addition to the 2i (48 h) followed by ESLIF (48 h) pre-culture condition described in detail in Subheading 3.1, other combinations of ESLIF and different pulse lengths of 2i medium have been shown to alter both the pluripotency state of mESCs and the consistency and efficiency of gastruloid formation [10]. Importantly, these effects are cell line-dependent; therefore, it is essential to optimize the conditions for the specific cell line used.

The medium conditions below are alternative examples to the mESC pre-culture conditions described in Subheading 3.1.

- *ESLIF medium only*: This condition was first described by Van den Brink and colleagues. They demonstrated that pre-culturing E14Tg2A mESCs in ESLIF medium leads to symmetry breaking and the formation of polarized structures [3, 19]. In previous work, we identified that this condition promotes naïve state pluripotency in B6 mESCs [10]. Additionally, reporter cell lines derived from CGR8, E14.1, and CK35 mESCs were reported to produce gastruloids efficiently in ESLIF medium [20].
- *2i medium only*: Sim et al. showed that J1 mESCs grown in 2i medium maintained high expression of core pluripotency markers [15]. In B6 mESCs, 2i medium resulted in transcriptomic enrichment for the ground-state pluripotency signature, but gastruloid formation efficiency was decreased [10]. In contrast, Cermola et al. reported enhanced success in gastruloid formation (95–98%) in TBV2 mESCs under 2i conditions [21].
- *ESLIF medium with a 2i pulse 24 h before aggregation*: After it was shown that E14Tg2A mESCs can form gastruloids in ESLIF medium [3, 19], Anlas et al. additionally demonstrated that a 2i pulse of 24 h before aggregation enhances gastruloid formation using E14Tg2A mESCs [20]. Reporter cell lines with a KH2 mESC background have also been shown to require a 24-h 2i pulse before aggregation [20]. Differently, we found that this condition resulted in low elongation efficiency for B6 mESCs, displaying the cell line-specific effect of this pre-culture on gastruloid formation success [10].

Beyond the combination of intervals of 2i and ESLIF-medium, alternative medium compositions such as ESLIF medium supplemented with the two inhibitors (2i; s2il) have been used to achieve efficient gastruloid formation in E14Tg2A mESCs, as reported by Girgin et al. and Vianello et al. [22, 23].

3.3.2 mESC Seeding Number for Aggregation

The number of mESCs used per aggregate is a critical factor for the efficiency of gastruloid formation. In our study, we tested 200–600 cells during the aggregation step in Subheading 3.2 *Day 0* and identified that 600 cells per aggregate yielded optimal results. Different cell lines may require varying starting cell numbers, for example, due to differences in their proliferation rate.

Previous research has shown that for E14-Tg2A cells, starting with a lower number of 300 ± 100 cells results in efficient elongation [3]. Additionally, Turner et al. optimized cell numbers for various cell lines in their study in 2017, finding that 200–400 cells were effective for gastruloid formation depending on the specific cell line [8]. More recently, it was shown that gastruloid growth dynamics vary with the number of cells used to start aggregation, highlighting the importance of this parameter on the absolute size and morphology of the gastruloids [16].

For uncharacterized cell lines, it is therefore recommended to always titrate the input cell number across a range of 200–600 cells to determine optimal conditions for reproducible gastruloid formation.

3.3.3 Pre-culture ESC Confluency and Morphology

The mESCs should be 80% confluent at **step 32** in Subheading 3.1. In order to reach the optimal confluency, it is essential to seed the cells at different dilutions during the final split before aggregation, as mentioned in **step 29** in Subheading 3.1. Afterward, the well that is closest to the target confluency is chosen to proceed with the rest of the experiment. This approach enables the establishment of a correlation between the initial seeding density and the resulting confluency after 2 days, providing information for the optimal dilution that can be used in future experiments.

In addition to assessing confluency, it is important to control the morphology of the cells at **step 32** in Subheading 3.1. Scanning the entire well and observing multiple positions help to establish an average confluency and allows for a comprehensive assessment of cell morphology. For example, cells cultured in pulses of 2i medium are generally more homogeneous, forming bright, white lines around dome-shaped colonies [13–15]. In contrast, cells maintained in ESLIF medium tend to be more heterogeneous, forming flatter and spread-out colonies [11–13]. Cells exposed to intermediate 2i pulses, as described in Subheading 3.3.1, display intermediate morphologies that should be evaluated to monitor reproducibility (Fig. 2). If the pre-culturing process is successful, comparable morphologies should be observed across the well and should be reproducible across experiments.

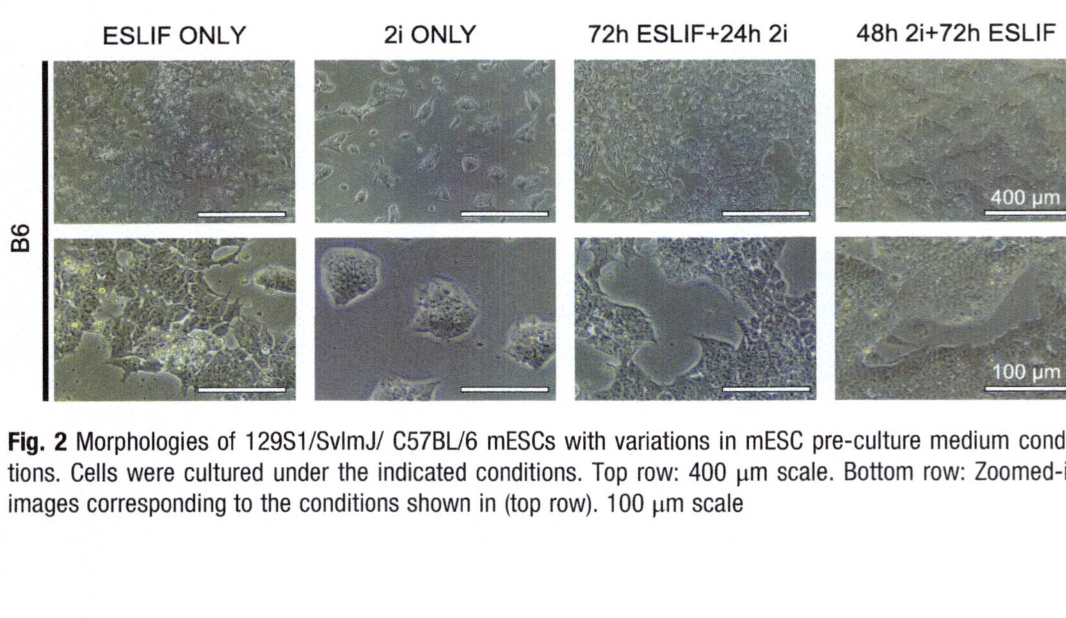

Fig. 2 Morphologies of 129S1/SvlmJ/ C57BL/6 mESCs with variations in mESC pre-culture medium conditions. Cells were cultured under the indicated conditions. Top row: 400 μm scale. Bottom row: Zoomed-in images corresponding to the conditions shown in (top row). 100 μm scale

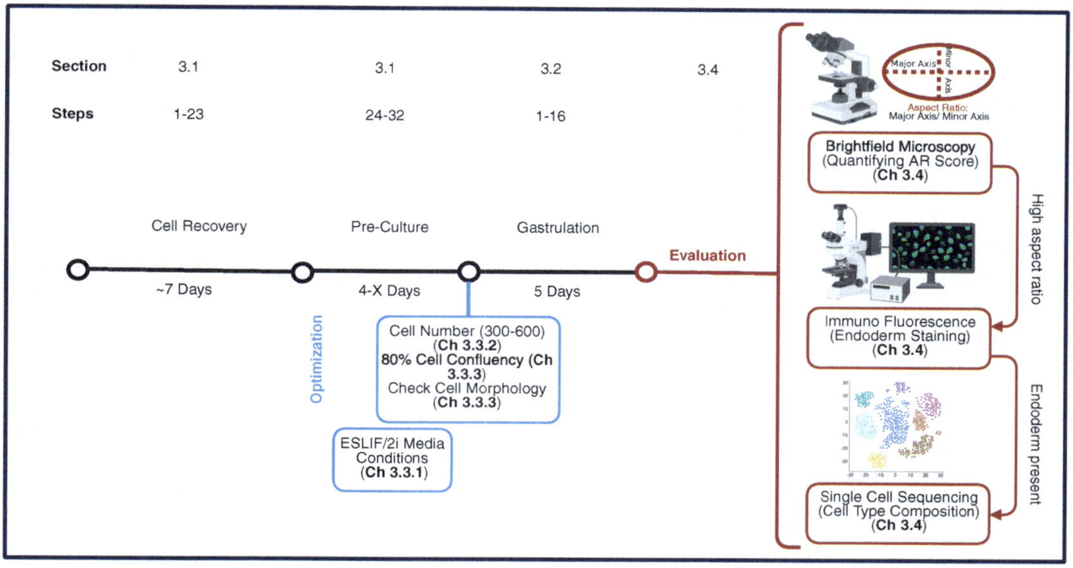

Fig. 3 Optimization of the pre-culture conditions and evaluation of the gastruloid formation process. (Created with Biorender.com)

3.4 Evaluation of the Gastruloid Formation Outcome

Previous sections have outlined how several parameters of the pre-culture procedure can be adjusted depending on the experimental design and the cell line. In this section, we focus on how to evaluate the results of the above-mentioned parameters in order to select the most suitable pre-culture condition to achieve reproducible gastruloid formation (Fig. 3). Gastruloids can be evaluated by brightfield microscopy, using the aspect ratio (AR) serving as an indicator for successful symmetry breaking. In addition, accurate generation of the desired cell types can be assessed through screening for the presence of key marker genes using immunofluorescence (IF) staining or performing single-cell RNA sequencing.

3.4.1 Brightfield Microscopy Analysis

Brightfield microscopy analysis can be used to quantify key morphological metrics, such as axis length and aspect ratio (AR), which serve as important indicators of gastruloid formation efficiency and is a straightforward approach to quantify efficient symmetry breaking [8, 17]. The aim is to achieve a stable higher aspect ratio, calculated as the ratio of the major axis to the minor axis length [24].

The following steps are provided as an example workflow using ImageJ Fiji and MOrgANA software (machine-learning based organoids analysis). Please refer to Gritti & Lim et al., 2021 for detailed instructions (*see* https://github.com/LabTrivedi/MOrgAna/tree/master) [24].

To acquire the images, a 10× microscope is required. In this example, we used the Leica THUNDER Imager Live Cell & 3D

Assay with a 10× dry lens for imaging B6 gastruloids at 120 h after aggregation. An advantage of this microscope is its ability to maintain cells in a stable CO_2 environment. For the image acquisition, a CO_2 and temperature-regulated microscope is not essential; however, such features are advantageous if imaging the gastruloids at multiple time points is desired.

1. Image gastruloids at 120 h after aggregation.
2. If Z-stacks of each gastruloid are taken, use the Stack Focuser plug-in in ImageJ data processing package Fiji (v.2.1.0/1.53c).
3. Segment images in MOrgAna using a subset of images to train the model.
4. During the quantification step, go to the "quantification" tab and select "morphology quantification." Choose "straightened" as the "type of mask."
5. Choose "major-axis length" and "minor axis length" parameters as "morphological parameter."
6. Results can be saved as csv file, and the aspect ratios can be calculated by dividing the major axis by the minor axis length.
7. We recommend the use of statistical tests to identify statistically significant differences in aspect ratio between the tested experimental conditions.

3.4.2 Immunofluorescence Staining

Assessment of lineage-specific marker genes provides an additional way to evaluate the outcome of gastruloid development. Proper organization of mesoderm and ectoderm is required for axial elongation; therefore, the presence of axial elongation can suggest these germ layers have been formed [3]. Endoderm, on the other hand, is not reflected in the AR and is considered the most challenging germ layer to generate due to its formation depending on multiple signaling pathways [25, 26].

Immunofluorescence (IF) staining for endoderm markers such as Sox17 and FoxA2 can confirm the presence of the endodermal germ layer [3, 4, 23]. The combination of AR and IF for endoderm markers can therefore be interpreted as evidence that gastruloid formation occurred successfully, with the presence of all germ layers [3, 23]. When performing IF for any cell type, we recommend the use of a panel of key marker genes, since co-staining of multiple markers is the most reliable way to identify the presence of the desired cell type.

3.4.3 Precise Evaluation with Single-Cell Sequencing

The most precise approach to evaluate gastruloid formation is by using single-cell RNA sequencing (scRNA-seq), which results in detailed gene expression profiles at the cellular level. In recent work, we applied the pre-culture conditions mentioned in Subheading 3.1 to B6 mESCs and performed scRNA-seq of aggregation-stage mESCs as well as of the resulting gastruloids 120 h after aggregation [10]. In mESCs at the time point of aggregation, this revealed a clear segregation between the pre-culture conditions ending with ESLIF, or with 2i [10]. However, we observed that even similar mESC states, for example those ending with ESLIF medium, showed considerable changes in cell type composition and reproducibility in the resulting gastruloids at 120 h after aggregation. These findings indicate that the transcriptional state of mESCs before aggregation is not sufficient to predict the outcome after gastruloid formation [10]. Therefore, it is essential to assess cell type composition using scRNA-seq approaches at the gastruloid stage. In addition to identifying the presence and abundance of the three germ layers using marker genes associated with endoderm (*Sox17* and *FoxA2*) [3, 4, 23], ectoderm (*Sox2 and Pax6*) [27, 28], and mesoderm (*Tbx6* [29] and *Mesp1* [27]), this approach also allows for a deeper analysis to identify more specific cell types using many additional marker genes.

Even though scRNA-seq provides a precise description of the gastruloid formation outcome, the high experimental costs may limit its use for large numbers of conditions. Therefore, we recommend initially screening a panel of different pre-culture conditions by assessing the degree of symmetry breaking in the resulting gastruloids using brightfield microscopy. Conditions that reproducibly lead to enhanced symmetry breaking can then be followed up by IF staining for endoderm markers. In addition, a limited number of conditions can be compared in detail through RNA-sequencing approaches. Together, this provides a comprehensive approach to optimize gastruloid formation in a cell line-specific manner (Fig. 3).

4 Notes

1. We used a vial containing B6 mESCs equivalent to one-third of a 6-well plate (~300-500 K cells per vial). The cells were thawed into two wells of a 6-well plate as described in Subheading 3.1 *Day -11*.
2. In this protocol, cells are thawed 1 week before the start of the pre-culture period to allow B6 mESCs sufficient time to recover from stress induced by thawing. However, the optimal recovery duration may vary between cell lines. During the recovery period, remember to split the cells every other day and refresh the medium daily.
3. The medium volume depends on the number of cells thawed. If the cryovial contains cells from one-sixth of a 6-well plate, resuspend in 2 mL of *ESLIF* medium and plate on one well.

If the cryovial contains one-third of a 6-well plate, resuspend in 4 mL of *ESLIF* medium and plate on two wells.

4. TrypLE volumes should be adjusted depending on the plate size. As a reference, we use 0.5 mL per well of a 6-well plate. For different plate sizes, please refer to *Useful Numbers for Cell Culture Guide* by Thermo Fisher Scientific [30].
5. The splitting ratio is cell line dependent. For B6 mESCs, a 1:5 ratio is used to achieve optimal confluency. Depending on the cell line properties, splitting ratios can range from 1:5 up to 1: 10.
6. As a guideline for weekly experimental planning: If B6 mESCs are thawed on Friday, splitting during the recovery week can be performed the next week Monday, Wednesday, Friday followed by a 1:2 split on Sunday. Pre-culturing then begins the following Monday.
7. Adhering to the specified time points is crucial to maintain the optimal pluripotency of the mESCs.

References

1. Tam PP, Behringer RR (1997) Mouse gastrulation: the formation of a mammalian body plan. Mech Dev 68(1–2):3–25. https://doi.org/10.1016/s0925-4773(97)00123-8
2. Beccari L, Moris N, Girgin M, Turner DA, Baillie-Johnson P, Cossy AC et al (2018) Multi-axial self-organization properties of mouse embryonic stem cells into gastruloids. Nature 562(7726):272–276. https://doi.org/10.1038/s41586-018-0578-0
3. van den Brink SC, Baillie-Johnson P, Balayo T, Hadjantonakis AK, Nowotschin S, Turner DA et al (2014) Symmetry breaking, germ layer specification and axial organisation in aggregates of mouse embryonic stem cells. Development 141(22):4231–4242. https://doi.org/10.1242/dev.113001
4. van den Brink SC, Alemany A, van Batenburg V, Moris N, Blotenburg M, Vivié J et al (2020) Single-cell and spatial transcriptomics reveal somitogenesis in gastruloids. Nature 582(7812):405–409. https://doi.org/10.1038/s41586-020-2024-3
5. Arias AM, Marikawa Y, Moris N (2022) Gastruloids: pluripotent stem cell models of mammalian gastrulation and embryo engineering. Dev Biol 488:35–46. https://doi.org/10.1016/j.ydbio.2022.05.002
6. Turner DA, Hayward PC, Baillie-Johnson P, Rué P, Broome R, Faunes F et al (2014) Wnt/β-catenin and FGF signalling direct the specification and maintenance of a neuromesodermal axial progenitor in ensembles of mouse embryonic stem cells. Development 141(22):4243–4253. https://doi.org/10.1242/dev.112979
7. Turner DA, Rué P, Mackenzie JP, Davies E, Martinez AA (2014) Brachyury cooperates with Wnt/β-catenin signalling to elicit primitive-streak-like behaviour in differentiating mouse embryonic stem cells. BMC Biol 12:63. https://doi.org/10.1186/s12915-014-0063-7
8. Turner DA, Girgin M, Alonso-Crisostomo L, Trivedi V, Baillie-Johnson P, Glodowski CR et al (2017) Anteroposterior polarity and elongation in the absence of extra-embryonic tissues and of spatially localised signalling in gastruloids: mammalian embryonic organoids. Development 144(21):3894–3906. https://doi.org/10.1242/dev.150391
9. Rosen LUSC, Argelaguet R, Barker CG, Yang A, Reik W, Marioni JC (2022) Inter-gastruloid heterogeneity revealed by single cell transcriptomics time course: implications for organoid based perturbation studies. bioRxiv. https://doi.org/10.1101/2022.09.27.509783
10. Blotenburg M, Suurenbroek L, Bax D, de Visser J, Bhardwaj V, Braccioli L et al (2025) Stem cell culture conditions affect in vitro differentiation potential and mouse gastruloid

formation. PLoS One 20(3):e0317309. https://doi.org/10.1371/journal.pone.0317309

11. Williams RL, Hilton DJ, Pease S, Willson TA, Stewart CL, Gearing DP et al (1988) Myeloid leukaemia inhibitory factor maintains the developmental potential of embryonic stem cells. Nature 336(6200):684–687. https://doi.org/10.1038/336684a0
12. Smith AG, Heath JK, Donaldson DD, Wong GG, Moreau J, Stahl M et al (1988) Inhibition of pluripotential embryonic stem cell differentiation by purified polypeptides. Nature 336(6200):688–690. https://doi.org/10.1038/336688a0
13. Kolodziejczyk AA, Kim JK, Tsang JC, Ilicic T, Henriksson J, Natarajan KN et al (2015) Single cell RNA-sequencing of pluripotent states unlocks modular transcriptional variation. Cell Stem Cell 17(4):471–485. https://doi.org/10.1016/j.stem.2015.09.011
14. Nichols J, Smith A (2009) Naive and primed pluripotent states. Cell Stem Cell 4(6):487–492. https://doi.org/10.1016/j.stem.2009.05.015
15. Sim YJ, Kim MS, Nayfeh A, Yun YJ, Kim SJ, Park KT et al (2017) 2i maintains a naive ground state in ESCs through two distinct epigenetic mechanisms. Stem Cell Rep 8(5):1312–1328. https://doi.org/10.1016/j.stemcr.2017.04.001
16. Merle M, Friedman L, Chureau C, Shoushtarizadeh A, Gregor T (2024) Precise and scalable self-organization in mammalian pseudo-embryos. Nat Struct Mol Biol 31(6):896–902. https://doi.org/10.1038/s41594-024-01251-4
17. Mantziou V, Baillie-Benson P, Jaklin M, Kustermann S, Arias AM, Moris N (2021) In vitro teratogenicity testing using a 3D, embryo-like gastruloid system. Reprod Toxicol 105:72–90. https://doi.org/10.1016/j.reprotox.2021.08.003
18. Vianello SD GM, Rossi G, Lutolf M (2020) Protocol to immunostain Gastruloids (LSCB, EPFL). Accessed
19. Baillie-Johnson P, van den Brink SC, Balayo T, Turner DA, Martinez AA (2015) Generation of aggregates of mouse embryonic stem cells that show symmetry breaking, polarization and emergent collective behaviour in vitro. J Vis Exp 105. https://doi.org/10.3791/53252
20. Anlas K, Baillie-Benson P, Arató K, Turner DA, Trivedi V (2021) Gastruloids: embryonic organoids from mouse embryonic stem cells to study patterning and development in early mammalian embryos. Methods Mol Biol 2258:131–147. https://doi.org/10.1007/978-1-0716-1174-6_10
21. Cermola F, D'Aniello C, Tatè R, De Cesare D, Martinez-Arias A, Minchiotti G et al (2021) Gastruloid development competence discriminates different states of pluripotency. Stem Cell Rep 16(2):354–369. https://doi.org/10.1016/j.stemcr.2020.12.013
22. Girgin MU, Broguiere N, Mattolini L, Lutolf MP (2021) Gastruloids generated without exogenous Wnt activation develop anterior neural tissues. Stem Cell Rep 16(5):1143–1155. https://doi.org/10.1016/j.stemcr.2021.03.017
23. Vianello SLM (2021) In vitro endoderm emergence and self-organisation in the absence of extraembryonic tissues and embryonic architecture. bioRxiv.:138883. https://doi.org/10.1101/2020.06.07.138883
24. Gritti N, Lim JL, Anlaş K, Pandya M, Aalderink G, Martínez-Ara G et al (2021) MOrgAna: accessible quantitative analysis of organoids with machine learning. Development 148(18). https://doi.org/10.1242/dev.199611
25. Nowotschin S, Hadjantonakis AK, Campbell K (2019) The endoderm: a divergent cell lineage with many commonalities. Development 146(11). https://doi.org/10.1242/dev.150920
26. Hashmi A, Tlili S, Perrin P, Lowndes M, Peradziryi H, Brickman JM et al (2022) Cell-state transitions and collective cell movement generate an endoderm-like region in gastruloids. elife:11. https://doi.org/10.7554/eLife.59371
27. Minn KT, Dietmann S, Waye SE, Morris SA, Solnica-Krezel L (2021) Gene expression dynamics underlying cell fate emergence in 2D micropatterned human embryonic stem cell gastruloids. Stem Cell Rep 16(5):1210–1227. https://doi.org/10.1016/j.stemcr.2021.03.031
28. Yang Y, Laterza C, Stuart HT, Michielin F, Gagliano O, Urciuolo A et al (2022) Human pluripotent stem cell-derived micropatterned ectoderm allows cell sorting of Meso-endoderm lineages. Front Bioeng Biotechnol 10:907159. https://doi.org/10.3389/fbioe.2022.907159
29. Chen B, Khan H, Yu Z, Yao L, Freeburne E, Jo K et al (2024) Extended culture of 2D gastruloids to model human mesoderm development. bioRxiv. https://doi.org/10.1101/2024.03.21.585753
30. Useful Numbers for Cell Culture. https://www.thermofisher.com/us/en/home/references/gibco-cell-culture-basics/cell-culture-protocols/cell-culture-useful-numbers.html. Accessed 2025

Methods in Molecular Biology (2026) 3048: 321–335
DOI 10.1007/7651_2025_639

Published online: 11 July 2025

Microfluidic Device Manufacturing by Light-Based 3D Printing for Organoid Vascularization

Rochelle Aubry, Idris Salmon, and Adrian Ranga

Abstract

3D printing by light-based vat polymerization enables the manufacturing of a variety of microfluidic devices which can be used to study growth, patterning, vascularization, and tissue interactions of stem cell-derived spheroids, organoids, and tissue explants. This technology allows to design and manufacture compartmentalized devices for precise seeding of cells and organoids, combined with the possibility to generate controlled media flow. Here, we detail the steps involved in the fabrication of such microfluidic devices, including the printing and post-processing stages using light-based 3D printing. We also give an example of how such a 3D printed microfluidic device can be used to culture and vascularize cerebral organoids. The use of 3D printing provides a rapid and inexpensive way to generate microfluidic devices without the need for cleanroom facilities and is therefore a technology accessible to every life science research lab. In addition, this high throughput method facilitates organoid studies in a more controlled environment, thereby representing a significant advancement in reproducibility for organoid research.

Key words 3D printing, Microfluidics, Stem cell-derived organoids, Tissue vascularization

1 Introduction

3D printing has emerged as a prominent technique for the manufacturing of microfluidic devices, offering significant advantages in terms of rapid prototyping, design flexibility, and the ability to create intricate architectures [1]. Widely used applications range from chemical analysis to diagnostics and drug delivery [2]. More recently, the advent of biocompatible materials has made this manufacturing technology compatible with cell and organoid culture [3, 4], where maintaining a controlled and physiologically relevant environment is essential for the growth and differentiation of cells. Notably, 3D-printed microfluidic devices also allow for co-cultures of multiple cell types and organoids, for example to vascularize stem cell-derived cerebral organoids [5].

Stereolithography (SLA) has become a leading technique for 3D printing microfluidic devices due to its flexibility and

Rochelle Aubry and Idris Salmon contributed equally.

accessibility. Unlike traditional photolithography, which requires expensive cleanroom facilities, advanced expertise, and limits design possibilities due to its architectural constraints, SLA offers a more versatile approach. One of its main advantages is the ability to construct complex shapes, including hollow structures such as micrometer-scale channels typically required in microfluidic devices, in a single build step. SLA uses the selective polymerization of light-sensitive resins to create highly detailed microstructures. In inverted SLA, a high-intensity UV laser light is specifically directed onto a vat of liquid resin, triggering a photochemical reaction that solidifies the resin in specific areas based on a digital design. Each pixel in the design corresponds to a precise point where polymerization occurs, enabling the 3D printing in high resolution, typically in the range of 100–300 μm [6]. This printing resolution largely depends on the laser spot size which is different between several generations and brands of 3D printers with the Form2 from Formlabs having a lasers spot size of 140 μm enabling a feature size of 300 μm. These high-resolution SLA 3D printers are small-sized desktop systems which can be used in standard laboratory settings without the need for specialized cleanroom requirements [7]. This significantly reduces infrastructure costs and reduces prototyping times to just a few hours. These factors broaden accessibility, making SLA 3D printing a cost-effective and practical option for labs interested in creating custom microfluidic devices to answer specific biological questions.

Here, we demonstrate how this technology can be applied to the manufacturing of a microfluidic device designed to support the growth and vascularization of hiPSC derived cerebral organoids on-chip. The main steps in this process are depicted in Fig. 1.

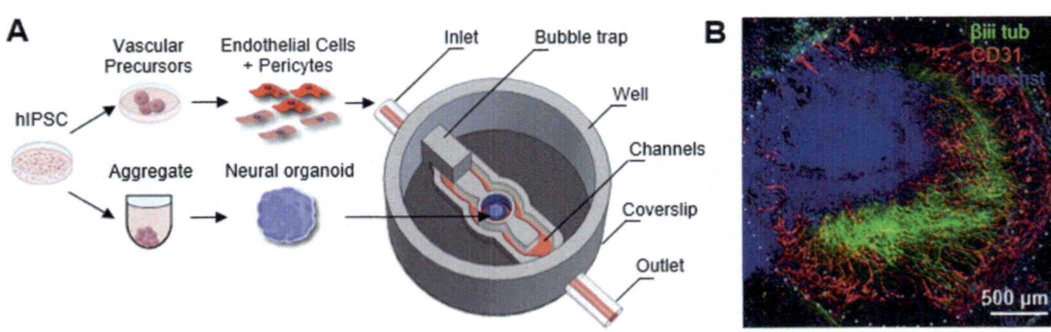

Fig. 1 3D printed microfluidic platform for vascularized organoid cultures on chip. (**a**) Schematic of co-culture: differentiation of hiPSC into early neural organoids in suspension and adherent vascular cells followed by seeding into 3D printed microfluidic; (**b**) Maximal intensity projection of a 400 μm z-stack from a day 30 neurovascular organoid on chip immunohistochemistry staining for βiii tubulin and CD31. Scale bar: 500 μm

2 Materials

1. SolidEdge CAD software (SolidEdge).
2. PreForm software (Formlabs).
3. Form 2 (Formlabs).
4. Form 2 Resin Tank LT (Formlabs, RT-F2-02).
5. Form Cure (Formlabs, FG-CU-01).
6. Dental SG resin (Formlabs).
7. Technicloth (Texwipe, TX609).
8. 99% Propanol-2 (IPA) (Chem-Lab, CL00.0901).
9. 100 mL glass beaker.
10. Ultrasonic cleaner (HBM machines, 01689).
11. UV filtered light.
12. P20 pipette and tips.
13. UV curable glue (Norland products, NOA86).
14. 13 mm round coverslips (VWR 631-0150).
15. Milli-Q water.
16. Compressed air.
17. Aluminum foil.
18. 20 mL syringes (Fisher scientific, 14955460).
19. Needle (BD, Microlance 3 21G 0.8X50mm).
20. Silikon tubing (Deutsch & Neumann, 3100525).
21. Prusa MK3S filament printer (Prusa).
22. Laser cutter (Trotec, Speedy100R).
23. Silicone tubing 0.5 mm ID × 2.5 mm OD (Deutsch & Neumann, 3100525).
24. Cutting mat (Carl Roth, KM07.1).
25. Cell culture hood aspirator.
26. Bleach 10% (Loda, 514471).
27. Scalpel (Swann Morton, 0503).
28. PTFE tubing 0.5 mm ID × 1 mm OD (Bohlender Z609730).
29. Petri dish Nunc (Thermo Scientific, 172931).
30. Parafilm.
31. SureOne P200 wide bore tips (FisherBrand, T118RS-FIS).
32. Growth Factor reduced Matrigel (Becton Dickinson 354230).
33. Needle BD Microlance 30G 0.30x13 (BD 304000).
34. Accutase (Gibco, A1110501).

35. DMEM-F12 (Gibco, 31330038).
36. VEGF-A (Peprotech 100-20).
37. Paraformaldehyde (Sigma-Aldrich).
38. Phosphate buffered saline (PBS) (Gibco, 10010015).
39. Bovine serum albumin (BSA) (Sigma Aldrich, A7906-100G).
40. Tx100 (PanReac Applichem).
41. NaN_3 (Thermo Scientific).
42. Orbital shaker.
43. Razor blade.

3 Methods

The design and manufacturing of a 3D-printed microfluidic chip plate involves a multistep process, encompassing design, 3D printing, post-processing, and final assembly of the microfluidic chip plate, followed by seeding of organoids and vascular cells (Fig. 2). In the initial design phase, specifications for the microfluidic architecture are established to ensure it meets all experimental requirements. The device is then fabricated through 3D printing, followed by a post-processing stage, which includes attaching a coverslip to the bottom and thoroughly cleaning and sterilizing the device to guarantee optimal functionality. Once post-processing is complete, the device is assembled into a chip plate. At this stage, the plate is ready for seeding cerebral organoids and vascular cells for co-culture and growth. Each stage is essential for producing a reliable and effective platform for advanced microfluidic applications.

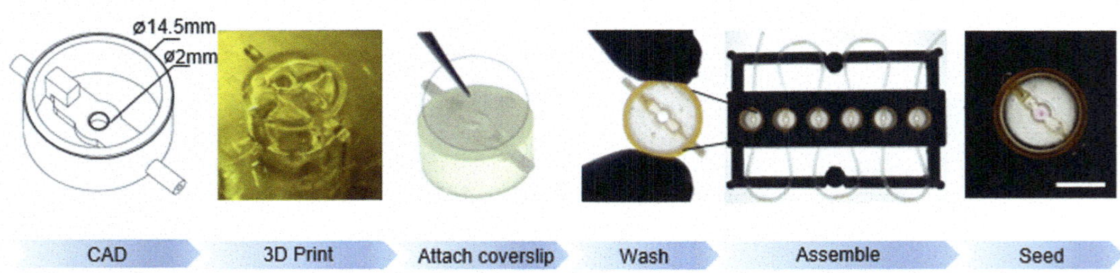

Fig. 2 Microfluidic chip design and manufacturing process. The microfluidic device design is generated in computer aided design (CAD) software and 3D printed with a FormLabs2 printer. A 13 mm round coverslip is glued onto the 3D-printed part and the chip is cured and extensively washed to ensure biocompatibility. Six microfluidic devices are inserted in a 3D-printed custom holder, and cells/organoids are seeded and incubated. Scale bar: 5 mm

3.1 Computer-Aided Design and Print Initiation of Microfluidic Devices

1. Design the microfluidic device in SolidEdge software and save the file both as a .par and .stl file (*see* **Note 1**).
2. Open the .stl file in the PreForm to create a new 3D print.
3. Assign print properties by opening the job setup tab and selecting the "Dental SG" resin with a material and 0.050 mm layer thickness and click apply.
4. Orient the design of the microfluidic device with the top of the design oriented to the build platform (Fig. 3 *left*).
5. Apply support structures manually having a touchpoint size of 0.40 mm to support overhanging design structures (Fig. 3 *middle*).
6. Duplicate (ctrl+D) up to 8 devices (Fig. 3 *right*) (*see* **Note 2**).
7. Save this as a .form file.

3.2 3D Printing of Microfluidic Devices

1. Connect a laptop via USB print cable to the Form2 3D printer (*see* **Note 3**).
2. Open the saved .form file and press "on click print" or Alt+P.
3. Install the Dental SG cartridge in the cartridge slot of the printer and a Form 2 resin tank LT, appointed to Dental SG resin, in the resin tank slot (*see* **Note 4**).
4. The 3D printer display will indicate once the print is uploaded from laptop to the 3D printer.
5. Disconnect the laptop from the printer.
6. Press "Print Now" to start printing .
7. Visually inspect whether the build platform is present and that no prints are sticking to it.

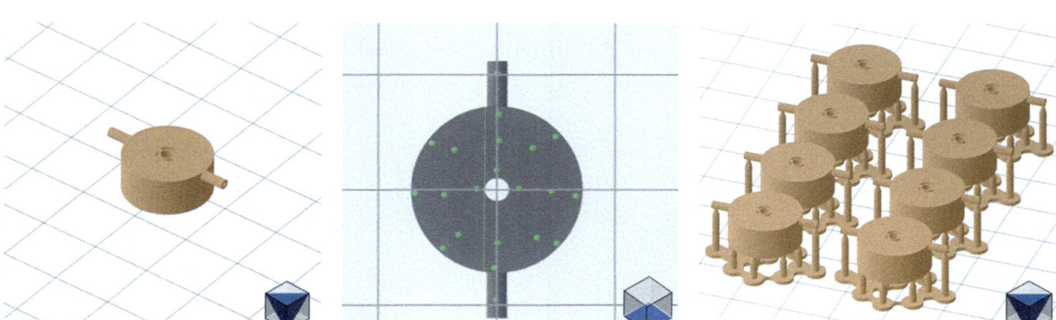

Fig. 3 Print preparation of the microfluidic device design in preform software. *Left:* screenshot of a microfluidic device, isometric view, with the top of the design oriented to the build platform without supports. *Middle:* screenshot of microfluidic device design, bottom view, in support edit mode with green dots depicting support attachment points having a touchpoint size of 0.40 mm. *Right:* screenshot of eight correctly oriented microfluidic devices, isometric view, with supports, with a touchpoint size of 0.40 mm

8. Press "Yes" on the display of the printer.
9. The print will be launched.

3.3 Post-processing

1. Turn on UV filtered light (*see* **Note 5**).
2. Remove the print platform from the printer.
3. Use a razor blade to cut of the supports positioned at the inlets and outlets.
4. Break off the other supports by twisting the microfluidic device, while keeping the support structures in place.
5. Clean the platform with a 99% isopropyl alcohol (IPA) moist cloth (*see* **Notes 6** and 7).
6. Immediately submerge the printed piece in 30–50 mL of IPA until fully covered in a 100 mL beaker.
7. Position the beaker in an ultrasonic cleaner for 2 min at room temperature. Set the power to 005 and frequency to H1.
8. Dispose of the used IPA in an appropriate solvent waste container and clean the beaker.
9. Fill a custom-enhanced 20 mL syringe with IPA and connect it to the inlet of the microfluidic device to wash out unpolymerized resin present inside the microfluidic channels (Fig. 4 *left*) (*see* **Note 8**).
10. Rinse the device at least 2 times more with 30–50 mL of clean IPA in a 100 mL beaker for a total amount of maximally 10 min (*see* **Notes 9** & **10**).
11. Allow the devices to air dry in a ventilated area for 10–30 min or use pressurized air to manually dry the devices.
12. Remove the rotating platform of the Form Cure station.
13. Preset the cure settings to 2 min without heat.
14. Leave the form cure station open.
15. Inspect if the bottom surface of the microfluidic device is flat and does not contain any printing debris (*see* **Note 11**).

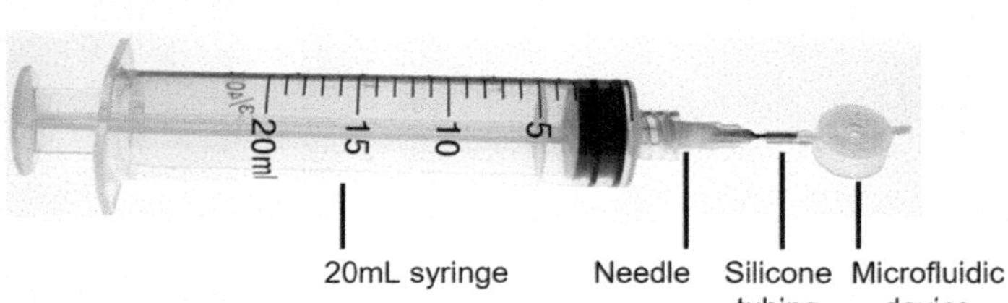

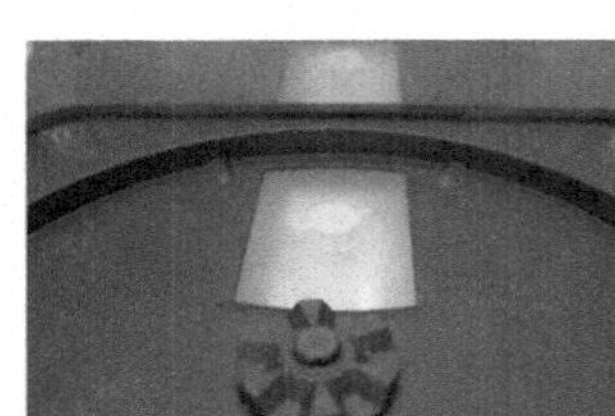

Fig. 4 Microfluidic device cleaning and coverslip attachment. *Left:* picture of custom-made syringe tool connected to microfluidic device used to clean microfluidic channels. *Right:* picture of coverslip-loaded microfluidic device in the preform curing station positioned directly above the LED strip

16. Pipette 5.5 μL of UV curable glue dropwise on the bottom of the microfluidic chip (*see* **Note 12**).
17. Carefully drop a 13 mm wide round coverslip on top of the glue using tweezers.
18. Inspect how the glue spreads (*see* **Note 13**).
19. Put the device directly on top of the LED strip located at the bottom of the Form Cure station (Fig. 4 *right*).
20. Close the Form Cure station and let the UV glue cure for 2 min.
21. Repeat the previous step for each microfluidic device.
22. Inspect the presence of glue between the microfluidic device and the coverslip (*see* **Note 14**).
23. Place the rotating platform in the Form Cure station.
24. Cure the microfluidic devices in the Form cure station for 2 h at 60 °C.
25. Place the cured microfluidic devices per 2 in a well of a six-well plate and submerge the cured devices in clean IPA until fully covered.
26. Leave the microfluidic devices on a shaker for 6 days with exchange of fresh IPA every other day.
27. At day 6, spray microfluidic devices with 70% ethanol.
28. Use compressed air to dry microfluidic devices.
29. Place microfluidic devices in a cell culture hood and turn UV on for 15 min to sterilize microfluidic devices.
30. Microfluidic devices are ready to be used for biological experiments directly or can be stored in tip boxes.

3.4 Manufacturing and Sterilization of Microfluidic Devices, Holder, and Tubing

1. A custom holder able to accommodate 6 microfluidic devices was designed in SolidEdge 2019.
2. The holder design is sliced at 150 μm and 3D printed with a Prusa MK3S filament printer equipped with polylactic acid (PLA).
3. Submerge the 3D printed holder in a 5% bleach solution for 15 min in order to sterilize the 3D printed part.
4. Spray the outside of the 3D printed holder with 70% ethanol solution and let the holder air dry in a cell culture hood.
5. Dried 3D printed holders are ready to be used for biological experiments directly or can be stored in an air-tight zip-lock bag.
6. Laser cut (Trotec Speedy 100R) a 2 mm thick plexiglass sheet into plexiglass lids (29 mm × 124 mm) used to cover the custom-made and 3D-printed holder (*see* **Note 15**).

7. UV-sterilize both sides of the plexiglass lids for 15 min (*see* **Note 16**).
8. Plexiglass lids are ready to be used for biological experiments directly or can be stored in an air-tight zip-lock bag.
9. Cut silicone tubing at a length of 42 cm (*see* **Note 17**).
10. Clean the outside of the cut pieces with 70% ethanol before placing the tubing onto a clean cutting mat within a cell culture hood.
11. Connect one side of the silicone tubing to the aspirator present in the cell culture hood.
12. Submerge the other side of the silicone tubing into a 50 mL falcon tube filled with a 5% bleach solution.
13. Turn on the aspirator and flow 5 mL of the 5% bleach solution through the tubing.
14. Repeat **steps 11–12** with a 70% ethanol solution.
15. Repeat **steps 11–12** with MilliQ water.
16. Keep the aspirator running for an additional 2 min without the tubing being submerged in any liquid in order to dry the silicone tubing.
17. Cut dry silicone tubing with a sterile scalpel in lengths of 6 cm.
18. UV sterilize the silicone tubing for 15 min by using the build in UV lamp of the cell culture hood.
19. Silicone tubing is ready to be used for biological experiments directly or can be stored in an empty tip box.
20. Cut small pieces (4–5 mm) of PTFE tubing, to be used to connect silicone tubing, on a petri dish in a cell culture hood.
21. UV-sterilize cut PTFE tubing for 15 min in a cell culture hood.
22. PTFE tubing is ready to be used for biological experiments directly or can be stored in an empty tip box.

3.5 Assembly of Microfluidic Chip Plate

1. On the day of organoid embedding and microfluidic device seeding, collect the sterilized, microfluidic devices, holder, plexiglass lid, silicone tubing, and PTFE tubing (*see* **Note 18**).
2. To assemble a microfluidic chip plate, 6 microfluidic devices, 12 silicone tubing pieces and 6 PTFE tubing pieces are needed per holder and plexiglass lid (Fig. 5 *left*).
3. Connect a silicone tube to the inlet and outlet of each microfluidic device.
4. Place the 6 microfluidic devices one by one by, while alternating top-down orientation of the inlet, in the holder by first inserting the tubing through the tubing holes (*see* **Note 19**).
5. Insert a PTFE tubing in the silicone tubing at each outlet side.

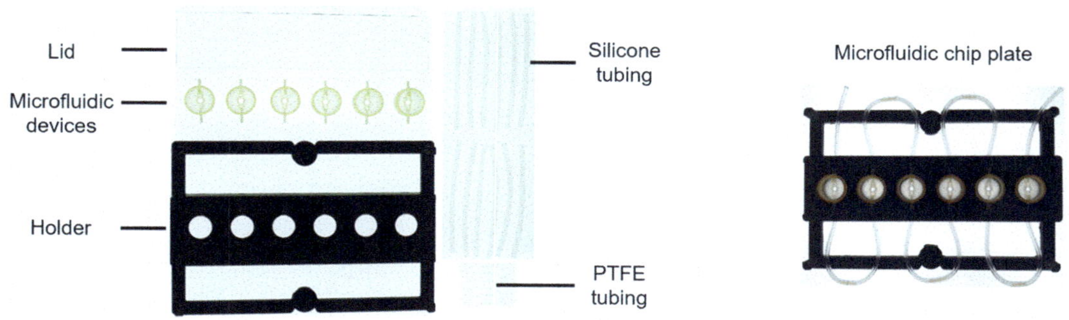

Fig. 5 Microfluidic chip plate assembly. *Left:* picture of parts needed to assemble the microfluidic chip plate comprising lid, microfluidic devices, holder, silicone, and PTFE tubing. *Right:* picture of assembled microfluidic chip plate

6. Slide the plexiglass lid in the microfluidic device holder (*see* **Note 20**).
7. The fully assembled microfluidic chip plate is ready to be seeded with vascular cells and cerebral organoids (Fig. 5 *right*).

3.6 Cerebral Organoid Seeding of the Microfluidic Chip Plate

1. Thaw an aliquot of growth factor-reduced Matrigel on ice.
2. Cut one parafilm square per microfluidic chip plate to be seeded.
3. Place the parafilm square inside a petri dish lid.
4. UV-sterilize the parafilm for 15 min in a cell culture hood.
5. At day 5 of cerebral organoid differentiation, transfer 6 organoids from the 96 U-bottom well plate to a square of parafilm using P200 wide bore tips (*see* **Note 21**).
6. Use a normal P200 tip to pipette away excess of medium around the organoids (*see* **Note 22**).
7. Use a P200 wide bore tip filled with 4.5 μL of Matrigel to lift up an organoid and pipette it in the central chamber of the first microfluidic device of the microfluidic chip plate (*see* **Note 23**).
8. Use a needle (0.3 mm outer diameter) to position the organoid in the middle of the central chamber (Fig. 6).
9. Repeat **steps 7–8** and until each microfluidic device of the microfluidic chip plate contains a Matrigel-embedded organoid.
10. Flip the microfluidic chip plate upside down and place it in the incubator for 10 min to allow gel polymerization.
11. Carefully pipette an additional 5 μL of growth factor-reduced Matrigel on top of the polymerized Matrigel to completely fill the central chamber with gel.
12. Put the plate back in the incubator for 15 min for the gel to polymerize.

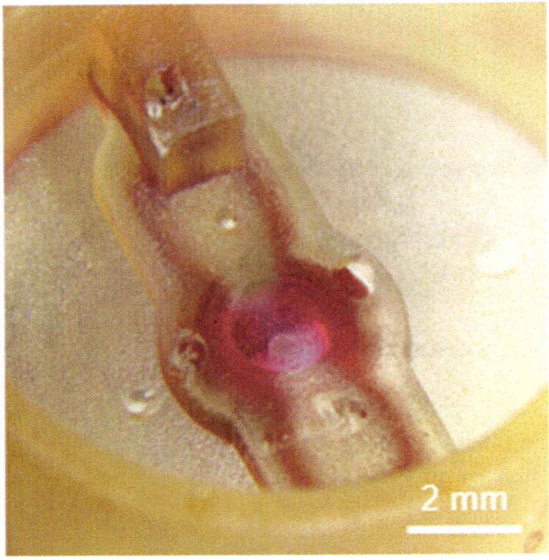

Fig. 6 Cerebral organoid on chip. Stereomicroscope image of the organoid on chip. Scale bar: 2 mm

13. Pipette 300 μL of cerebral organoid differentiation into each well (*see* **Note 24**).
14. Slide the plexiglass lid to close the top of the plate.
15. Repeat **steps 5–14** to seed additional microfluidic chip plates with cerebral organoids.

3.7 Vascular Cell Mix Seeding of the Microfluidic Chip Plate and Co-culture Maintenance

1. At day 6 of vascular cell differentiation, make a single cell suspension of vascular cells by pipetting 0.700 mL of Accutase to a well of a six-well plate (*see* **Note 25**).
2. Collect cells in an 1.5 mL Eppendorf.
3. Rinse the well with 0.7 mL of DMEM-F12 to collect remaining cells.
4. Count the amount of cells present in the Eppendorf.
5. Spin the cells for 5 min at 300 g.
6. Resuspend the cells in vascular differentiation medium containing 5% growth factor-reduced Matrigel at a concentration of six million cells per mL (*see* **Note 25**).
7. Pipette 60 μL of cell suspension with a P200 tip.
8. Connect the tip with the silicone tubing at the inlet side of the microfluidic device.
9. Carefully pipette out the 60 μL to load the channels of the microfluidic device until a droplet of medium is hanging out of the outlet tubing.
10. Repeat **steps 7–9** to load each microfluidic device individually.

11. Connect the outlet silicone tubing, containing the PTFE tubing, of the first microfluidic device with the inlet silicone tubing of the second microfluidic tubing by pushing the silicone tubing of the second microfluidic device over the PTFE connector tubing.
12. Subsequently connect the outlet silicone tube of the second microfluidic device to the inlet silicone tube of the third device.
13. Continue in the same manner until all the microfluidic devices are linked.
14. Connect the silicone outlet tube of the sixth microfluidic device with the silicone inlet tube of the first microfluidic device, thereby fully closing the fluidic circuitry.
15. Change the medium of the central well with organoid differentiation medium containing vitamin A and 100 ng mL^{-1} VEGF-A.
16. Change the medium in the central well every other day until the end of the experiment at D30.

3.8 Immunohistochemistry

1. At D30 of differentiation, disconnect the tubing and transfer microfluidic devices from the microfluidic chip plate to a six-well plate.
2. Fix vascularized cerebral organoids on-chip using 4% paraformaldehyde overnight at 4 °C in the fridge.
3. Wash microfluidic devices 3 times with PBS.
4. Break the inlet and outlet port of each microfluidic device.
5. Wrap parafilm around the outside of the microfluidic device to prevent leakage from the inlet and outlet.
6. Block and permeabilize the samples by adding 250 μL of a PBS-based solution containing 5% BSA, 2% Tx100, and 0.05% NaN_3, termed blocking solution, for 2 days at room temperature on an orbital shaker.
7. Pipette 250 μL of primary antibody mix, containing a 1:300 dilution of βiii TUBULIN (rabbit, biolegend 802001), and a 1:75 dilution of CD31 (mouse, Agilent M082329) in blocking solution per chip and leave on the shaker at RT for 2 days.
8. Wash off primary antibody mix during 2 days with PBS containing 0.05% NaN_3 with and refresh the PBS mix twice a day.
9. Pipette 250 μL of secondary antibody mix, containig a 1:500 dilution of donkey anti rabbit Alexa fluor 488 (invitrogen), 1:500 dilution of donkey anti mouse Alexa fluor 555 (invitrogen), and 1:500 dilution of hoechtst in blocking solution, per chip and leave on the shaker at RT for 2 days.

10. Wash off secondary antibody mix during 2 days with PBS containing 0.05% NaN_3 and refresh the PBS mix twice a day.
11. Samples are ready to be imaged.

4 Notes

1. We made use of SolidEdge 2019 CAD software to design our microfluidic device. Other CAD software packages such as Rhino 3D software or AutoCad can be used as long as the software supports export of the design as an .stl file.
2. Up to 8 devices can be printed at once without loss of print quality.
3. In addition to the Form2 printer, we successfully 3D printed microfluidic devices with both a Form3 equipped with Dental SG resin and an Asiga Max X UV27 equipped with Asiga Denta Guide resin (Asiga, 04504), and printodent GR-10 guide resin (pro3dure, D1000800).
4. The Dental SG (Formlabs) resin got discontinued and replaced by the Surgical guide resin (Formlabs, RS-F2-SGAM-01) which supports cerebral organoid growth and vascularization.
5. Resins used for vat polymerization 3D printing are sensitive to UV light due to presence of UV-specific photoinitiators. In order to avoid unwanted and random resin polymerization, especially within the microfluidic channels, opening of the printer and washing of the not-fully cured 3D printed microfluidic devices are done under UV-filtered light. In practice either yellow LED light or filtered white light can be used to achieve this.
6. It is highly recommendable to use wipes with zero or minimal fiber release in order to avoid fiber contamination of the resin vat.
7. Sticky unpolymerized resin is toxic. Avoid contact with eyes and skin, wear appropriate PPE such as laboratory coat, gloves, and glasses. Resin-contaminated areas can be easily cleaned with IPA moist cloths.
8. In order to efficiently wash out the internal channels of the microfluidic device, a custom syringe tool was developed making use of a 20 mL syringe, BD microlance needle, and silicone tubing. First, tweezers were used to break off the sharp end of BD microlance needle. This modified needle is attached to the syringe using the Luer lock system. A short piece of 2 cm silicone tubing was fit over the metal part of the needle. The flexible tubing of this syringe easily fits over the in- and outlets of the microfluidic device allowing to easily clean the channels.

9. IPA washes have to be thorough in order to avoid unpolymerized resin remaining on or in the microfluidic channels of the device. Nonetheless care should be taken in order to avoid breaking fine details on the device. Devices can be placed under a stereo microscope to better inspect for cleanliness and whether microfluidic channels still contain unpolymerized resin.
10. IPA can affect not fully cured resin structures and therefore the contact of the freshly 3D printed parts is limited to maximum 15 min in addition to the initial 2 min of ultrasonic cleaning. It is advised not to wash more than 8 devices in parallel in order to prevent part degradation due to prolonged IPA exposure.
11. The bottom surface of the microfluidic device, glue surface, needs to be flat to allow equal spreading of the UV curable glue upon placement of the coverslip. Equal spreading of glue assures water-tight sealing of the microfluidic device by the coverslip. Debris can be easily removed by wiping over the bottom surface with a finger. In addition, printing errors can result in an uneven bottom surface of the microfluidic device which are not usable.
12. The NOA86 UV curable glue is viscous. P20 pipette tips can be precut to increase tip diameter and improve glue handling.
13. In order to guarantee a proper water-tight seal, UV curable glue has to spread uniformly between the coverslip and the bottom of the microfluidic device. This can be visually checked by looking at how light reflects. The presence of air pockets is only problematic if it stretches from the central organoid culture chamber all the way to the edge. In this case, the coverslip might be glued to the device but it is not water tight. The presence of small isolated air pockets is not problematic.
14. If there is glue in the inner ring of the microfluidic device, endothelial cells loaded in the microfluidic channel cannot sprout to the organoid culture chamber. Microfluidic devices with glue in the inner circle cannot be used for biological experiments.
15. Inkscape software package is recommended to be used for the design of plexiglass lids which communicates well with the JobControl software from Trotec.
16. Laser cut plexiglass has the tendency to crack when exposed to ethanol; therefore, UV sterilization is preferred over ethanol sterilization.
17. The length of silicone tubing to be cut depends on the amount of prepared microfluidic devices. Each microfluidic chip has one inlet and one outlet each designed to be linked to a 6 cm

long silicone tube, with a total of 2 piece of 6 cm long silicone tubing needed per microfluidic device.

18. Work underneath a cell culture hood during the assembly of the microfluidic devices plate in order to maintain sterility as much as possible.
19. It is important to appoint a top and bottom of the plate in order to properly orient the microfluidic devices. At the first slot of the microfluidic devices holder, the inlet of the microfluidic device is pointed toward the top. In the second slot, the inlet of the second microfluidic device is oriented toward the bottom. Keep alternating between top and bottom until the holder is filled with microfluidic devices. Note that the inlet of the devices is different from the outlet due to the bubble trap feature immediately next to the inlet.
20. The assembled microfluidic plate can be UV sterilized for 15 min under a cell culture hood before inserting the plexiglass lid.
21. Cerebral organoids were generated based on the protocol published by Lancaster and Knoblich [8]. However, slight modifications were made to the timing of medium delivery. Additionally, the aggregation medium needed to form embryoid bodies was slightly different and is described in Salmon et al. [5].
22. Be careful not to pipette the organoid in the pipette tip because this can break the organoid in pieces. Avoid aspirating all the medium, always leave a minimum of medium, because dry organoids tend to stick to the parafilm and are therefore more difficult to pipette.
23. Do not fully pipette out the 4.5 μL of Matrigel containing the organoid in order to avoid the presence of air bubbles in the microfluidic device.
24. Cerebral organoid differentiation medium is prepared exactly the same as described in Lancaster and Knoblich [8].
25. The generation of vascular cells from hIPSC used in this protocol is based on Wimmer et al. [9] with the main difference that the vascular differentiation is performed in 2D in six-well plates, as described in Salmon et al. [5].

Acknowledgments

These studies were supported by Research Foundation Flanders grants G087018N, G086622N, G0ACA24N, G0AQK25N; KU Leuven grants I009718N, IDN/20/007, IDN/22/012, C14/22/108; Allen Distinguished Investigator Award a Paul G. Allen Frontiers Group advised grant of the Paul G. Allen Family Foundation.

References

1. Jiang T et al (2023) Study of forming performance and characterization of DLP 3D printed parts. Materials 16(10):3847. https://doi.org/10.3390/MA16103847
2. Su R, Wang F, Mcalpine MC (2023) 3D printed microfluidics: advances in strategies, integration, and applications. Lab Chip 23:1279. https://doi.org/10.1039/d2lc01177h
3. Piironen K, Haapala M, Talman V, Järvinen P, Sikanen T (2020) Cell adhesion and proliferation on common 3D printing materials used in stereolithography of microfluidic devices. Lab Chip 20(13):2372–2382. https://doi.org/10.1039/D0LC00114G
4. Khan I, Prabhakar A, Delepine C, Tsang H, Pham V, Sur M (2021) A low-cost 3D printed microfluidic bioreactor and imaging chamber for live-organoid imaging. Biomicrofluidics 15(2). https://doi.org/10.1063/5.0041027
5. Salmon I et al (2022) Engineering neurovascular organoids with 3D printed microfluidic chips. Lab Chip 22:1615–1629. https://doi.org/10.1039/d1lc00535a
6. Sagot M et al (2024) Functionality integration in stereolithography 3D printed microfluidics using a 'print-pause-print' strategy. Lab Chip 24(14):3508. https://doi.org/10.1039/D4LC00147H
7. "Formlabs Professional 3D Printers | Formlabs." Accessed: 24 Sept 2024. [Online]. Available: https://formlabs.com/eu/3d-printers/catalog/
8. Lancaster MA, Knoblich JA (2014) Generation of cerebral organoids from human pluripotent stem cells. Nat Protoc 9(10):2329–2340. https://doi.org/10.1038/nprot.2014.158
9. Wimmer RA, Leopoldi A, Aichinger M, Kerjaschki D, Penninger JM (2019) Generation of blood vessel organoids from human pluripotent stem cells. Nat Protoc 14(11):3082–3100. https://doi.org/10.1038/s41596-019-0213-z

Methods in Molecular Biology (2026) 3048: 337–348
DOI 10.1007/7651_2025_640

Published online: 22 May 2025

Engineering Synthetic PEG Hydrogels to Model Peri-implantation Epiblast Morphogenesis with Tunable Biophysical Properties

Michael Patrick Seitz, Zhen Ma, and Era Jain

Abstract

Implantation triggers critical morphological transformations in the embryo, where the epiblast transitions from a cluster of unpolarized cells into a highly organized, polarized epithelium characterized by a central lumen. Human pluripotent stem cells (hPSCs) are valuable models for studying this process, but conventional matrices like Matrigel have significant limitations, including variability and poor control over mechanical properties. To overcome these challenges, we developed a synthetic polyethylene glycol (PEG) hydrogel system with tunable mechanical stiffness to model peri-implantation epiblast morphogenesis.

Our platform enables hPSCs to form unpolarized 3D aggregates that undergo stiffness-dependent transformation into lumen-forming, apicobasal-polarized structures resembling epiblast morphogenesis during peri-implantation. Unlike natural ECMs, PEG hydrogels maintain hPSC pluripotency for extended periods and support trilineage differentiation upon induction. The modular hydrogel design facilitates targeted mechanistic studies on the biophysical and biochemical regulation of cell morphogenesis. We present a comprehensive protocol for fabricating PEG hydrogels, encapsulating hPSCs, and assessing cell polarity, lumen formation, and pluripotency using immunostaining and RT-PCR. This platform provides a robust, cost-effective, and versatile tool for advancing developmental biology and regenerative medicine.

Key words hiPSC, Hydrogels, In vitro embryo model, Mechanobiology, Pluripotency, Stem cells

1 Introduction

Implantation triggers dramatic changes in the embryo, transforming the epiblast from a cluster of unpolarized cells into a highly organized, polarized epithelium characterized by a central lumen [1–3]. Human pluripotent stem cells (hPSCs) have emerged as valuable models for studying this morphogenetic transformation when grown within Matrigel, a tumor-derived extracellular matrix (ECM) [4–7]. However, Matrigel poses significant limitations due to its undefined composition, batch-to-batch variability, and lack of control over critical physical properties such as mechanical stiffness [8, 9] Moreover, hPSCs cultured in natural ECM matrices like Matrigel often undergo differentiation within a short time, restricting their utility for investigating how matrix mechanical properties influence the morphogenesis of undifferentiated hPSCs mimicking the epiblast stage of development [8].

While the role of matrix stiffness in stem cell differentiation is well documented [10], its influence on epithelial morphogenesis—particularly the transformation of hPSCs into epithelial-like cells—remains poorly understood. Capturing peri-implantation stages of human embryo development in vitro is critical, as these events are inaccessible for direct observation in vivo due to the embryo's implantation in the uterine wall. A synthetic matrix with tunable biophysical properties is therefore essential for advancing mechanistic studies on how matrix stiffness regulates cell morphogenesis and polarization during this pivotal developmental stage [8, 9, 11].

Here we introduce a simple and inexpensive platform for producing 3D hPSC aggregates to model the peri-implantation epiblast using synthetic polyethylene glycol (PEG) hydrogels. Our PEG hydrogels feature tunable mechanical properties and present a minimalistic, modular system for mimicking the biophysical aspects of ECM, particularly mechanical stiffness. We have demonstrated that single hPSCs encapsulated within the hydrogel can reliably proliferate and generate unpolarized 3D aggregates of cells. Upon fine-tuning the hydrogel stiffness, the unpolarized 3D aggregates form a central lumen with radially organized, apicobasal-polarized cells, reminiscent of epiblast morphogenetic changes during peri-implantation and bearing the hallmarks of the epiblast 10–12 days post-fertilization (dpf). Unlike natural ECMs, our PEG hydrogels maintain hPSC pluripotency for at least 6 days, and the aggregates remain capable of trilineage differentiation upon induction. This system provides unprecedented control over the mechanical environment, enabling precise investigations of how stiffness drives morphogenetic transformations and cell polarization in epiblast. Further, the modular nature of the synthetic matrix allows for controlled integration of adhesion ligands and can facilitate targeted studies on the role of specific biochemical cues. The PEG hydrogel system described here represents a significant advancement in modeling human epiblast morphogenesis. By providing a simple, cost-effective, and customizable platform, it offers new opportunities for mechanistic studies of ECM stiffness in morphogenesis and has the potential to serve as a valuable tool in developmental biology and regenerative medicine [12].

This protocol outlines the methods for fabricating PEG hydrogels with varied mechanical stiffness, encapsulating and culturing hPSCs within these hydrogels, and probing the constructs using immunostaining for cell polarity markers and lumen morphology (Fig. 1). Additionally, we detail protocols for assessing pluripotency maintenance and trilineage differentiation via RT-PCR and immunostaining.

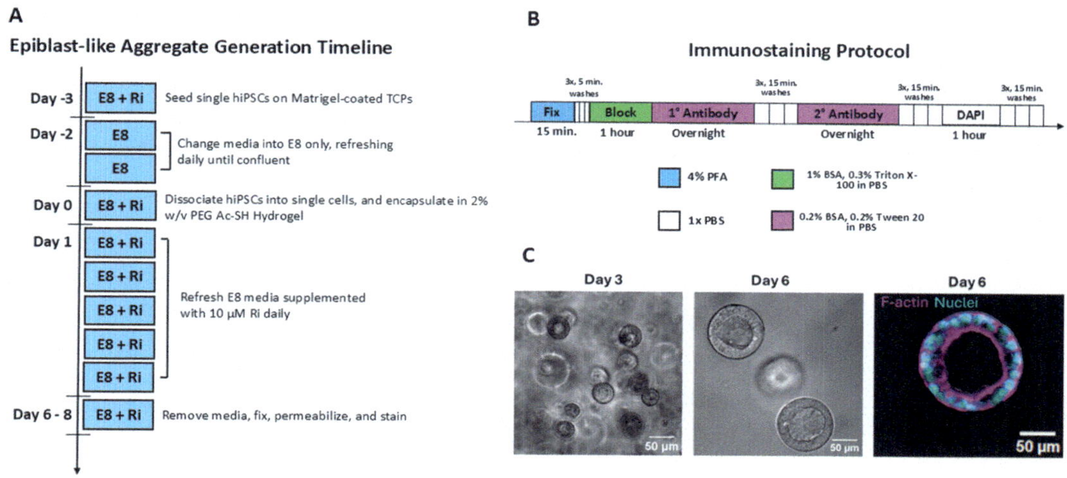

Fig. 1 Timeline of experiments and expected results. (**a**) Overview of the timeline of experiment for encapsulating single hPSCs in the hydrogel and obtaining apicobasal polarized cells enclosing a central lumen. (**b**) Overview of timeline for immunofluorescent analysis of resulting aggregates. (**c**) Single hPSC upon encapsulation in 2% PEG hydrogels form unpolarized aggregate and transform to form radially organized, polarized cells enclosing a central lumen. Bright field image of the aggregates after 3 and 6 days of culture. Note the lumen expands in size from day 3 to 6. Immunostained image of a hPSC aggregate culture in PEG hydrogels showing polarized accumulation of F-actin in apical domain

2 Materials

2.1 PEG Hydrogel Precursors

1. Reaction Buffer: 100 mM HEPES, pH 8. Add 45 mL of DNase, RNase-free water to a sterile, 50 mL centrifuge tube. Weigh 1.19 g of HEPES sodium salt and transfer to the same tube. Mix and adjust the pH to 8 with NaOH. Bring the solution to a total volume of 50 mL with DNase, RNase-free water. Store at 4 °C (*see* **Note 1**).
2. 8-arm PEG Acrylate 20 kDa: store at −20 °C (Jenkem Technology).
3. 4-arm PEG Thiol: store at −20 °C (Laysan Bio).

2.2 Cell Culture

1. Biosafety Cabinet (BSC).
2. 37 °C, 5% CO_2 cell culture incubator.
3. WTC-11 hiPSCs.
4. Essential 8™ Medium.
5. Accutase.
6. Matrigel.
7. 10 mM Y-27632 Rho-associated protein kinase inhibitor (ROCK inhibitor).
8. 1× Phosphate buffer saline (PBS).

9. Glass slides.
10. Parafilm.
11. 600 μm thick silicone spacers.
12. 60 mm petri dish.
13. Methanol.
14. 6-well tissue culture plate.
15. 48-well tissue culture plate.
16. Trypan Blue.
17. Sterile microcentrifuge tubes.
18. Small metal spatula.
19. Light microscope.

2.3 Fixing and Staining the Aggregates

1. PBS, DPBS with calcium and magnesium (Gibco 14040141).
2. 4% paraformaldehyde.
3. Blocking buffer: 1% Bovine Serum Albumin (BSA) and 0.3% Triton X-100 diluted in PBS.
4. PBST blocking buffer: 0.2% Tween 20 and 0.2% BSA diluted in PBS.
5. Primary antibody (antibodies used are listed in Table 1) diluted in either blocking buffer or PBST blocking buffer.
6. Fluorescently labeled secondary antibody diluted in blocking buffer.
7. DAPI (to visualize nucleus).

Table 1
List of primary antibodies used in routine assessment of model epiblast pluripotency, polarization, and differentiation capacity

Antigen	Marker	Isotype	Dilution	Vendor	Product #
OCT4	Pluripotency	Rabbit IgG	1:50	Abcam	ab19857
SOX2	Pluripotency	Rabbit IgG	1:200	Abcam	ab92494
NANOG	Pluripotency	Rabbit IgG	1:50	Abcam	ab21624
ZO-1	Apical membrane	Rabbit IgG	1:100	Thermo Fisher	402200
E-cadherin	Cell adhesion	Mouse IgG	1:50	Abcam	ab1416
T-Brachyury	Mesoderm	Goat IgG	1:40	Thermo Fisher	PA5-46984
PAX-6	Ectoderm	Mouse IgG2a	1:50	Biolegend	862001
SOX17	Endoderm	Mouse IgG1	1:100	Biolegend	628151
Isotype control	NA	Rabbit IgG	Assay dependent	Abcam	ab37415
Isotype control	NA	Mouse IgG2a	Assay dependent	Thermo Fisher	02-6100

8. Phalloidin (to visualize the actin cytoskeleton).
9. Pipettes for washes.
10. Aspirator for washes.
11. Fluorescence microscope suitably equipped for imaging 3D aggregates via optical sectioning (e.g., laser scanning confocal microscope).

3 Methods

3.1 Hydrogel Mold Preparation

1. Wrap a thin layer of parafilm around two glass microscope slides (*see* **Note 2**).
2. Wash both slides with methanol (*see* **Note 3**).
3. Cut two small squares (5 × 5 mm approximately) out of the silicone spacer sheet. Remove the adhesive and place at opposite ends of one microscope slide.
4. Place both slides in a petri dish, and transfer to a BSC, followed by UV sterilization for at least 15 min.

3.2 Hydrogel Precursor Preparation

Before attempting any experiments, inspect 2D-cultured iPSCs for health and quality (*see* **Note 4**).

The following recipe makes 100 μL of hydrogel precursor solution; however, these values can be scaled up or down to match experimental demands:

1. Remove dry PEG acrylate and PEG thiol macromers from the freezer and allow them to equilibrate to room temperature.
2. Weigh at least 1.1 mg of PEG acrylate and PEG thiol macromers individually into two separate sterile microcentrifuge tubes (*see* **Note 5**).
3. After spraying the microcentrifuge tubes with 70% ethanol, place them in a BSC. Open the lids of both tubes and UV sterilize the powder for at least 15 min.
4. Add 5 μL of reaction buffer to each tube to make 20% w/v stock solutions.
5. Vortex PEG macromer stocks for 30 s before centrifugation at 5000 RPM for 2 min. This step is essential to collect all liquid to the bottom of the tube.
6. Transfer 20% stocks back to the BSC for later use.

3.3 Fabricating Polyethylene Glycol Hydrogels: Hydrogel Recipe

1. Mix the reaction buffer and the PEG macromer stocks in the indicated amounts to get the hydrogels of desired stiffness. Note each formulation results in 100 μL of the hydrogel precursor solution.

Table 2
PEG hydrogel composition. Amount of various components and reaction buffer for making hydrogels of desired stiffness

Polymer concentration	20% PEG acrylate (μL)	20% PEG thiol (μL)	Reaction buffer (μL)	Cell suspension 2.77×10^6 cells/mL (μL)	Y-27632 (10 mM) (μL)	Expected Young's Modulus (E kPa)
1.5	3.75	3.75	47.5	45	0.1	0.5
2	5	5	45	45	0.1	1
3	7.5	7.5	40	45	0.1	1.5

2. Mix the hydrogels components and cell suspension in indicated amounts and order and incubate for indicated times as described in 3.4 and Table 2 (*see* **Note 6**). Please note highest cell polarization and consistent lumen formation are seen when cells are cultured in hydrogel with 2% polymer concentration.

3.4 Routine Cell Maintenance and Cell Encapsulation in Hydrogels

All directed reagent volumes are assuming that the hPSCs are being cultured on standard 6-well tissue culture plates, as this yields an appropriate number of cells. Values must be adjusted if substrates with different surface areas are used. Routine hiPSC culture using E8 media is well established and described in the manufacturer's protocol (www.thermofisher.com/order/catalog/product/A1517001) as well as in published papers [13]. hiPSC-aggregates are derived from a singly dissociated hPSC monolayer (*see* **Note 7**). The process of hydrogel fabrication and cell encapsulation is depicted in Schematic 1.

1. Plate cells at a density that will reach 70–80% confluency in 4 to 5 days.
2. Inspect cell on days 3–4 and remove any differentiated cells (*see* **Note 8**).
3. On days −3 through −1, check the entire well and remove any differentiated cells.
4. On the day of encapsulation or when colonies are 70–80% confluent, remove E8 medium, and rinse the well twice with 1 mL of PBS.
5. Add 1 mL of Accutase and incubate for 10 min in a standard 37 °C CO_2 incubator (*see* **Note 9**).
6. Quench with 1 mL of E8 medium, and rinse the plate at least 5 times with suspension to break up clumps.
7. Transfer the suspension to a 15 mL centrifuge tube and centrifuge at 500 RPM for 3 min.

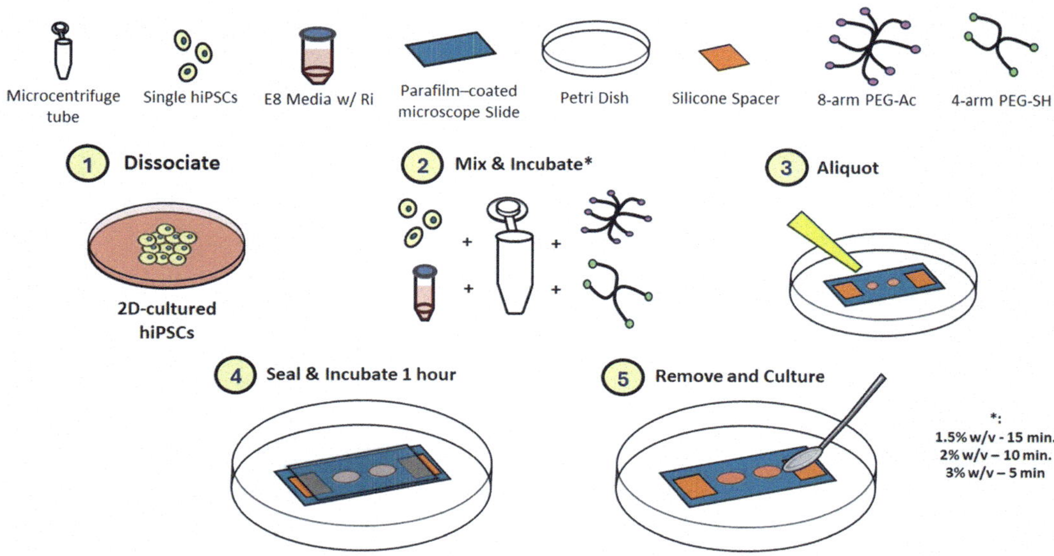

Schematic 1 Schematic of the procedure for encapsulating single hiPSCs in PEG hydrogels

8. Aspirate the supernatant and resuspend the pellet with 1 mL of E8 medium.
9. Count the cells using trypan blue and adjust the cell density to 2.77×10^6 cells/mL (*see* **Note 10**).
10. To a sterile microcentrifuge tube, add 45 μL of reaction buffer.
11. Add 0.1 μL of 10 mM Y-27632.
12. Add 45 μL of cell suspension.
13. Separately, add 5 μL 20% PEG-acrylate and PEG-thiol stock solutions (*see* **Note 11**).
14. After mixing thoroughly, place the gel precursor solution in a 37 °C incubator for 10 min (*see* **Note 12**).
15. On the parafilm-coated glass slide with the spacers from Subheading 3.1, aliquot five, 20 μL droplets of gel precursor solution, pipetting up and down before each aliquot (*see* **Note 13**).
16. Gently place the other microscope slide on top of the first (*see* **Note 14**).
17. Place the lid back on the petri dish and incubate at 37 °C for 1 h.
18. Prepare 2 mL of E8 media, supplemented with 10 μM Y-27632.
19. In a 48-well plate, add 300 μL of this media to 5 wells.
20. After gelation, transfer the petri dish to the BSC.

21. Wet the end of a metal spatula with a small amount of sterile PBS, and use it to gently pry open the microscope slides (*see* **Note 15**).
22. Gently scoop each gel into the wells of the 48-well plate filled with media.
23. Transfer the plate to a 37 °C incubator.
24. Add fresh E8 supplemented with 10 M Y-27632 (*see* **Note 16**).
25. To ensure proper culture, the size and morphology of cells must be assessed every day (*see* **Note 17**).
26. Every 24 h post-encapsulation, aspirate media and replace with fresh E8 supplemented with 10 μM Y-27632 (*see* **Note 17**).

3.5 Viability Staining

1. Wash hydrogel samples once with 1× PBS.
2. Staining: Incubate samples with a calcein AM/ethidium homodimer working solution (LIVE/DEAD™ Viability/Cytotoxicity Kit) for 30 min at 37 °C in 5% CO_2.
3. Wash samples once with 1× PBS.
4. Image samples using a Leica Thunder DM4 B microscope (10× objective). (Note: An inverted fluorescent microscope can be used too to capture images.)
5. Capture at least three images per sample.
6. Analyze using Image J plugin using standard procedure.

3.6 Determining the Pluripotency of hPSCs Aggregates

Pluripotency of the aggregates can be determined using a standard qPCR or via immunostaining for typical markers NANOG, SOX2, and OCT4. For PCR analysis, prepare the samples as follows. For immunostaining, follow the procedure as listed in the Subheading 3.7. Sample preparation for qPCR:

1. Wash the hydrogels with PBS buffer.
2. Remove the hydrogels from the cell culture plates and place in a microcentrifuge tube.
3. Prepare the sample to be processed by a bead beater to degrade the hydrogels.

3.7 Immunostaining the hiPSCs Aggregate Cultured in Hydrogels

1. Fix gels with 4% paraformaldehyde (PFA) at room temperature for 15 min.
2. Wash gels three times with PBS for 5 min each.
3. Incubate gels in blocking buffer for 1 h.
4. Incubate gels with primary antibody (Table 1) in PBST overnight at 4 °C.
5. Include appropriate isotype controls for each experiment.

6. Wash gels three times for 15 min each with PBS at room temperature.
7. Incubate gels with secondary antibody in PBST overnight at 4 ° C.
8. Wash gels three times for 15 min each with PBS.
9. Stain gels with 100 μL of 300 nM DAPI for 1 h.
10. Wash gels three times for 15 min each in PBS.
11. Image samples using a confocal lasing scanning microscope (*see* **Note 18**).

3.8 Determination of Cell Polarity and Lumen Imaging

Cell polarity can be determined by staining for apical F-actin and tight junctions (ZO-1) (*see* **Note 19**). From a 1000× Phalloidin-iFluor reagent stock (Abcam), prepare a 1× working solution using 1% BSA in PBS.

1. Stain gels with 100 μL of phalloidin working solution for 1 h.
2. Remove the phalloidin working solution and counterstain with 100 μL of 300 nM DAPI for 1 h.
3. Wash gels three times for 15 min each in PBS.
4. Perform imaging using a laser scanning confocal microscope. A z-stack step size of at most 5 μm is recommended for proper morphological analysis.
5. Follow the immunostaining procedure outlined in Subheading 3.7 to stain for apical marker ZO-1.

3.9 Trilineage Differentiation: Differentiating Cells to Different Lineages

1. Grow encapsulated hiPSCs under self-renewing conditions using E8 media for 4 days.
2. *Endoderm Differentiation*: Switch E8 medium to Stemdiff™ Trilineage Endoderm medium supplemented with 10 μM Y-27632. Incubate for 4 days, with daily media changes. Follow manufacture instructions (Stemcell Technologies # 05230).
3. *Ectoderm Differentiation*: Switch E8 medium to Stemdiff™ Trilineage Ectoderm medium supplemented with 10 μM Y-27632. Incubate for 6 days, with daily media changes (Stemcell Technologies # 05230).
4. *Mesoderm Differentiation*: Switch E8 medium to RPMI 1640 with B27 supplement minus insulin, 6 μM CHIR-99021, and 10 μM Y-27632. Incubate for 2 days. (Adapted from Arkenberg et al. (2022)).
5. Prepare samples for immunostaining and imaging as described in the viability staining protocol.

4 Notes

1. To protect cell viability, either DNase-free water or ultrapure water should be used to make the HEPES buffer. From our experience, using water from DI taps has had a negative effect on cell survival during encapsulation.
2. Ensure the parafilm layer is flat, as this will affect the gels' geometry. The same set of parafilm-coated slides may be reused multiple times if washed with 70% ethanol and dried, so this does not need to be done every time.
3. PEG is hydrophilic and will stick to many surfaces. A hydrophobic coating like methanol makes removal from the mold easier. Rain-x may also be used.
4. For encapsulations, hiPSCs are best used when cultured for 3 days or less. Longer culture times make dissociation into single hiPSCs more difficult, requiring longer incubation times in Accutase.
5. When preparing PEG stocks, it is advised to weigh at least 10% extra to account for reagent loss during transfer steps. Purging PEG macromer powders with nitrogen before transferring back to the freezer also helps extend lifespan of the reagent.
6. To test hydrogel mechanical stiffness, make hydrogels without cells by replacing the volume of the cell suspension with reaction buffer or cell culture media. Hydrate hydrogels for 2 h in PBS buffer and measure using a standard rheometer with plate geometry and diameter of 8 mm. Method to measure hydrogels modulus can be found in various publications [12, 14].
7. It is critical to use properly maintained hPSC monolayers free of spontaneously differentiated cells. Any differentiated cells will rapidly disseminate throughout the culture dish, complicating imaging and analysis. Therefore, it is important to start with samples containing a few differentiated cells. Cells should be passaged as aggregates into a six-well plate or 35 mm dish 4–5 days before making single-cell suspension. Passaging can be performed with enzymatic, or enzyme-free methods.
8. Differentiated cells are apparent as colonies with opaque raised centers, or colonies with gaps between adjacent cells, or colonies containing cells with fibroblast morphologies. Remove the entire colony by scraping with a pipette tip.
9. Incubate in Accutase until colonies are freely floating, to make dissociation into single cells easier. Formation of size-controlled aggregates requires single cells, making this a critical step.

10. This seeding density yields ~25,000 cells per 20ul aliquot, which assures that resulting aggregates will be separated enough to not eventually overlap.
11. While the order of macromer addition does not matter, it is imperative that they are mixed separately, as they will begin gelling immediately after being brought in contact. Mix thoroughly after adding each macromer, as uneven mixing can lead to premature gelation.
12. As the gelation time of 2% w/v hydrogels are ~15 min, this step ensures the cells will remain suspended. Immediate transfer into the mold would cause cells to sink to the bottom.
13. It is important to leave appropriate space between aliquots, and to conduct the aliquoting on a level surface, as drops can bleed into each other once they are sandwiched between the slides. Alternatively, a 96-well plate can also be used.
14. Gels should become noticeably disc shaped once the top slide is added.
15. While a hydrophobic coating will help, the gels may still stick to the surface of the slides. As such, it is important to be gentle when removing them. Applying a little pressure and wedging the spatula underneath the gel can help pry them off.
16. After the first 24 h of culture, cells should start growing as large, circular aggregates. The size of these aggregates should become progressively larger, reaching ~50 um by day 3, and ~100 um by day 6. Lumens should be noticeable by day 3.
17. The gels are soft enough to be sucked into pipettes, so it is advised to place the pipette tip against the bottom wall of the well while aspirating the old media. If the media begins to turn progressively yellow after the first few days, the amount of media added should be increased.
18. Since optical sections are needed to image the 3D aggregates within the hydrogel, a confocal is best suited for this application.
19. We recommend including a second stain for tight junction protein ZO-1 to ensure accurate determination of cell polarity.

Acknowledgments

This work was supported by the NIH [R01HD101130] and the NSF [CMMI-2130192 and CBET-1943798] to ZM and Bioinspired Seed Funds by Bioinspired Syracuse: Institute for Material and Living System, Syracuse University to EJ and ZM. We also acknowledge Blatt BioImaging Center for use of the LSM 980 or LSM 710, which was supported by NIH S10 OD026946–01A1.

References

1. Shahbazi MN et al (2016) Self-organization of the human embryo in the absence of maternal tissues. Nat Cell Biol 18(6):700–708
2. Bedzhov I, Zernicka-Goetz M (2014) Self-organizing properties of mouse pluripotent cells initiate morphogenesis upon implantation. Cell 156(5):1032–1044
3. Sozen B et al (2021) Reconstructing aspects of human embryogenesis with pluripotent stem cells. Nat Commun 12(1):5550
4. Deglincerti A et al (2016) Self-organization of human embryonic stem cells on micropatterns. Nat Protoc 11(11):2223–2232
5. Durdu S et al (2014) Luminal signalling links cell communication to tissue architecture during organogenesis. Nature 515(7525): 120–124
6. Romero-Morales AI, Ortolano NA, Gama V (2017) Apical polarization and lumenogenesis: the apicosome sheds new light. J Cell Biol 216(12):3891–3893
7. Taniguchi K et al (2015) Lumen formation is an intrinsic property of isolated human pluripotent stem cells. Stem Cell Rep 5(6):954–962
8. Aisenbrey EA, Murphy WL (2020) Synthetic alternatives to Matrigel. Nat Rev Mater 5(7): 539–551
9. Gjorevski N, Lutolf MP (2017) Synthesis and characterization of well-defined hydrogel matrices and their application to intestinal stem cell and organoid culture. Nat Protoc 12(11):2263–2274
10. Evans ND et al (2009) Substrate stiffness affects early differentiation events in embryonic stem cells. Eur Cell Mater 18:1–13; discussion 13-4
11. Srivastava P et al (2023) Defined microenvironments trigger in vitro gastrulation in human pluripotent stem cells. Adv Sci (Weinh) 10(5): e2203614
12. Seitz MP et al (2024) Soft polyethylene glycol hydrogels support human PSC pluripotency and morphogenesis. ACS Biomater Sci Eng 10(7):4525–4540
13. Castro-Vinuelas R et al (2021) Tips and tricks for successfully culturing and adapting human induced pluripotent stem cells. Mol Ther Methods Clin Dev 23:569–581
14. Jain E et al (2017) Control of gelation, degradation and physical properties of polyethylene glycol hydrogels through the chemical and physical identity of the crosslinker. J Mater Chem B 5(14):2679–2691

The manufacturer's authorised representative in the EU is Springer Nature Customer Service Centre GmbH, Europaplatz 3, 69115 Heidelberg, Germany. If you have any concerns regarding our products, please contact ProductSafety@springernature.com

Printed and bound by CPI Group (UK) Ltd, Croydon, CR0 4YY
15/09/2026
02217004-0002